AF753103

Physics of Asymmetric Continuum: Extreme and Fracture Processes

Roman Teisseyre · Hiroyuki Nagahama · Eugeniusz Majewski (Eds.)

Physics of Asymmetric Continuum: Extreme and Fracture Processes

Earthquake Rotation and Soliton Waves

Springer

Roman Teisseyre
Polish Academy of Sciences
Institute of Geophysics
ul. Ksiecia Janusza 64
01-452 Warszawa
Poland

Hiroyuki Nagahama
Tohoku University
Graduate School of Science
Inst. Geology & Paleontology
Sendai 980-8578
Japan

Eugeniusz Majewski
Polish Academy of Sciences
Institute of Geophysics
ul. Ksiecia Janusza 64
01-452 Warszawa
Poland

ISBN: 978-3-540-68354-4 e-ISBN: 978-3-540-68360-5

Library of Congress Control Number: 2008928850

© 2008 Springer-Verlag Berlin Heidelberg

This work is subject to copyright. All rights are reserved, whether the whole or part of the material is concerned, specifically the rights of translation, reprinting, reuse of illustrations, recitation, broadcasting, reproduction on microfilm or in any other way, and storage in data banks. Duplication of this publication or parts thereof is permitted only under the provisions of the German Copyright Law of September 9, 1965, in its current version, and permission for use must always be obtained from Springer. Violations are liable to prosecution under the German Copyright Law.

The use of general descriptive names, registered names, trademarks, etc. in this publication does not imply, even in the absence of a specific statement, that such names are exempt from the relevant protective laws and regulations and therefore free for general use.

Cover design: deblik, Berlin

Typesetting: Institute of Geophysics, Polish Academy of Sciences, Warszawa

Printed on acid-free paper

9 8 7 6 5 4 3 2 1

springer.com

Introduction

Our new monograph has been inspired by the former one, *Earthquake Source Asymmetry, Structural Media, and Rotation Effects* (R. Teisseyre, M. Takeo, and E. Majewski, eds, Springer 2006). Some problems, concerned primarily but not exclusively with the basic theoretical nature, have appeared to us as worthy of further analysis. Thus, in the present monograph we intend to develop new theoretical approaches to the theory of continua that go far beyond the traditional seismological applications. We also try to present the links between the experimental data, the observed rotational seismic waves, and their theoretical evaluation and description.

In addition, we consider the basic point motions and deformations, and we intend to find the invariant forms to describe such point motions. We believe that there must exist the basic equations for all point motions and deformations, and we derive such relations within a frame of a continuum theory. Thus, in the considered standard asymmetric theory, we include relations not only for the displacement velocities but also for a spin motion and basic point deformations as well. We include here the axial point deformation and twist point deformation represented by the string-string and string-membrane motions. A twist vector is defined here as a vector perpendicular to the string-string plane and representing its magnitude. It becomes an important counterpart to spin and a key to the presented theory. We show in the forthcoming chapters that the twist motion describes the oscillations of shear axes.

Practically all motions and deformations can be described by the displacement field, but this is not true for the point-related spin, the string-string vector, and point axial motions, which become independent fields related to different source processes. This approach not only widens our possibilities for the description of material deformations but also leads us to its projection onto some basic problems of the classical relativistic theory. Furthermore, we include considerations of an advanced approach to space geometry and deformations in asymmetric continua. Applications of differential geometry are extremely important for the non-linear processes. Finally, we arrive at some analogies leading us up to the general relativity problems.

The monograph is presented in three parts. In the first part, *Introduction to Asymmetric Continuum and Experimental Evidence of Rotation Motions*, we consider the basic deformations in a continuum and discuss the full system of the point deformations and motions that need to be independently described by an appropriate system of basic equations. We present an important evaluation of the experimental data and methods used to derive the rotation wave part. Our considerations include the necessity of simultaneous recording of the translational and rotational earthquake motions and the strains—at least the deviatoric strains. We believe that the related data will provide new insights for earthquake engineering studies as well.

The new achievements in experimental studies related to rotation motions involve some important relations with engineering seismology and with interaction effects, including some reverse actions. For instance, a strong vibration of buildings erected on flexible soil can induce, through soil-structure interaction, additional rotational motions in the ground. Thus, we argue that the vibration of some objects on the ground surface may interact back on the ground motions.

In the next part, *Continuum with Defect Densities and Asymmetry of Fields*, we develop the standard asymmetric continuum theory. We discuss the fundamental problems related to the continuum theories and the need for a new approach that includes asymmetric fields, and we present a review of achievements of the asymmetric continua. When considering continua contaminated by a field of defects and interaction nuclei, we include the different material responses to stresses or other active fields, with these responses most frequently being described by an appropriate choice of constitutive relations. The other approach—the Kröner approach—relies on suitable modifications of the equation of motion so as to include a physical interaction involving the defects and internal nuclei (thermal, electric, rotation nuclei, and others).

Instead of the above-mentioned approaches, we introduce the standard asymmetric continuum theory that brings us more directly to an interactive nature of defects and internal nuclei. In a frame of this new approach, we discuss the basic deformations and the related aspects of the constitutive bonds between stress moments and the rotation of particles in a continuum. Stress moments, when introduced into a frame of classical continuum theory, require that at least one discrete element be introduced into the continuum—i.e., an internal length unit. This inconsistency is usually overcome by the concepts of the micropolar or micromorphic theories, in which the point-grains are subjected to deformation. Such an apparent contradiction disappears in the asymmetric continuum theory when the internal rotations become connected to antisymmetric stresses in a constitutive way. Therefore, we present an equivalence between this new treatise with the antisymmetric stresses and the one that uses the stress moments.

Many aspects of continuum theories have their origin in problems of the crystal lattice defects. The densities of such defects have their description in the continuum theories. Thus, in developing the standard asymmetric continuum theory we include a counterpart of dislocation density fields together with the stress-dislocation relations in a concise manner. This theory of continua, with rotational motions (spin and twist together), with the inner central motion (e.g., the thermal one), and with the other defect fields, can be projected onto the intrinsic properties of the Riemannian or non-Riemannian space. Moreover, based on the notion of the antisymmetric stresses (introduced, among others, by Shimbo 1975, 1995), we may extend the theory to include a case of a homogeneous continuum. The standard asymmetric continuum is, thus, the theory that includes symmetric and antisymmetric stresses. The additional constitutive laws introduced form a basis for the existence of an antisymmetric stress field, with such stresses arising as an elastic response to the rotational deformations of bonds in a lattice network.

Furthermore, we introduce the deviatoric potentials, which help us to define the twist motion as the counterpart to the spin motion and to provide better understanding of the symmetry properties of fields. This theory permits us to present a new insight into fracture processes with a counterpart of rotational motions, spin and twist, and also an axial deformation when a confining load is included. Consequently, we discuss processes leading to material granulation and fragmentation. The hypothesis of fracture synchronization processes is based on the special solution for the spin and twist fields.

An important problem in seismology, and especially in the search for earthquake precursory phenomena, concerns the interaction between the mechanical and electric fields. We present a frame for the interaction asymmetric continuum theory and discuss the numerous approaches to the study the electric field and electric current related precursory phenomena.

We extend our considerations for non-linearity in terms of the theory of fluids and extreme deformations, including the soliton waves.

Continuing with the problem of the non-linear processes, we give examples concerning the motion of dislocations and, strictly speaking, its density field. We provide examples of the numerical simulation of some fracture sequences of the in-plane (edge dislocation) and anti-plane (screw dislocation) motions. Furthermore, we present the most characteristic examples of soliton applications to problems of fracture physics. We then end this part with a canonical approach to the theory of asymmetric continua. Our considerations, starting with Hamilton's Principle, lead us to the equations of motion formulated in terms of the spin structure and spin rates.

In the last part, entitled *Deformations in Riemannian Geometry*, we pass to the more advanced descriptions for the continua, basing the discussion on a number of theories presented in the Riemannian and other generalized spaces, which may include the intrinsic asymmetric properties. This approach can be treated as a natural introduction to nonlinear physics. We review possible analogies between an elastic continuum with a continuum defect distribution and the Einstein-Cartan theory, paying the most attention to the analogies between dynamical concentrations of continuous fields (disclinations and dislocations) and the tensors of curvature and torsion in the Riemannian spaces extended by torsion. We attempt to relate the defect densities to spinors in the spin-spaces, arriving at the disclination and dislocation spinors. Then, we employ twistors to describe spin and twist solitons, and we formulate Maxwell-like equations for spin and twist motions in terms of spinors. Finally, we consider how the antisymmetric potentials for the spin and twist can lead us to formulate a complex description of motions, with the asymmetric perturbations applied only to the metric tensor. This approach permits us to present the relevant relations in the frame of the classical Riemannian space.

We rely also on the numerous analogies between a continuum theory approach and General Relativity. With regard to such analogies, we recall, first of all, the Cartan works (Cartan 1923, 1924, 1925) as being influenced by the work of the Cosserat brothers (1909) on a generalized continuum in which a moment stress tensor is included. Therefore, we dare to extend the obtained results for the electromagnetic potentials. In this way we define the antisymmetric perturbations to metric tensor, and then we come to the relevant relations for the Complex General Relativity in the frame of the classical Riemannian space.

On a final note, our treatise in the first and second parts is presented in a rather accessible form and suggests some possible future applications. The related considerations and results may be recommended for researchers in physics, geophysics, material sciences, and for advanced students as well. A more advanced approach is used in the last part; however, we point out that the applications of differential geometries in physics have become extremely enlightening for the fundamental understanding of nonlinear processes.

Roman Teisseyre

Contents

Contributors

Wojciech BORATYŃSKI
Institute of Geophysics, Polish Academy of Sciences
ul. Księcia Janusza 64, 01-452 Warszawa, Poland
and
Faculty of Mathematics and Information Science
Warsaw University of Technology
Plac Politechniki 1, 00-661 Warszawa, Poland

Marek GÓRSKI
Institute of Geophysics, Polish Academy of Sciences
ul. Księcia Janusza 64, 01-452 Warszawa, Poland

Vladimir GRAIZER
California Division of Mines and Geology
Strong Motion Instrumentation Program
801 K Street, MS13-35, Sacramento, California 95814, USA

Leszek R. JAROSZEWICZ
Institute of Applied Physics, Military University of Technology
ul. gen. Kaliskiego 2, 01-908 Warszawa, Poland

Eugeniusz MAJEWSKI
Institute of Geophysics, Polish Academy of Sciences
ul. Księcia Janusza 64, 01-452 Warszawa, Poland

Hiroyuki NAGAHAMA
Department of Geoenvironmental Sciences
Graduate School of Science
Tohoku University, Aoba-ku, 980-8578, Japan

Krzysztof P. TEISSEYRE
Institute of Geophysics, Polish Academy of Sciences
ul. Księcia Janusza 64, 01-452 Warszawa, Poland

Roman TEISSEYRE
Institute of Geophysics, Polish Academy of Sciences
ul. Księcia Janusza 64, 01-452 Warszawa, Poland

Mihailo D. TRIFUNAC
Department of Civil Engineering,
University of Southern California
Los Angeles, California 90089-2531, USA

Jan WISZNIOWSKI
Institute of Geophysics, Polish Academy of Sciences
ul. Księcia Janusza 64, 01-452 Warszawa, Poland

PART I

INTRODUCTION TO ASYMMETRIC CONTINUUM AND EXPERIMENTAL EVIDENCE OF ROTATION MOTIONS

1 Introduction to Asymmetric Continuum: Fundamental Point Deformations

Roman Teisseyre and Marek Górski

Institute of Geophysics, Polish Academy of Sciences
ul. Księcia Janusza 64, 01-452 Warszawa, Poland
e-mail: rt@igf.edu.pl

1.1 Introduction

A more general approach to continuum mechanics is inspired by the asymmetric theory; therefore, we first consider some arguments forming the basis to constitute such an asymmetric continuum theory based on asymmetric stresses:

I. Already when studying the elastic field of an edge dislocation we find some asymmetry in its components; in confrontation with symmetry of shears, this fact results in the asymmetry of stresses for a continuous distribution of dislocations (for screw dislocations such a contradiction does not exist). Therefore, a direct differential relation between any density of dislocations and the related stresses cannot be adequately found in a symmetric continuum.

II. Fracture reveals usually its asymmetric pattern relative to the main slip plane; we believe that the premonitory processes, as described by deformations in a continuum with defects, develop also in an asymmetric way.

III. In the classical continuum, the balance of angular momentum holds only if the stresses are symmetric; the angular motions can be introduced only artificially with the help of a length element and a reference rotation point. This classical theory has also many other limitations and therefore many trials have been undertaken to generalize it. The asymmetric theory of elasticity with asymmetric stresses and couple-stresses has been founded by Nowacki (1986). However, a first generalization to include the moments in a continuum is due to Voigt (1887) and a complete theory including asymmetry of stress and strain is that known as the Cosserat theory of elasticity with the displacement vector and rotation vector (E. Cosserat and F. Cosserat 1909). Micropolar and micromorhic elastic theories have been developed by Eringen and his co-workers (see: Eringen

1999). Teisseyre (2005, 2006) proposed a simpler version of the asymmetric theory which includes the asymmetric stresses, strains and rotations, but in which the equations for the antisymmetric stresses differ from that of couple moments in Nowacki's theory; their roles are interchanged, but both systems remain almost equivalent.

IV. Usually, when searching for the fault slip solutions we rely on classical elasticity with the friction constitutive laws introduced according to the experimental data. The obtained equations do not theoretically apply to elastic continuum that includes any distribution of defects as objects with their own stress field; nevertheless, the obtained results well explain the observational data. We consider an elastic continuum with asymmetric stresses and defects. This approach permits to study the defect interactions and elastodynamic solutions describing slip propagation along a fault.

V. The asymmetry of fields follows also from the notion of antisymmetric stresses introduced by Shimbo (1975, 1995) and Shimbo and Kawaguchi (1976) and related to the friction processes and rotation of grains. Fracture processes develop usually along the main fault plane; hence, there appears the initial asymmetry of the fracture pattern (Teisseyre and Kozak 2003); due to friction, the rotation of grains adjacent to the main slip plane gives rise to the antisymmetric part of stresses and twist deformations. According to Shimbo (1975, 1995) we introduced the constitutive law joining the antisymmetric stresses with the rotation nuclei (self-rotation field); without such a constitutive law, any theory reduces the rotation motions and waves (except the displacement rotation) to zero.

VI. In the asymmetric continuum, defined as that including both the symmetric stresses and the antisymmetric stresses, there appear also the rotational motions/deformations which split into the pure spin and twist motions, the latter relating to the shear deformations of grains; when considering the point like nuclei, the twist deformation passes to 3D space torsion (Riemannian space).

VII. Experimental evidence discovering an appearance of the spin and twist motions in a seismic field is based on the records of seismic rotation fields obtained both with the help of the ring laser or fiber optics interferometers, based on the Sagnac principle (Cochard et al. 2006, Schreiber et al. 2006, Takeo 2006, Jaroszewicz et al. 2006) and by the rotation seismographs (Moriya and Teisseyre 2006, Wiszniowski 2006). For spin motion we shall be aware of the fact that the recorded rotation contains two elements: rotation of displacements and independent spin motion. Both these elements, according to experimental evidence, appear in different proportions depending on the source processes and propagation conditions. New observations have been obtained due to the modern instrumentation techniques; the results are in the form of the rotation wave seismograms. The first one is probably that achieved in an indirect way (Teissyere 1973); the azimuth array of horizontal seismographs, installed in one of coal mines in Upper Silesia to record the very near tremors, permitted to deduce the rotational component of motions. The really first rotation seismograms were obtained at the two fundamental geodetic stations in Wetzell, Germany, and in Australia (Cochard et al. 2006, Schreiber et al. 2006) equipped with the ring-laser interferometers based on the Sagnac principle. Later on, sensors of another type – the fiber-optic interferometers – were used by Takeo (2006), specifically for the seismic observations; one of the versions of his sensors included the tri-axial system. Jaroszewicz et al. (2006) applied this system of interferometers for the study of Silesian seismic events. Using a more traditional approach, Moriya (see: Moriya and Teisseyre 2006) has constructed the first rotation seismograph system consisting of a pair of anti-parallel seismographs; such a system with the common suspension of the anti-parallel seismographs was subsequently used in later constructions (Wiszniowski 2006).

Data collected by the above-mentioned recording systems brought at least two important results:

- records of different events in the very near field indicate that some events, e.g., shallow volcanic and those of explosion type, differ from the common characteristics by their extremely small rotation components (Teisseyre et al. 2003);
- correlations between the rotation seismograms obtained from the ring-laser system and the rotation motions, rot **u**, derived from the array of seimometers (located at the same site) show almost perfect fit (Cochard et al. 2006).

The independent rotation field, e.g., the rotation related to grains or points of continuum, has been introduced by Shimbo (1975, 1995) in his considerations on the friction and fracturing processes. In such a way, the constitutive law joining the antisymmetric stresses and rotation was introduced

$$S_{[ks]} = 2\mu^{*} \omega_{ks} ,$$

which relates directly to the internal friction μ^{*}, called the rotation rigidity, as kind of material resistance.

Such a constitutive law is requested by the asymmetric theory; this constitutive law concerns any deformation in which the internal friction plays any role.

The Shimbo law has been later generalized for the spin and twist rotation motions (Teisseyre et al. 2006).

Thus, we cannot deny the existence of independent rotation waves; and the question is reduced to the magnitude of the rotation rigidity μ^*; when μ^* exceeds the modulus of rigidity, $\mu^* \gg \mu$, we may, of course, neglect these rotation waves, but in the reverse case, $\mu^* \ll \mu$, the pure rotation motions would prevail. In our considerations (see: Teisseyre 2006) we make the natural assumption $\mu^* \equiv \mu$, as the rigidity modulus relates to the transversal motions which are connected to rotations.

To summarize, we may note that the independent, pure rotation, $\omega_{[s]}$, can be presented by means of the potentials u_k^{micro} :

$$\in_{nsk} \omega_{[k]} = \frac{1}{2}\left[\frac{\partial}{x_n} u_s^{micro} - \frac{\partial}{x_s} u_n^{micro}\right].$$

Such potentials contribute to displacement field, as the rotations contribute to the displacement derivatives; reversely, the displacements may produce the rotation field; such a statement may appear as a far-going one as the values of displacement u_k^{micro} could be extremely small. However, we shall note that in the asymmetric theory the rotation motions would exist even in the case when we neglect displacement fields – such a case can be called a degenerated continuum.

Therefore, basing on the above reasoning, showing that the rotations may contribute to displacement derivatives and displacements may contribute to rotations, we can state that these motions are interrelated; this statement is empirically supported by the above mentioned almost perfect fit between the derived rotations, rot u, and the rotations obtained from the ring-laser system data. Therefore, we think that the problem related to the existence of rotation waves appears irrelevant. However, we shall stress that the displacements and rotation motions differ in general, especially when considering their origin and effects. Instead of the problem whether the rotation waves exist or not, we propose to consider the classification of rotation motions from the point of view of their origin, scale, and the effects produced.

We propose the following classification:

- **the micro-rotations** or rotations, ω, as related to the wave motions based on the internal friction processes (rotation rigidity), as well as to slip motions with friction/fracture processes;

- **the meso-rotations** related to material granulation and formation of the mylonite zones under the shear load fracturing processes;
- **the total rotations**, ω^T (the nomenclature introduced by Kröner 1981) related to the displacement field, **u**;
- **the macro-rotations** as related to fragmentation of material at the fracturing under compression load;
- **the mega-rotation effects** related to the ground tilts and tilting of high objects on the ground.

The important role of rotational processes in mechanics of fracturing and the related energy release has been discussed in our former papers (Teisseyre 2006, Teisseyre et al. 2006). Both under confining pressure and under external shears the role of micro-fracturing in the bond breaking process is similar; however, we observe here the essential differences for rotations in larger scales. The confining condition leads to the formation of induced opposite arrays of dislocations, resulting in fragmentation processes and related macro-rotations. On the other hand, shear load leads to more concentrated fracturing along some planes. We have underlined, in the cited papers, that macro-rotation processes and related energy release is more effective for the fracturing under confining pressure.

These considerations lead us to the question of classification of basic motions.

Worth mentioning here is the fundamental contribution to the rotation motions as caused by earthquakes, presented in the paper by Cochard et al. (2006); the described simulations and comparison of the rotation of displacement field, as computed from the array seismometer system, excellently fit to the spin motion measured by means of modern ring laser technology. We shall note that this well proved agreement has been obtained using the pass band filtering, 0.03-0.08 Hz, limiting results to the surface wave trails. However, the near source effects, including friction processes at cracks, and an influence of medium structure and its defects, may contribute to the formation of an independent rotation field; the question how big such effects could be remains still open.

For seismic source processes, the independent rotation, the spin motion, clearly overpasses rotation of displacements, however we shall also note some exceptions related to events of explosion nature or for some volcanic near surface events; here, the situation is reverse and both the pure spin motion and rotation of displacement are highly reduced. For strong motions which include a tilting component the rotation of displacements can reach high magnitudes. Same is observed in engineering seismology: the rotation of displacements is very high and this effect is due to the magnification of horizontal rotation components and to the appearance of rocking/tilting component caused by the geometry of construction, especially for high buildings.

1.2 Self-Field Nuclei: Deviations from Classical Elasticity

Any continuum could be described using the Kröner approach based on the concept of internal fields excited by a density of defects and internal nuclei; stresses and strains are related by the unique constitutive law, that for the ideal elasticity. This approach is equivalent to the other one in which we change the constitutive law in the way appropriate to describe the plastic, viscous and relaxation effects.

In the Kröner continuum with a density of internal nuclei, also treated as points, the elastic strains, rotations and stresses can be expressed as differences between total and self-fields (Kröner 1982). These elastic fields shall obey the constitutive laws valid in the ideal elasticity.

Frequently, the deviations from ideal elasticity are taken into account by the appriopriate elasto-plastic constitutive laws. However, following the Kröner approach (Kröner 1982), we maintain the ideal elastic law, and we introduce the self/inner stresses, strains and rotations as related to the internal nuclei or defects. The Kröner method allows to include the deviations from symmetry properties for stresses, strains and rotations. We also introduce additional constitutive laws joining the antisymmetric stresses with rotations; in such a way we include a role of the antisymmetric stresses which replace that of stress moments.

The asymmetric continuum includes the displacements and also rotations; the related balance equations split into those related to symmetric and antisymmetric parts (Teisseyre 2005).

In the present approach we rely on the standard asymmetric continuum theory in which we assume that strains remain symmetric and rotations antisymmetric for the symmetric and antisymmetric fields:

$$S_{kl} = S_{(kl)} + S_{[kl]}, \qquad E_{kl} = E_{(kl)}, \qquad \omega_{kl} = \omega_{[kl]} \tag{1.1}$$

and we can join these deformation fields, in an independent way, with displacement motions:

$$E_{kl} = \frac{1}{2}\left(\frac{\partial u_l}{\partial x_k} + \frac{\partial u_k}{\partial x_l}\right), \qquad \omega_{kl} = \chi^0 \frac{1}{2}\left(\frac{\partial u_l}{\partial x_k} - \frac{\partial u_k}{\partial x_l}\right) \tag{1.2}$$

where the structural index, x^0, is introduced to account for the phase shifts between shear and spin motions; $|x^0| = 1$.

We recall that any symmetric tensor can be split in a given coordinate system into the axial and deviatoric parts: the deviatoric tensor, $S^D_{(ik)}$, having zero value of its trace, and axial tensor, $S^A_{(ik)}$, are defined as

$$\mathbf{S}^D = \mathbf{S} - \frac{1}{3}\mathbf{I}\,\mathrm{Tr}\,(\mathbf{S}), \qquad \mathbf{S}^A = \frac{1}{3}\mathbf{I}\,\mathrm{Tr}\,(\mathbf{S}), \qquad \mathrm{Tr}\,(\mathbf{S}^D) = 0. \tag{1.3}$$

For the symmetric part of stresses we can take the classical constitutive relation (including its deviatoric form):

$$S_{(kl)} = \lambda\delta_{kl}E_{ss} + 2\mu E_{kl}, \qquad S^D_{(ik)} = 2\mu E^D_{ik}, \tag{1.4}$$

and we supplement it with the Shimbo law for the antisymmetric part

$$S_{[ik]} = 2\mu\omega_{ik}. \tag{1.5}$$

There is no problem with including here the appropriate linear deviations related to visco-plastic effects. In the classic elastic continuum, only displacement motions are taken into account, while any independent rotation motions are excluded because of the lack of appropriate constitutive laws supporting the existence of elastic response to the rotational deformations of bonds in a lattice network. The Shimbo law joins the friction related rotations with the antisymmetric stresses. We may also note that the rotation rigidity constant introduced, in the Shimbo law and related to bonds (inner friction) has been taken as equal to the rigidity modulus.

Finally, we shall underline that the body deformation is given by strain and rotation.

The experimental evidence for independent spin and twist motions appeared when analyzing the records of seismic rotation fields. The obtained experimental data clearly indicate that the independent spin and twist motion are detectable. As mentioned above, the twist deformations represent the grain deformations caused by elastic strain; however, when considering the grains as the "rigid" points of continuum, such a twist deformation converts to a kind of 3D space curvature (Teisseyre et al. 2005, Teisseyre et al. 2006).

The proper definition of the twist motion we will be given further on. Here, we can state that the antisymmetric stresses relate to rotational deformation and form the internal rotation nuclei, as, e.g., those appearing when micro-cracks are formed.

Antisymmetric stresses relate to an internal rotation motion; these stresses become important in materials with higher dislocation densities, under high stresses, or in zones where microfracture processes nucleate; in such zones, we can expect the presence of rotation nuclei.

The continuum mechanics found its basis in considerations on deformations caused by displacement field, including the moment of momentum and angular deformation. Only later, in the micropolar and micromorphic theories with the infinitesimally small nuclei or in other advanced continuum treatments, the spin motions, or say the angular fields, appear as independent variables. In the asymmetric continuum theories, the displacements and rotations may appear as equally and similarly treated independent fields. Starting with such theories as the basis, we can define the degenerated continuum as that in which the displacements vanish – do not exist; we remain confined to the rotational deformations only. Rotational deformations split into the pure rotation and twist/shear oscillations; in the micromorphic theories the latter can be also related to the shear deformations of the point-nuclei. In our generalized approach, there appear also the axial fields, with a structure similar to that of a thermal field, and meaning the axial deformations: expansion/compression. Deformations related to thermal field, and deformations of the internal nuclei appear in the micromorphic theories.

1.3 Basic Deformations and Simple Motions in an Asymmetric Continuum

Basic and simple motions could be defined as those which may be reduced to the 3D point motion in the Cartesian or Riemannian spaces or the deformations conceived as the respective curvatures.

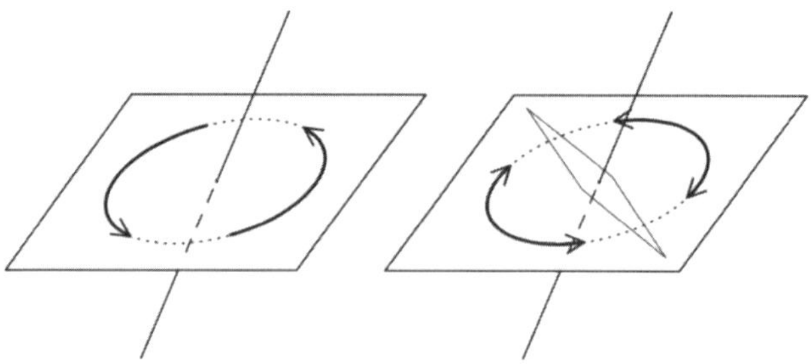

Fig. 1.1 Rotational motions: spin and twist (for explanation, see further on)

Considering basic motions we can distinguish here the simple motions, first the translations described by vector **u** and next the independent rotation, further on called spin; in a non-homogeneous continuum, the grains having different material parameters can rotate due to interaction of a displacement field with the moment of inertia of grains.

Then, we shall pass to the tensorial motions/deformations:

Any **antisymmetric tensor** $\omega_{[ks]}$ can be related to the vectorial field $\omega_{[i]} = \in_{iks} \omega_{[ks]}$, e.g., to spin motion; thus, we come again to the equivalent vector field. Theoretically, this simple spin motion, $\omega_{[i]}$, could be treated as independent of the displacement rotation, however both contribute to the total spin field as observed, e.g., in seismology,

$$\omega^{0}_{[ks]} = \frac{1}{2}\left(\frac{\partial u_s}{\partial x_k} - \frac{\partial u_k}{\partial x_s}\right) + \omega_{[ks]}. \tag{1.6}$$

Any **symmetric tensor** can be split into the axial and deviatoric tensors (Eq. 1.3); the axial field represents the point deformation (compression/dilatation nucleus, e.g., related to thermal anomaly), see Fig. 1.2. These axial oscillation motions include the equal translation motions which, for both elastic or thermoelastic continua, may be directly related to the incompressibility moduli K, or to thermal expansion coefficients. With the axial formations and for the point-like continuum, we obtain, respectively, the Riemannian curvature and, for the last case, the more complicated Riemannian torsion tensor.

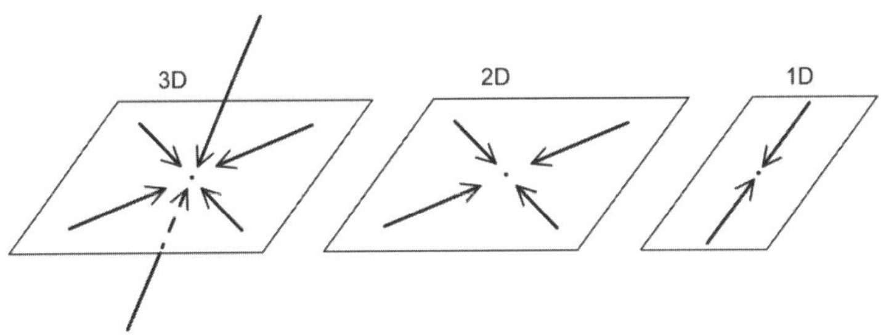

Fig. 1.2 Axial basic deformations (3D, 2D and 1D)

There remains the deviatoric field; in the limit related to the point-like deviatoric deformations presented in the system of off-diagonal axes we will have the string-string type (Fig. 1.3) and the string-membrane type (Fig. 1.4) deformations.

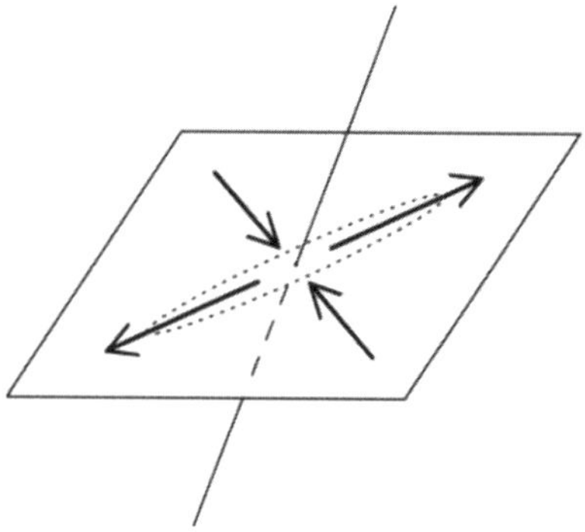

Fig. 1.3 String-string basic deformation

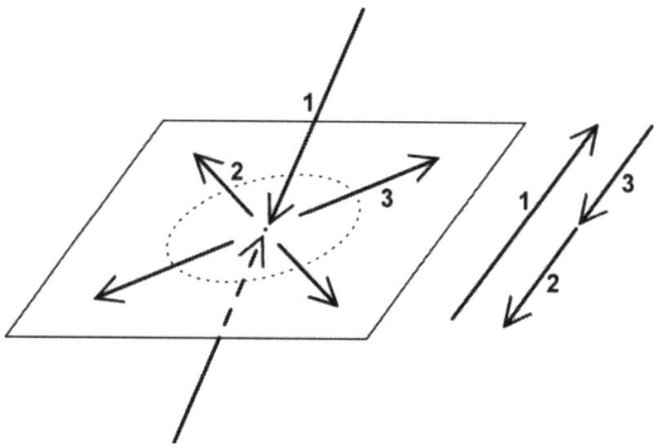

Fig. 1.4 String-membrane basic deformation

However, it is possible to show that these deviatoric deformations may be used to define a new antisymmetric tensor related to the simple deformations representing another kind of the rotations motion – the twist deformation, $\omega_{(i)}$. Such a deformation can be related to pure shear oscillations (Fig. 1.5) related to the oscillation of the off-diagonal shear axes as perturbated by dynamical and fracture processes. In a system related to the off-diagonal axes, we can present a shear deformation in the form:

$$S_{(ik)}^{D} = 2\mu E_{ik}^{D}, \qquad E_{(ks)}^{D} = \in_{ksi} \omega_{(i)}. \tag{1.7}$$

In the next chapters (Chaps. 6 and 7) we will discuss an invariant definition of the twist motion $\omega_{(i)}$.

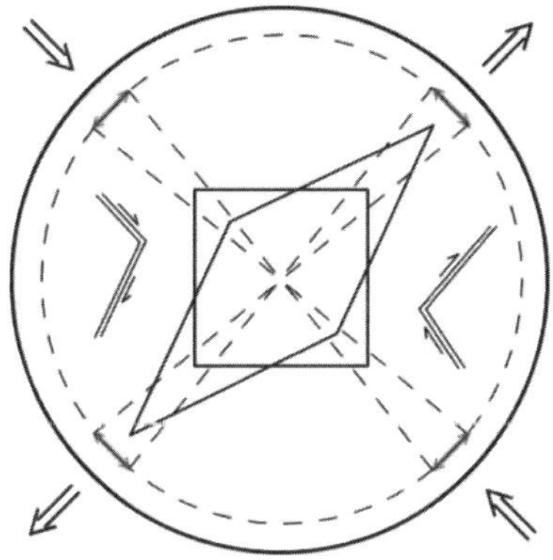

Fig. 1.5 Twist motion: rotational oscillations of the off-diagonal shear axes and internal fracturings as sources of perturbations

The spin and twist fields contribute to the outcoming asymmetric rotation field:

$$\omega_s = \omega_{[s]} + \omega_{(s)}. \tag{1.8}$$

We may notice that in the above formulae the displacement field might be presented by the equivalent vector potential field:

$$u_k = \lambda^2 \in_{ksn} \frac{1}{2}\left(\frac{\partial \tilde{\omega}_n}{\partial x_s} - \frac{\partial \tilde{\omega}_s}{\partial x_n}\right)$$

where λ will be the intrinsic length parameter of the continuum; in the extreme case it is equal to the Planck length unit.

1.4 Conclusions

This chapter is based on an advanced approach to the theory of degenerated continuum, and relates to its sources in our former studies, as well as in the recent monograph "Earthquake Source Asymmetry, Structural Media and Rotation Effects" (Springer 2006) and to our other former papers; however, we introduce here a new theoretical approach based on the standard continuum theory.

We could consider the mechanics based exclusively on rotation motions as the exactly opposite case to the classic elastic continuum in which only displacement motions are taken into account, while any independent rotation motions are excluded. In fact, the independent rotation motions are excluded only due to the lack of appropriate constitutive laws supporting the existence of an elastic response to the rotational deformations of bonds in a lattice network. In the asymmetric continuum we included the bonds joining asymmetric stresses with spin and twist motions.

The twist motions were defined as those related to oscillations of shear type; we shall remember that the notion of twist field appeared from an analysis of the records of seismic rotation fields.

In the degenerated mechanics, in which we confine ourselves to rotation motions, excluding that related to displacements, we may describe any process, phenomenon, body property, or shift and position, which is related to some length element l, by the help of the "arm spin" θ defined as:

$$\mathrm{tg}\,\theta = \frac{l - l_0}{l_0}$$

where by this expression we can reduce any length element l to the angle θ, "arm spin" and to a certain basic reference unit l_0 (e.g., for point-like continuum to the Planck length unit).

We shall note that the seismic spin and twist motions can be measured with the help of the ring laser or fiber optics interferometers, based on the Sagnac principle and by the strainmeters, or partly by the rotation seismographs. The obtained experimental data clearly indicate that the independent spin and twist motions are detectable, forming the counterparts of displacement rotations and shear deformations.

As mentioned above, the shear deformations can be presented in the main axes system (point-grain deformation) or in the off-diagonal system (twist motion); however, when considering the grains as "rigid" points of continuum, such deformations convert the Euclidean space into a kind of a 3D space torsion curvature. We may present such geometrical features related to the Riemannian geometry aspects in a schematic way as shown in Fig. 1.6.

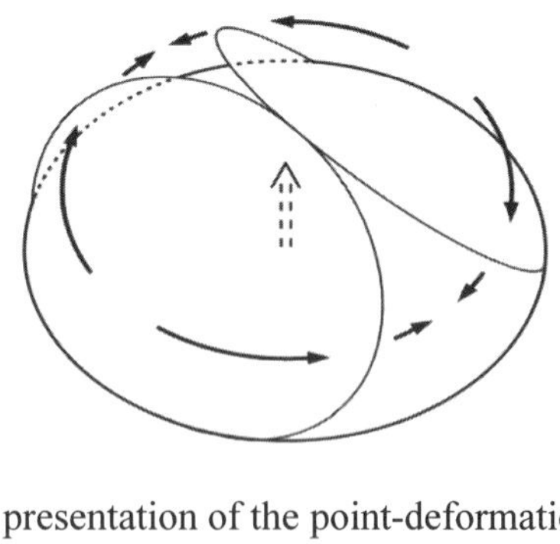

Fig. 1.6 Schematic presentation of the point-deformations converting the Euclidean space into the Riemannian space with 3D space torsion and curvature

In the above considerations, we have defined the simple basic motions and point-like deformations: translation was supplemented with independent spin and twist motion and, furthermore, the complex pure axial motions.

Concluding, we defined two basic independent motions: displacement u_i and spin $\omega_{[i]}$ and the basic deformations: twist/shear $\omega_{(i)}$ and the axial deformations E^A_{kk}.

In this chapter we have considered a total of the ten basic motions: u_n, $\omega_{[n]}$, $\omega_{(m)}$ and dilatation/compression as given by traces of deformation.

References

Cochard A, Igel H, Schuberth B, Suryanto W, Velikoseltsev A, Schreiber U, Wassermann J, Scherbaum F, Vollmer D (2006) Rotational motions in seismology: theory, observation, simulation. In: Teisseyre R, Takeo M, Majewski E (eds) Earthquake source asymmetry, structural media and rotation effects. Springer, Berlin, pp 391-411

Cosserat E , Cosserat F (1909) Theorie des corps déformables. A. Hermann et Fils, Paris

Eringen AC (1999) Microcontinuum field theories. Springer, Berlin

Jaroszewicz LR, Krajewski Z, Solarz L (2006) Absolute rotation measurement based on the Sagnac effect. In: Teisseyre R, Takeo M, Majewski E (eds) Earthquake source asymmetry, structural media and rotation effects. Springer, Berlin, pp 413-438

Krőner E (1981) Continuum theory of defects. In: Balian R, Kléman M, Poirer JP (eds) Physics of defects (Les Houches, Session XXXV, 1980). North Holland, Amsterdam, pp 217-315

Moriya T, Teisseyre R (2006) Design of rotation seismometer and non-linear behaviour of rotation components of earthquakes. In: Teisseyre R, Takeo M, Majewski E (eds) Earthquake source asymmetry, structural media and rotation effects. Springer, Berlin, pp 439-450

Nowacki W (1986) Theory of asymmetric elasticity. PWN-Pergamon Press, Warszawa, Oxford

Schreiber KU, Stedman GE, Igel H, Flaws A (2006) Ring laser gyroscopes as rotation sensors for seismic wave studies. In: Teisseyre R, Takeo M, Majewski E (eds) Earthquake source asymmetry, structural media and rotation effects. Springer, Berlin, pp 377-390

Shimbo M (1975) A geometrical formulation of asymmetric features in plasticity. Bull Fac Eng, Hokkaido Univ **77**: 155-159

Shimbo M (1995) Non-Riemannian geometrical approach to deformation and friction. In: Teisseyre R (ed) Theory of earthquake premonitory and fracture processes. PWN, Warszawa, pp 520-528

Shimbo M, Kawaguchi M (1976) A note on the asymmetric fields. Bull Fac Eng, Hokkaido Univ **80**: 75-79 (in Japanese)

Takeo M (2006) Ground rotational motions recorded in near-source region of earthquakes. In: Teisseyre R, Takeo M, Majewski E (eds) Earthquake source asymmetry, structural media and rotation effects. Springer, Berlin, pp 157-167

Teisseyre R (1973) Earthquake processes in a micromorphic continuum. Pure Appl Geophys **102**: 1, 15-28

Teisseyre R (2005) Asymmetric continuum mechanics: deviations from elasticity and symmetry. Acta Geophys Pol **53**: 2, 115-126

Teisseyre R (2006) Asymetric continuum and anisotropy. Acta Geophys **54**: 3, 225-238

Teisseyre R, Białecki M, Górski M (2005) Degenerated mechanics in homogenous continuum: Potentials for spin and twist. Acta Geophys Pol **53**: 3, 219-231

Teisseyre R, Białecki M, Górski M (2006) Degenerated asymmetric continuum theory. In: Teisseyre R, Takeo M, Majewski E (eds) Earthquake source asymmetry, structural media and rotation effects. Springer, Berlin, pp 43-56

Teisseyre R, Górski M, Teisseyre KP (2006) Fracture-band geometry and rotation energy release. In: Teisseyre R, Takeo M, Majewski E (eds) Earthquake source asymmetry, structural media and rotation effects. Springer, Berlin, pp 169-184

Teisseyre R, Kozak J (2003) Considerations on the seismic rotation effects. Acta Geophys Pol **51**: 3, 243-256

Teisseyre R, Suchcicki J, Teisseyre KP, Wiszniowski J, Palangio P (2003) Seismic rotation waves: basic elements of theory and recording. Ann Geophys **46**: 4, 671-685

Voigt W (1887) Theoretische Studien über die Elasticitätsverhältnisse der Krystalle. Abh Ges Wiss, Gottingen, **34**

Wiszniowski J (2006) Rotation and twist motion recording – couple pendulum and rigid seismometers system. In: Teisseyre R, Takeo M, Majewski E (eds) Earthquake source asymmetry, structural media and rotation effects. Springer, Berlin, pp 451-470

2 Measurement of Short-Period Weak Rotation Signals

Leszek R. Jaroszewicz[1] and Jan Wiszniowski[2]

[1]Institute of Applied Physics, Military University of Technology
ul. gen. S. Kaliskiego 2, 01-908 Warszawa, Poland
e-mail: jarosz@wat.edu.pl

[2]Institute of Geophysics, Polish Academy of Sciences
ul. Księcia Janusza 64, 01-452 Warszawa, Poland
e-mail: jwisz@igf.edu.pl

2.1 Definition of Rotation and a Review of the Measurement Methods

The term rotation has several meanings and relates to various topics. Generally, it is used to mean (in three-dimensional space) the rotation movement of a rigid body in such a way that any given point of that body remains at a constant distance from some fixed point.

In seismology, rotation means mainly a curl of a spatial vector field of displacements **u**. Hence, it can be defined as the limit of a ratio of the surface integral (over a close surface S) of the cross product of **u** with the normal **n** of *S*, to the volume *V* enclosed by the surface *S*, as the volume goes to zero:

$$\operatorname{curl}\mathbf{u} = \lim_{V\to 0}\left(\frac{1}{V}\int_{\Sigma}\mathbf{n}\times\mathbf{u}\,d\mathbf{S}\right). \tag{2.1}$$

In the Cartesian coordinates *x*, *y*, *z*, rotation is given by the following formula:

$$\operatorname{curl}\mathbf{u} = \left(\frac{\partial u_z}{\partial y}-\frac{\partial u_y}{\partial z}\right)\mathbf{e_x} + \left(\frac{\partial u_x}{\partial z}-\frac{\partial u_z}{\partial x}\right)\mathbf{e_y} + \left(\frac{\partial u_y}{\partial x}-\frac{\partial u_x}{\partial y}\right)\mathbf{e_z}\,, \tag{2.2}$$

where $\mathbf{e}_i$ are unit vectors of each coordinate.

We can also say that the component of rotation of **u** in the direction of unit vector **n** is the limit of a line integral per unit area of the surface *S* over a closed curve *C* which encloses surface *S*, where **n** is the normal of *S*:

$$(\operatorname{curl}\mathbf{u},\mathbf{n}) = \lim_{S\to 0}\frac{1}{S}\int_C \mathbf{u}d\mathbf{r}. \tag{2.3}$$

The notation $(\mathbf{a}, \mathbf{b})$ means a scalar product of vectors **a** and **b**. The Stokes theorem is related to rotation. It says that

$$\oint_C \mathbf{u}d\mathbf{r} = \int_\Sigma \operatorname{curl}\mathbf{u}\times d\mathbf{S}. \tag{2.4}$$

Formulas (2.2) and (2.3) suggest two approaches to the rotation measurement. In the first approach, we need to determine partial differences of movement as an approximation of partial derivatives. Components of the rotation vector can be then computed by subtraction of proper partial derivatives. The second approach is based on measurement of an integral or sum of projections of movements along a closed curve.

There are two ways to apply the gradient method. The most widely applied procedure is based on the measurement of the ground motion displacement[1] in various but specified points and in specified direction; the ratio of differences in signals recorded simultaneously from appropriate seismometers to the distance between the seismometers is then computed. To estimate one component of rotation, e.g. vertical, four seismometers are required. In we have more seismometers than required, the rotation can be calculated directly from recorded signals. Saito (1968) suggests to estimate strain and vertical rotation as a weighted sum of horizontal components, where coefficients in the sum will be determined by a set of polynomials. The measurement of rotation from arrays of seismometers was made by Saito (1968), Gomberg et al. (1999), Bodin et al. (1997), Suryanto et al. (2006), and Huang (2003). In most experiments, seismometers cover an area, but there were also such experiments in which the seismometers were put in line in the direction of the event. In this case, the assessment was simple, because the event was an explosion.

The second approach to gradient measurement is to transmit the displacement in one point to a second point by a rigid bar (Aki and Richards 1980, Smith and Kasahara 1969) or by a laser beam (Duncan 1986). Then

[1] In most cases, instead of displacement, the measurement is made of velocity of displacement (by a seismometer) or acceleration of displacement (by an accelerometer). Similarly, an application of the Sagnac effect gives the velocity of rotation. However, we use the terms displacement and rotation, because the velocity measurement does not change the essence of meaning. However, while describing specific measurements the real measured quantities will be given.

the gradient is measured as a distance between the end of the bar and the second point. This method is applied mainly to record strain.

The measurement of the rotation based on Eq. (2.3) was made by seismometers placed on the circle or other closed curve. The seismometers were oriented towards the recorded movement, tangential to the curve. (Droste and Teisseyre 1976). Then the integral is approximated by the sum of the signal from the seismometers. The gradient methods can also be treated as an approach based on Eq. (2.3). They need the assumption of the strain tensor. It can change slightly on segments that connect sensors (Wiszniowski 2006). For the measurement of rotation it is enough to have three seismometers in a triangle; however, in order to ensure the fulfillment of constant strain tensor between seismometers and because of differences in the seismometer's responses, it is better to use a greater number of seismometers distributed on a regular polygon. When the distribution is arbitrary, we must use scaling coefficients.

A next group of sensors to measure rotations based on Eq. (2.3) are magnetohydrodynamic sensors, MHD (Nigbor et al. 2007), which can be also categorized as measuring rotation based on Eq. (2.3).

The sensors based on Sagnac effect (Sagnac 1913) are ideal because they measure the real integral over a closed curve. We distinguish two such systems: ring- laser (Schreiber et al. 2001), and fibre-optic (Jaroszewicz et al. 2006). In both systems, the optical path length difference, ΔL, experienced by light propagating in opposite directions along the closed path is detected (Post 1967):

$$\Delta L = \frac{4}{c_0}(\mathbf{A}, \mathbf{\Omega}), \tag{2.5}$$

where **A** is the vector of the geometrical area enclosed by the wave path, c_0 is the velocity of light in vacuum, $\mathbf{\Omega}$ is the rotation vector. The investigation of the above formula leads to the two important conclusions. The first is that for a given resolution of measurement of the optical path length difference ΔL, a method for detecting very small values of the rotational speed $\mathbf{\Omega}$ is to enlarge the geometrical area **A**. The other conclusion is connected with scalar product of two vectors $(\mathbf{A}, \mathbf{\Omega})$ which shows that the system detects rotational component with axes perpendicular to the geometrical area enclosed by the wave path, and this axis can be positioned freely over this area.

Usually, the distance ΔL generated by the Sagnac effect is extremely small; for instance, the Earth rotation rate equal to 0.26 rad/h gives magnitude of ΔL equal to $9.7 \cdot 10^{-13}$ cm for $A = 100\ \text{cm}^2$. Hence, the above-mentioned ring laser and fibre-optic systems are a technical implementation of the loop interferometer for appropriate detection of so small or even smaller distances.

The ring-laser setup for measurement of ΔL, shown in Fig. 2.1a, is the loop interferometer (with triangular or square shape of the loop) and includes an optical amplifier within the resonator (Rosenthal 1962, Macek and Davis 1963, Killpatrick 1966). Such an amplifier enables to produce laser oscillation at f^q along the (q = +) and also (q = –) directions within the resonator (bottom part of Fig. 2.1a).

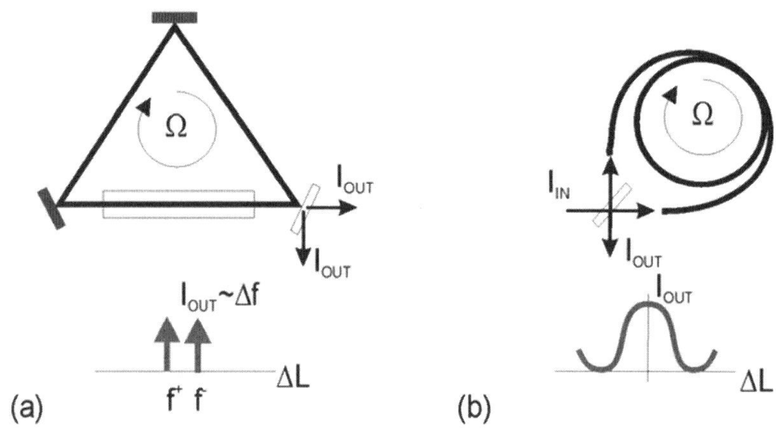

Fig. 2.1 Interferometric systems for Sagnac effect detection: (a) active method in ring-laser approach, (b) passive method in two-beam interferometer approach. Parameters I_{IN} and I_{OUT} are the intensities of input and output beams, respectively

In the presence of rotation Ω, we get the frequency difference, Δf, given by

$$\Delta f = f^{+} - f^{-} = \frac{4A}{\lambda P}(\mathbf{n}, \mathbf{\Omega}), \tag{2.6}$$

where λ is the optical wavelength of the laser oscillator, $\mathbf{n}$ is the normal vector to the laser beam plane and P is the perimeter enclosed by the beam path.

The advantage of the ring-laser method is that no external means are needed to measure Δf, since f^{+} and f^{-} are automatically generated within the ring laser and may be coupled out through one of the mirrors. To obtain Δf, one simply beats the f^{+} and f^{-} outputs outside the ring laser.

The ring-laser approach using a He-Ne amplifier (Aronowitz 1971) was the first successful optical gyroscope and is now being used in a number of civilian and military inertial navigation systems. The implementation of such a system for seismological research has been discussed by Schreiber (2006) which presents system G with square laser resonator ($A = 16$ m^2) and sensor resolution of $9 \cdot 10^{-11}$ rad/s$^{1/2}$ installed in the Geodetic Observatory Wettzell.

The fibre-optic version, named fibre-optic rotational seismometer (FORS) (Jaroszewicz et al. 2003), which uses the two-beam interferometer method, applies the fibre loop interferometer configuration (Vali and Shorthill 1976) with a 3 dB fibre coupler as input-output gate for optical beam (Fig. 2.1b). In such a system, a phase shift $\Delta\phi$ is produced between clockwise (*cw*) and counterclockwise (*ccw*) propagating light, given by

$$\Delta\phi = \frac{2\pi}{\lambda_0}\Delta L = \frac{8\pi A}{\lambda_0 c_0}(\mathbf{n}, \mathbf{\Omega}). \tag{2.7}$$

where λ_0 is the wavelength on the light in vacuum and $\mathbf{n}$ is the normal vector to the fibre loop plane.

The bottom part of Fig. 2.1b shows the cosinusoidal variation of the output intensity from this interferometer, I_{OUT}, as a function of Ω. Therefore, to measure Ω, we need to measure the change in I_{OUT}. In the case of a fibre interferometer, however, it is possible to loop the fibre many times (Vali and Shorthill 1976), say N times, before returning to the fibre coupler. In this case, Δt as well as ΔL become N times longer and the corresponding $\Delta\phi$ becomes

$$\Delta\phi = \frac{2\pi}{\lambda_0}\Delta L \cdot N = \frac{8\pi A \cdot N}{\lambda_0 c_0}(\mathbf{n}, \mathbf{\Omega}). \tag{2.8}$$

For a fibre of length L wound in a coil of diameter D, we have $A = \pi D^2/4$ and $N = L/\pi D$, so at last we get

$$\Delta\phi = \frac{2\pi LD}{\lambda_0 c_0}\Omega, \tag{2.9}$$

where Ω is rotation component in the axis perpendicular to fibre-optic loop. In other words, the sensitivity of the Sagnac interferometer in this approach is enhanced not only by increasing the physical sensor loop diameter but also by increasing the total length of the fibre used.

The fibre-optic approach using a classical fibre-optic gyroscope (Takeo et al. 2002, Jaroszewicz et al. 2003) were the first successful applications of such a system for seismological research. The next generation of this system, seismometer FORS-II installed in the Ojcow Observatory, Poland (Jaroszewicz et al. 2006) for the rotational events investigation had a resolution of $9.5 \cdot 10^{-9}$ rad/s$^{1/2}$, as a result of optimization of its sensor loop radius and the optical fibre length.

The presented approaches based on Eqs. (2.2) and (2.3) are equivalent providing that distances are as small as possible. Unfortunately, decreasing the area of surface in (2.3) lead to worsening of the signal to noise ratio for ring laser and fibre-optic systems. The area of fibre-optic can be enlarged by increasing the length L of the optical fibre but this can also increase the noise. Taking into consideration the fibre-optic system operation limited by short noise, the expected minimum value of detectable rotation, the so-called resolution, is (Ostrzyżek 1989):

$$\Omega_{\min} = 5.66 \frac{\lambda_0 \cdot c_0}{\pi \cdot D \cdot L} \cdot \frac{10^{\alpha L + \sigma} \sqrt{\frac{V_A^2}{R_o^2} + \frac{1}{8} eSP \cdot 10^{-\sigma - \alpha L}(1+X) + \frac{4kT}{R_o} + (i_A)^2}}{S \cdot J_1[2\phi_o \sin(\pi \cdot f_m \cdot \tau)]P}, \tag{2.10}$$

where k is the Boltzman constant, e the elementary charge, T the temperature, i_A and V_A are the amplifier noises (for current and voltage input sources, respectively), R_o the resistance of loaded photodiode, S the sensitivity of photodiode, α the optical fibre attenuation, σ the total optical losses without sensor loop losses, P the optical power of source, X describes the non-coherent source overflows noise, ϕ_o and f_m are amplitude and frequency of phase modulation, respectively, τ is the time delay of light during propagation through the loop, and J_1 is the Bessel function of the first kind.

As one can see, the maximum sensitivity of the system requires maximization of such parameters as: radius R of the loop, optical power P, length L of the used fibre; it also depends on wavelength λ and total losses of optical path σ. It should be noticed that the sensor loop length has the main influence on sensitivity. However, because with growing fibre length, the losses increase too, the optimum length is evaluated at about 12-15 km for the standard single-mode optical fibre at 1285 nm, as shown in Fig. 2.2 (Krajewski et al. 2005).

The distances measured by rigid bars become immeasurable and the differences, if the velocities are measured by seismometers, become lower than the inaccuracy of recording.

On the other hand, by increasing the area of surface in the Sagnac-effect sensors, we measure the mean value in the area not the rotation in a point. The question is how much the rotation we can averaged.

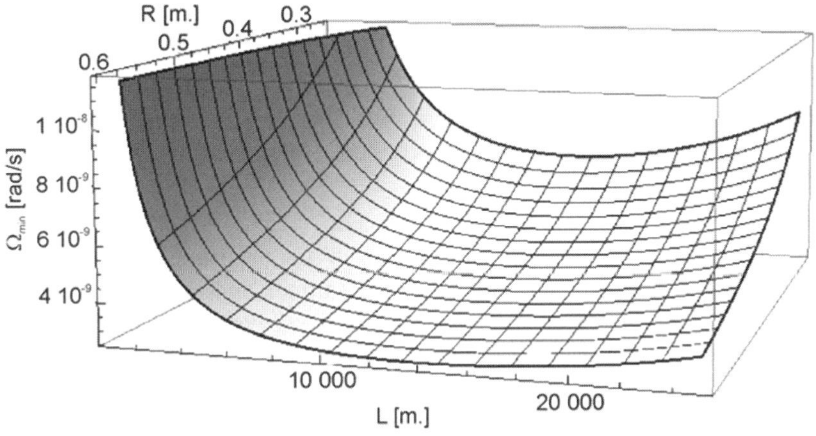

Fig. 2.2 Fibre-optic system resolution versus total optical length L and loop radius R in 1 Hz detection band. Parameters for simulation: $\lambda = 1285$ nm, $\alpha = 0.45$ dB/km, $\sigma = 15$ dB, P = 10 mW

The computation of rotation based on Eq. (2.2) assumes a small error of approximation of the partial derivatives by differences, and consequently small or possible to specify changes of strain tensor in the path between points and seismometers. The measurements of rotation were performed on the arrays of seismometers located at distances of: a few kilometers (Saito 1968); about 4 m and more for seismometers L28, and about 600 m for seismometers STS-2 (Gomberg et al. 1999); 72-139 m (Bodin et al. 1997); or about 500-1000 m (Suryanto et al. 2006). The frequency band of the rotational signal is limited by the distance between the seismometers. The shorter the distances between the seismometers, the higher the frequency of the signal that can be measured. Gomberg et al. (1999) wanted to measure a signal in band up to 8 Hz, whereas Suryanto et al. (2006) compared rotation recorded by ring laser to array-derived rotation in band 0.03-0.3 Hz.

This consideration concerned the continuous medium. Only this approach to rotation can be applied to the seismometers array. In many cases, rotation can be treated as a rotation of a rigid body. This is correct for small-size and small-distance sensors where we can assume that displacement and rotation are locally constant. In fact, almost all measurement instruments, except of the array of seismometers, measure the rotation of their own rigid chassis. The sensors that measure rotation of a rigid body can be categorized basing on the method of measurement as follows: fibre-optic gyroscopes, ring laser gyroscopes, piezoelectric gyroscopes (Nigbor 1994), hemispherical resonators gyros, tuning fork gyroscopes, vibrating wheel gyroscopes (Huang 1963, Farrell 1969), MHD sensors (Nigbor et al. 2007), and balanced pendulum sensors (Smith and Kasahara 1969, Ferrari 2006).

In order to determine rotation of a rigid body, it is enough to measure the displacement in two points, or even the displacement in one point if the body is the balanced pendulum.

The gradient measurement of rotation (Eq. 2.2) and the methods based on Sagnac effect (Eq. 2.3) was made also in a rigid trunk. Teisseyre et al. (2003), Moriya and Marumo (1998) and Bradner and Reichle (1973) applied seismometers placed in a rigid trunk. Teisseyre et al. (2002) used a rotational-seismograph system with two oppositely oriented independent seismometers, having pendulums suspended on a common axis, to record small earthquakes at Ojców Observatory, Poland, and L'Aquila Observatory, Italy. The structure of sensor and its coupling points was elongated. It makes conjectures that depend on the direction of foundation of the sensor. The measurement of vertical rotation in Ojców Observatory, Poland, and L'Aquila Observatory was performed by two sensors mounted perpendicularly – for seismometers in common. Bradner and Reichle (1973) presented a Sem and Lear instrument employing one normal and inverted pendulum as well as two back-to-back vertical seismometers to separate tilt from vertical displacement.

Nigbor (1994) measured rotations of ground during an underground chemical explosion experiment with a solid-state rotational velocity sensor based on Coriolis effect. The resolution of that sensor made in MEMS technology was not so good (about 0.1 mrad/s) and cannot be applied to week rotational signals in practice. Takeo (1998) recorded an earthquake swarm on Izu peninsula in Japan by sensors of the same type.

A next group of sensors exploited the principle of simple balanced pendulum. First, very simple instruments of this type were made by Jean de Hautefeuille in 1703, Nicola Cirillo in 1731, Andrea Bina in 1751 and at last Filippo Cecchi, who obtained the first records of rotation on smoked paper (see Ferrari 2006). The idea of balanced pendulum is still applicable to record rotations. The inertial rotation meter proposed by Smith and Kasahra (1969) is a balanced cross-shaped pendulum where rotation can be measured in four points at the ends of all arms of the cross. The rigid seismometer (Wiszniowski et al. 2003) is also a balanced pendulum designed as arms of two seismometers rigidly joined with each other. Both sensors can measure rotation in a few points although one point looks sufficient. However, measurement in many points allows us to eliminate such effects like oscillation of the axis of pendulum or springy vibration of the pendulum (Zadro and Braitenberg 1999).

The ratio of the inertia mass of the pendulum to the moment of inertia should be as low as possible to prevent the influence of displacement on the pendulum (see the sensor made by Filippo Cecchi in Ferrari 2006). Gyroscopic seismometers (Huang 1963, Farrell 1969) increase the moment of inertia by a built-in gyro. The noise of such a sensor was rather high. The noise of gyroscopic seismometer made by Farrell (1969) was equivalent to 5 μrad of ground tilt and 0.1 cm/s of ground velocity.

It has been noted in many articles (Droste and Teisseyre 1976, Bradner and Reichle 1973, Teisseyre et al. 2003, Trifunac and Todorowska 2001, Graizer 2006) that a typical unbalanced pendulum seismometer records the displacement and rotation simultaneously. The equation of motion of horizontal pendulum is:

$$\ddot{y} + 2\alpha\omega_0\dot{y} + \omega_0^2 y = -\ddot{u} + l_0\ddot{\phi} + \xi\ddot{u}_\perp, \tag{2.11}$$

where ω_0 is the circular frequency of free vibrations, α is the damping coefficient, ξ is an angle of deflection of the pendulum from its equilibrium position, ϕ is the vertical rotation, l_0 is the reduced length of the pendulum, u is the horizontal displacement of ground in seismometer direction whereas $u_\perp$ the horizontal displacement of ground orthogonal to u.

Wiszniowski et al. (2003) presented another approach to the signal recording by the pendulum. They show that it is possible to present a formula for recording the displacement component alone, without rotation. The pendulum seismometer is then equivalent to a seismometer with straight-line movement of inertial mass placed in the centre of inertia of a simple pendulum. This approach is better than the multi-seismometers recording of rotation.

The problem is with an internal deformation of the instrument and how the body of sensor is attached to the elastic medium of the earth. The rigidity of the base and its coupling to the earth is critical for such instruments (Smith and Kasahara 1969).

2.2 Classification of Rotation Measurements and Requirements for Recording Instruments

The measurement of rotations involves a wide range of problems:

(a) Near source rotational ground motions: Bouchon and Aki (1982) measured natural earthquake strike-slip by the stations put 1-20 km away from the fault strike with epicenter distance of 1-50 km. The recorded signal amplitude was about 0.1-1.2 mrad/s. Huang (2003) showed rotation with an amplitude of about 40-200 μrad/s recorded at a distance of 6 km from earthquake. Takeo recorded, in the near field, a rotation with amplitude of about 30 μrad/s (Takeo 1998) and 26 mrad/s (Takeo 2006). Recording of rotation in the near field allows us to learn more about the mechanism of a seismic event.

(b) Rotations connected with volcanoes eruptions. The amplitude of rotations recorded close to a volcano was tens of μrad/s (Moriya and Teisseyre 2006).

(c) Rotation measured during chemical explosion (Nigbor 1994) had rather big amplitude. Rotational signal recorded 1 km away from a 1 kton explosion had amplitude of about 138 mrad/s. The same recording system has not recorded a natural earthquake with $M = 3.5$ at a distance of 8 km from the hypocenter because the rotational signal did not exceed the noise of instruments.

(d) Engineering seismology (Zembaty 2006) is interested in recording rotations in the range of mrad and more.

(e) Measurement of tilt (Bradner and Reichle 1973, Graizer 2006, Bodin et al. 1997) recorded tilts with an amplitude of 5 μrad of waves from an earthquake with $Mw = 6.7$ at a distance of 311 km.

(f) Measurement of rotation of teleseismic waves (Pancha et al. 2000, Igiel et al. 2003, Schreiber et al. 2006). The recorded amplitudes of rotations are small, from nrad/s up to 400 nrad/s. These signals were measured by ring lasers.

(g) Measurement of rotations for identification and separation of waves enables better and more unique interpretation and identification of P versus SV versus SH wave components (Smith and Kasahara 1969) as well as separate Love from Rayleigh waves.

(h) Research into self-rotations in micromorfic continuum (Teisseyre and Nagahama 1999).

Based on the recording conditions, the measurements of rotation can be grouped into recording of strong rotations, as listed in points a-d (tens of μrad/s and more) and recording of very weak rotations and very small ratios of rotation to movement, as listed in points f-h. The measurement of rotations needs sensors with sensitivity less than 10^{-9} rad/s. Ring laser sensors (resolution of $9 \cdot 10^{-11}$ rad/s$^{1/2}$, Schreiber et al. 2006), fibre-optic sensors (resolution of $9.5 \cdot 10^{-9}$ rad/s$^{1/2}$, Jaroszewicz et al. 2006), and MHD sensors (resolution of $6 \cdot 10^{-6}$ rad/s$^{1/2}$, PMD Scientific Inc.) can record very low rotation signals. Unlike seismometers, these sensors are not sensitive to linear motions. The sensitivity of seismometers is the best, but there is a problem with separating the recording of rotation from linear motion because of different responses. The problem of the discrepancy of response of seismometers and homogeneity of the Earth's crust beneath the seismometers seems to be negligible at the long period of the signal (Saito 1968) but at the short period of interest to us a similar approach was unsuccessful (Smith 1966). Besides, some signals, like rotation waves, may be observed by the seismometers close enough to each other (Moriya and Teisseyre 1999). The quoted paper describes further the errors of recording in this situation and the way to reduce the errors.

2.3 The Influence of Recording Error on the Computed Rotation Signal

The most widely applied method of seismic rotation and strain waves and motions depends on the measurements of the ground motion velocity in various but closely situated points; then the signals recorded simultaneously from many seismometers are compared (Moriya and Teisseyre 1999, Teisseyre et al. 2003a). In optimal case, the responses of all seismometers to equal stimuli are equal, so the recorded differences correspond only to differences in the ground motions. However, this happens only in theory. In practice, there is a hidden equipment response non-equality in the differences between simultaneous records. In the seismic far field, differences in ground motion at various points are much smaller than this motion, thus any non-identity of responses of channels seriously spoils the results.

The strain tensor is

$$\delta\mathbf{u} = \left(\delta\mathbf{x}\cdot\nabla\right)\mathbf{u}, \tag{2.12}$$

where $\mathbf{x}$ is coordinate vector and $\mathbf{u}$ is vector of displacement. It consists of symmetric and antisymmetric parts (Aki and Richards 1980):

$$\delta\mathbf{u} = \mathbf{e}\,\delta\mathbf{x} + \frac{1}{2}\left(\operatorname{curl}\mathbf{u}\times\delta\mathbf{x}\right), \tag{2.13}$$

where $\mathbf{e}$ jest a symmetric strain tensor. The antisymmetric tensor describes the rotation, whereas deviatoric part of the tensor $\mathbf{e}$ represents twist.

Let $\mathbf{v}_i$ be a displacement recorded by the i-th seismometer

$$\mathbf{v}_i = (\mathbf{n}_i, \mathbf{u}_i)\mathbf{n}_i, \tag{2.14}$$

where $\mathbf{n}_i$ is the direction of i-th seismometer movement, whereas $\mathbf{u}_i$ is a displacement in the i-th seismometer site.

Because the scalar value is $v_i = (\mathbf{n}_i,\mathbf{u})$, the displacement difference for two seismometers, i and j, is

$$\delta v_{ij} = (\mathbf{u}_i, \mathbf{n}_i) - (\mathbf{u}_j, \mathbf{n}_j). \tag{2.15}$$

Additionally, it was so far assumed that

$$\mathbf{n}_i = \mathbf{n}_j, \tag{2.16}$$

(or equivalently for rotation $\mathbf{n}_i = -\mathbf{n}_j$). This allowed us to describe the measured difference as

$$\delta v_{ij} = (\mathbf{u}_i \pm \mathbf{u}_j, \mathbf{n}_i). \tag{2.17}$$

When the seismometers are placed perpendicular to the segment connecting them, as in the case of rotation measurement, we have (Teisseyre et al. 2003):

$$\mathbf{v} \perp \delta\mathbf{x}, \quad \text{or} \quad (\mathbf{v}, \delta\mathbf{x}) = 0. \tag{2.18}$$

We cannot measure the changes of volume, because when we equate **u** to **v**, then

$$\mathrm{Tr}\,\mathbf{e}\delta\mathbf{x} = 0. \tag{2.19}$$

When the seismometers are placed parallel to the segment connecting them, as in the case of strain measurement, we get

$$\mathbf{v} \parallel \delta\mathbf{x}, \quad \text{or} \quad \mathbf{v} \times \delta\mathbf{x} = 0, \tag{2.20}$$

which means that second term in (2.13) is always zero.

The recorded signal of displacement is processed by the response of seismometer and a recording device. The signal recorded by the *i*-th seismometer can be described in the Laplace domain by the formula

$$v_i(s) = \left(\mathbf{G}_i(s), \mathbf{u}_i(s)\right), \tag{2.21}$$

where $\mathbf{G}_i$ is the tensor of response of the *i*-th seismometer to displacement vector $\mathbf{u}_i$. Usually, the response is assumed as one-dimensional. Then

$$\mathbf{G}_i(s) = G_i(s)\mathbf{n}_i. \tag{2.22}$$

The STS-2 seismometer (Streckeisen 1995) is an example of tree-axial sensor whose response is multidimensional and formula (2.22) is not fulfilled. But the departure from that is minimal. Assuming a small difference in the recording conditions of the two seismometers, the difference of the two seismometers can be presented in the form

$$v_{ij}(s) = \bar{G}_{ij}(s)\left(\bar{\mathbf{n}}_{ij}, \delta\mathbf{u}_{ij}(s)\right) + \delta G_{ij}(s)\left(\bar{\mathbf{n}}_{ij}, \bar{\mathbf{u}}_{ij}(s)\right) + \bar{G}_{ij}(s)\left(\delta\mathbf{n}_{ij}, \bar{\mathbf{u}}_{ij}(s)\right), \tag{2.23}$$

where $\bar{G}_{ij}(s)$, $\bar{\mathbf{n}}_{ij}$, $\bar{\mathbf{u}}_{ij}(s)$ are the mean values, whereas $\delta G_{ij}(s)$, $\delta\mathbf{n}_{ij}$, $\delta\mathbf{u}_{ij}(s)$ are differences of the two values or vectors. The first term is a gradient of displacement, whereas the remaining two terms are a linear motion recorded due to the discrepancy of responses and positions of seismometers. Adopting the assumption (2.23), we can separately deal with the error of position and the error of response of the seismometer.

2.3.1 The error of the seismometers position

The directions of movement of two seismometers differs slightly.

$$\mathbf{n}_i = \mathbf{n}_j + \mathbf{c}_{ji}, \tag{2.24}$$

where $\mathbf{c}_{ji}$ is the position difference. For small values of $\mathbf{c}_{ji}$

$$\mathbf{n}_i \perp \mathbf{c}_{ij}, \quad \text{and} \quad \mathbf{n}_j \perp \mathbf{c}_{ij}. \tag{2.25}$$

The difference of recording of the same displacement signals by two seismometers ($\mathbf{u}_i = \mathbf{u}_j = \mathbf{u}$) will be

$$\delta v_{ij} = (\mathbf{u}, c_{ij}). \tag{2.26}$$

and will be correlated with the displacement recorded by perpendicular seismometers. This effect, named the cross-axis sensitivity, was previously shown by Graizer (2006), and Trifunac and Todorovska (2001). In their measurements they neglected the cross-axis sensitivity as relatively low.

The experiment to measure rotation and strain took place in observatory Książ, Poland, in 2006. Three horizontal seismometers, labeled 1, 2 and 3, were put in line at distances of 192 and 263 cm. The direction of movement of the seismometers pendulums agrees with the direction of the line and was perpendicular to the direction towards the expected waves from the Lubin Copper Mine Region, Poland. Additional couple of seismometers on a rigid basis, labeled 5 and 6, recorded signals in the wave direction. The two seismometers recorded the signal for estimation of rotation. The third seismometer recorded the signal for estimation of the strain perpendicular to the wave direction.

During the measurement we noticed that the difference of signals from two seismometers depends on a signal from the perpendicular seismometer (see Fig. 2.3). The relationship of two signals, s_j and s_i, can be described by the self correlation coefficient

$$\rho_{ij} = \frac{\sum s_i s_j}{\sqrt{\sum s_i^2 \sum s_j^2}}. \tag{2.27}$$

The correlation coefficient for the difference signal from seismometers 1 and 2 and the signal from seismometer 5 for an examined earthquake from the Lubin Copper Basin is shown in Fig. 2.3c. It can be explained by the difference of direction of seismometers 1 and 2. The differential signal equals 0.02 of perpendicular signal (Fig. 2.3d). It correspond to the angle error of ~1°.

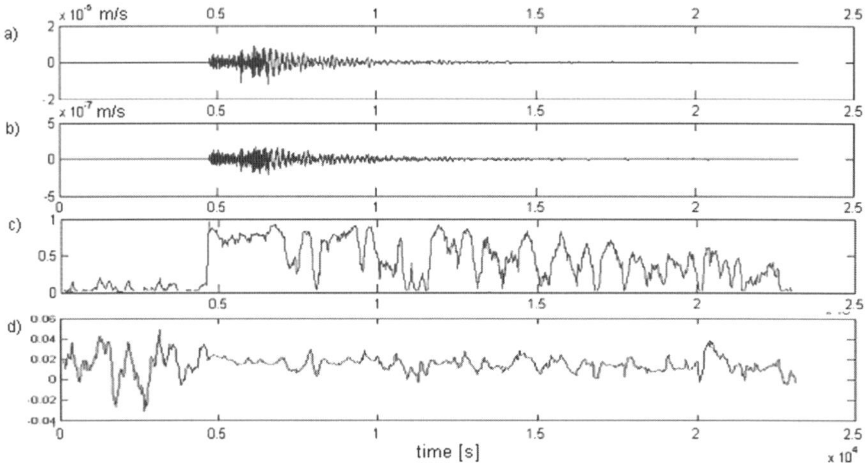

Fig. 2.3 Earthquake from the Lubin Copper Basin, 2006.07.28, 15:44, M = 3.3, recorded by a group of seismometers installed in the Książ seismic station: (a) velocity recorded by the seismometer installed in radial direction; (b) difference of velocity recorded by horizontal seismometers (1 and 2) installed in-line in transversal direction; (c) correlation between signals (a) and (b); (d) ratio of signal (b) to (a)

The signals were recorded by pendulum seismometers SM-3. The equilibrium position of such seismometer moves in time. This means that the recording of displacement direction changes during the measurement, and the position error cannot be corrected by more precise installation. The position of pendulum has to be systematically tested and corrected.

This problem does not occur when seismometers with strait movement of mass are applied. The force-balance seismometers do not have this problem either, because the zero position of the pendulum is forced by the electronic feedback. Muramatsu et al. (2001) describe a similar problem. They suggest to solve it by applying the coupled pendulum connected by a crossing wire. The difficulties with positioning the seismometers to work in the same direction with the accuracy less than 1° still remain.

It is easy to eliminate this correlation owing to the correlation with perpendicular movement. The problem is how to discriminate between the position error and the real strain or rotation. The correlation between the rotation or strain and the movement does exist. An example is a compressional wave along a thin rod.

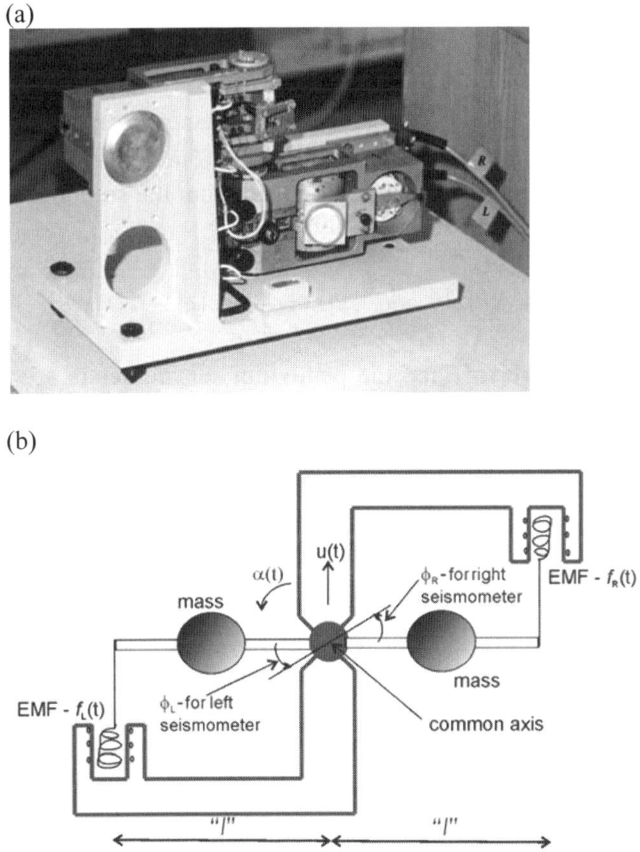

Fig. 2.4 The TAPS system – two antiparallel pendulum seismometers: (a) general view, (b) schematic

2.3.2 Error of seismometer responses and methods of correcting it

As a classical example of the above situation, let us consider an application of the rotational seismometer named TAPS (Teisseyre and Nagahama 1999). It is a set of two antiparallel pendulum seismometer (named left – L and right – R) situated on a common axis and connected in parallel, but with opposite orientations, as shown in Fig. 2.4.

In the case of ground motion containing displacements $u(t)$ and rotation $\alpha(t)$, the electromotive force EMF recorded by each simple seismometer, f(t), contains a component of displacement $\pm u$ and the rotation motion α multiplied by a proper length of pendulum l (Moriya and Teisseyre 1999):

$$f_{L,R}(t) = \pm u(t) + l \cdot \alpha(t), \tag{2.28}$$

where signs "+" and "–" are for R and L seismometers, respectively.

As one can see, in the case of two identical seismometers the rotational and translational components can be obtained from the sum and difference of the two recorded signals respectively as

$$\alpha(t) = \frac{1}{2l}\left[f_R(t) + f_L(t)\right] \quad \text{and} \quad u(t) = \frac{1}{2l}\left[f_R(t) - f_L(t)\right]. \tag{2.29}$$

Because, as a matter of fact, the pendulum seismometers are different, the special TAPS channels equalization algorithm for a clear rotation detection (Suchcicki et al. 2001) has been applied originally. Unfortunately, this procedure can be ineffective, especially if the TAPS seismometer components have different attenuation characteristics. In such a situation the existing finite sensitivities related to the signal sampling procedure used during the data recording generate errors in the signal (Jaroszewicz et al. 2003), as shown in the simulation presented in Fig. 2.5. In this simulation, the difference between the left and right seismometers attenuation $|\beta_L - \beta_R| = 0.05$ has been assumed. Moreover, the two seismometers should also be considered as elements with a different noise level.

As one can see, the main error signal exist in the region where the rotational events have small amplitude in comparison to the displacement. Because, in fact, it is the expected region of the rotational seismic event, the method of TAPS calibration is a crucial problem for credibility of its operation. Moreover, the extremely high sensitivity of the translational motions of the seismometers taken into account in their construction can limit the accuracy of such devices, too.

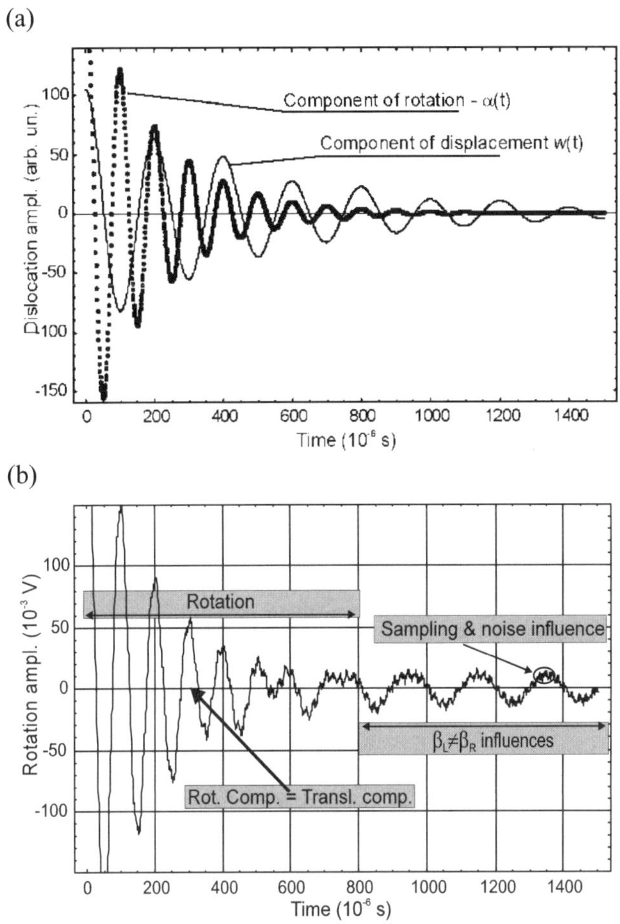

Fig. 2.5 (a) Simulated rotational and displacement components of a seismic event, and (b) rotation signal detected by the TAPS system (Jaroszewicz et al. 2003)

The experimental verification of the above consideration was a joint application of the TAPS and FORS-I systems where the latter is a fibre-optic rotational seismometer with sensitivity equal to 2.3×10^{-6} rad/s (for 2σ, where σ is the standard deviation of measured noise level) in the used 20 Hz detection band. The results presented in Fig. 2.6 show that the rotational signal obtained from the TAPS is fuzzed, whereas the signal from the FORS-I is very smooth. This results show, in the first, the advantage of direct method of rotation measurement by the FORS in comparison to the differential method realized by the TAPS. Secondly, obtained results suggest the necessity of searching for other methods of improving the TAPS performance.

(a)

(b)

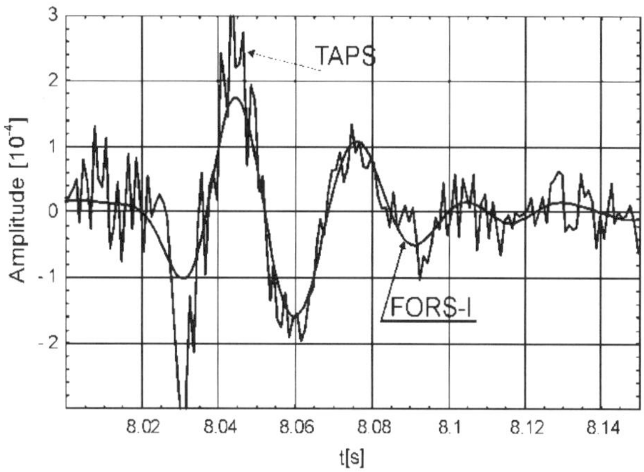

Fig. 2.6 (a) The rotation table with the TAPS (bottom box) and the FORS-I system (top box), and (b) output signals from the FORS-I and the TAPS after proper numerical processing (Jaroszewicz and Krajewski 2002)

One of the possible approaches is to apply the filtering procedure in the FFT domain (Teisseyre et al. 2002) or the time-domain (Nowożyński and Teisseyre 2003). The precise estimation of the filter is important and difficult. Parameters of such filters are estimated on the basis of some data recorded previously from the same seismometers but these methods use the so-called test positioning of TAPS (the seismographs of the system are turned so as to make them situated in the parallel-parallel position), that generally changes the conditions of the TAPS operation. The other procedure of the recorded data processing proposed by Solarz et al. (2004) based on smoothing by the spline functions (Kojdecki 2002, Eubank 2000).

The recorded digital data $\mathbf{Y} = \{Y_i,\ i = 0, \ldots, N\}$ with sampling at Δt is smoothed by the spline function:

$$S(t) = a_j\tau^3 + b_j\tau^2 + c_j\tau + d_j, \qquad j\Delta t \le t \le (j+1)\Delta t, \qquad \tau = t - j\Delta t, \qquad j = 0, \ldots, N-1. \tag{2.30}$$

In this way, the functional:

$$F[S] = p\int_0^{N\Delta t}\left[S''(t)\right]^2 dt + \sum_{i=0}^{N} p_i\left[S(i\Delta t) - Y_i\right]^2, \qquad p \ge 0, \qquad p_i > 0, \tag{2.31}$$

reaches its minimum. It should be emphasized that there exists a relation between parameter p of the above functional and mean-square error ε (Kojdecki 2002) defined as

$$\varepsilon = \sqrt{\frac{1}{N+1}\sum_{i=0}^{N} p_i\left[Y_i - S(i\Delta t)\right]^2} \Bigg/ \sqrt{\frac{1}{N+1}\sum_{i=0}^{N} p_i Y_i^2}\ . \tag{2.32}$$

This relation calculated for $p_i = 1$ $\{i = 0, \ldots, N\}$ (Kojdecki 2002) by implementation of the falsi method (Flannery 1998) is shown in Fig. 2.7. As one can see, the smoothing procedure generates an error by one order of magnitude greater for TAPS than for FORS-I.

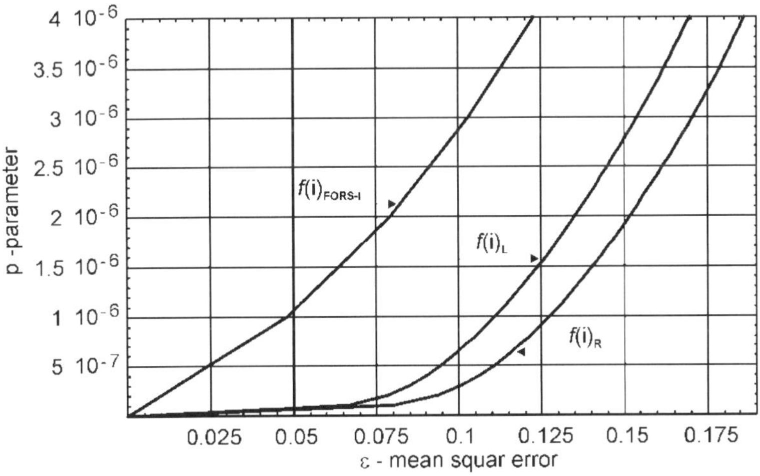

Fig. 2.7 Dependence between the mean square error ε and parameter p for TAPS and FORS-I systems

The effectiveness of this method for improving the recording of rotation events by TAPS (in comparison with the method presented in Fig. 2.6b) is shown in Fig. 2.8a. For the spline function, the parameter p equal to 5×10^{-6} has been chosen as optimum for smoothing. As Solarz et al. (2004) have shown, such a value is high enough for rotational component smoothing without reducing the really existing displacement component (see Fig. 2.8b).

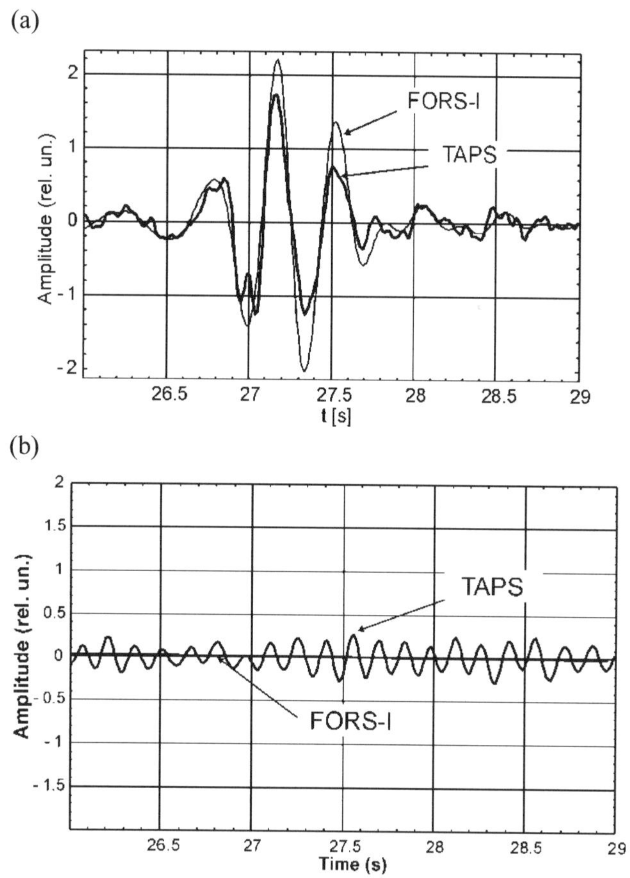

Fig. 2.8 (a) The rotational component recorded during the test presented in Fig. 2.6b after smoothing, and (b) additional displacement effect recorded by TAPS

The error of response of seismometers and the error of position can also be reduced by increasing the number of seismometers in an array. This enhances also the sensitivity and accuracy of measurement.

2.4 Direct Detection of the Rotational Component

Because the system based on the Sagnac effect realizes the absolute rotation measurement, it is probably the best solution for rotational events recording. That is why the new system named FORS-II using the classical gyroscope configuration (see Fig. 2.9) has been proposed by Jaroszewicz et al. (2005). The application of a standard single-mode fibre with length L equal to 11130 m in 0.63 m diameter sensor loop, high optical power source (10 mW superluminescence diode, $\lambda = 1285$ nm) and total optical loss equal to 21 dB give the theoretical sensitivity of 4.4×10^{-9} rad/s$^{1/2}$.

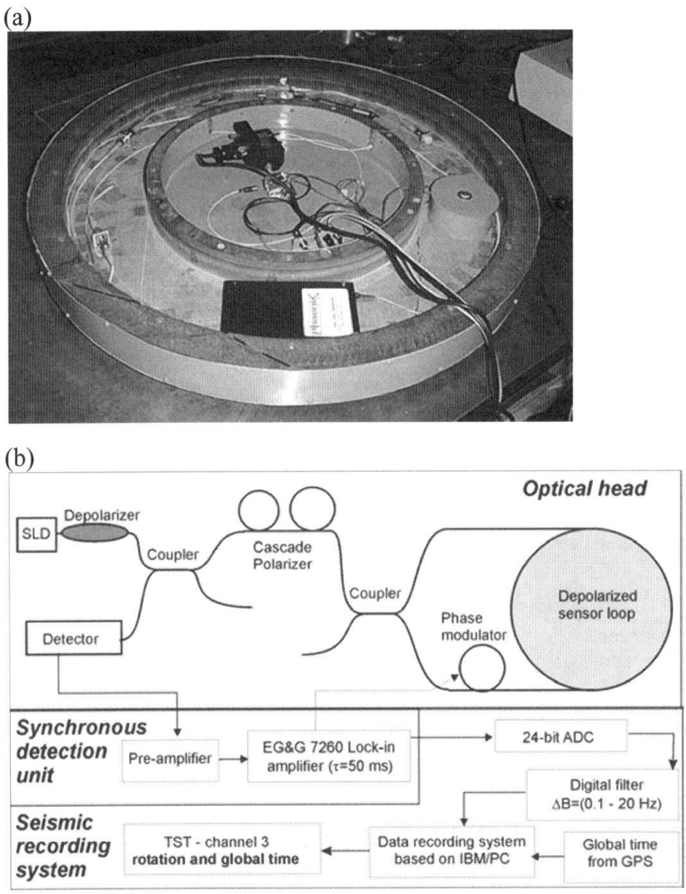

Fig. 2.9 (a) The view of the FORS-II optical part, and (b) the general scheme of the system

Moreover, the system uses the cascade of two fibre-optic polarizers with a high extinction ratio. To provide slow drift in a long period of time, the FORS-II generally operates applying depolarized light. For this reason, the set of two fibre depolarizers have been used. One of them is placed behind the source. The second one is the sensor loop whose work is equivalent to the depolarizer for the used wide-band source (Krajewski et al. 2005). The detection unit realizes synchronic detection with optimization for frequency equal to 9.0 kHz (Jaroszewicz et al. 2005). Additionally, a standard seismic recording station (KST) has been used for the data processing; its ADC samples a signal with frequency 1 kHz and after re-sampling stores it with frequency 100 Hz. The system calibration basing on the Earth rotation shows that the estimated FORS-II resolution is 4.3×10^{-8} rad/s (for 2σ where σ is standard deviation of measured noise level) in the 20 Hz detection band used (Jaroszewicz et al. 2006).

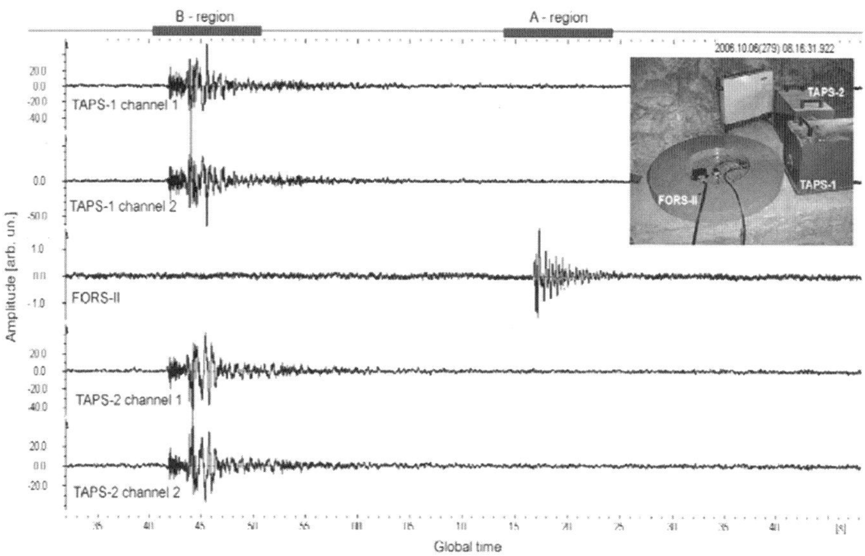

Fig. 2.10 (a) Seismograms of seismic events recorded on 6 October 2006 at 8:16 and general view of rotational seismometers (upper right window)

The first results obtained by application the FORS-II together with a set of two TAPS for rotational events investigation in Ojców Observatory, Poland (see the upper-right window in Fig. 2.10) provided new interesting results which should be underlined first of all. The data shown in Fig. 2.10 present an example of the seismic events recorded on 2006.10.06 at 8^h16^m; the first two plots are seismograms from two channels of TAPS-1, the third one presents data from FORS-II and last two are the seismograms from two channels of TAPS-2. As initial information it should be underlined that data obtained for TAPS systems present only linear motion described as $f_L(t)$ and $f_R(t)$ (see Eq. 2.29) and rotational components must be calculated by suitable method. Additionally, the initial impact test during the systems installation showed that all the electronic channels of the seismic recording system KST give the same time delay (Jaroszewicz et al. 2005).

Of utmost interest is the fact that the FORS-II has registered rotation with the time delay to linear motion characteristic of this earthquake registered by channels of TAPS systems (in region A instead of region B – see Fig. 2.10). The final results of the numerical processing with data correction by spline function approximation with $\varepsilon = 0.3$ (Solarz et al. 2004) applied for the data presented in Fig. 2.10 designed for calculation of the rotational component is shown in Fig. 2.11.

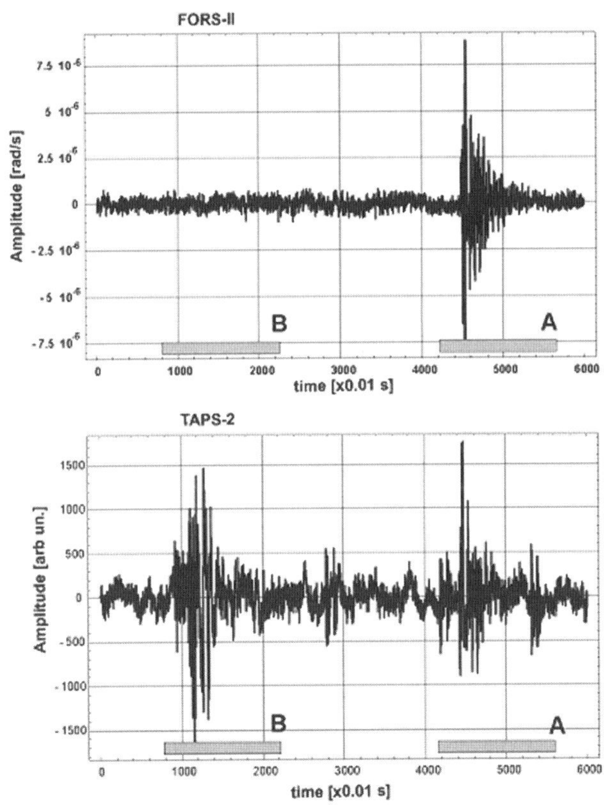

Fig. 2.11 Recognition of seismic rotational components by FORS-II and TAPS-2 from data presented in Fig. 2.10

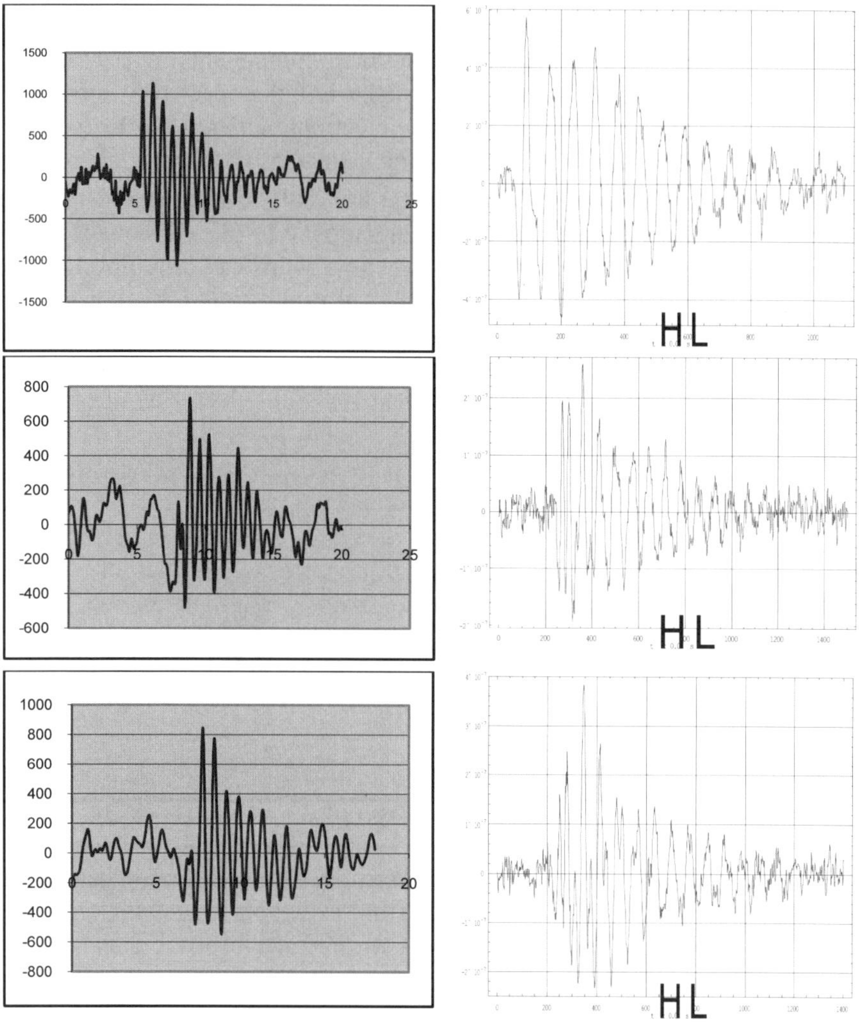

Fig. 2.12 The comparison of the rotational components recorded by TAPS (left column) with FORS-II (right column) obtained during the earthquakes recorded on 22.10.04 at 8^h16^m, and the two events of 21.10.04 at 11^h42^m, respectively

The analysis of these data, based on the comparison of the translational and rotational components registered by TAPS as well as the spectrum of rotational components registered by two types of seismometers, show that real rotational components exist only in the A region (see Fig. 2.11), whereas other rotation components recorded by TAPS (region B at Fig. 2.11) are probably erroneous due to the fact that the characteristics of its two channels are not identical, as it has been mentioned at the beginning of this chapter. In region A, the translational component does not exist. Moreover, the rotational characteristics recorded by TAPS and FORS-II are the same and their amplitudes are twice smaller than expected previously.

Figure 2.12 presents a comparison of different rotational components obtained from four seismic events previously recorded by TAPS and FORS-II. It is easy to recognize a similarity of the characteristics recorded by each of the above rotational seismometers. Moreover, the FORS-II calibration procedure gives additional information about the absolute amplitudes of these events which are in the range from 1.5·10-6 rad/s to 2·10-7 rad/s.

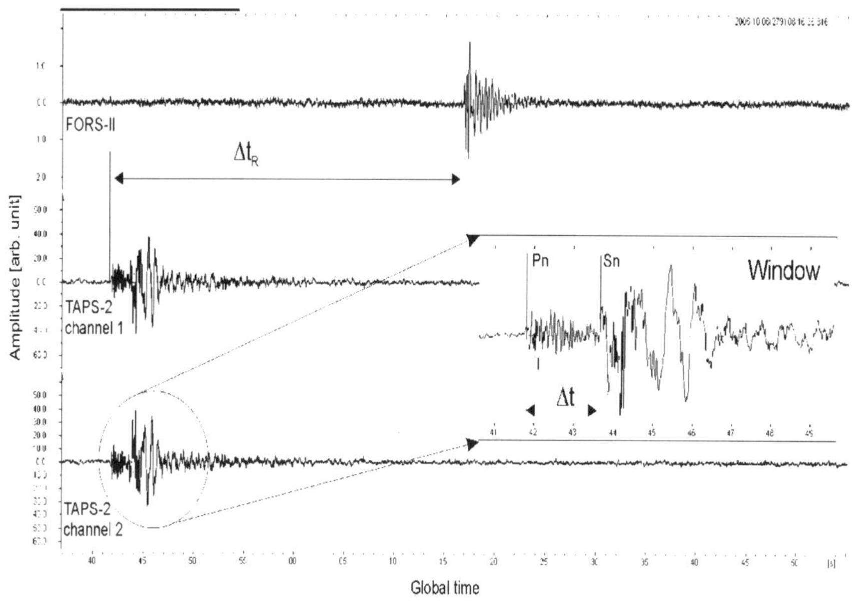

Fig. 2.13 Time relation between signals registered by FORS-II and TAPS-2 system as well as between P-waves and S-waves registered by second channel of TAPS-2 (in the window) of the seismic events recorded on 2006.10.2006, at 8^h16^m

If the recorded rotational components are related to the seismic rotational waves, SRW, the main conclusion to be drawn from the observed time delay ΔtR (see Fig. 2.13) is that the SRW are the seismic waves which propagate with velocities different from the classical longitudinal or transversal ones. Because the seismic S-waves have higher velocities than the P-waves and both of them have different attenuation and frequency characteristics, the delay time between them (Δt – see window in Fig. 2.13) can be used for calculation of the distance from the seismic events epicentre as $L = 7.86 \cdot \Delta t$ (Teisseyre et al. 2001). Additionally, for each of the recorded seismic events the time delay ΔtR between the P-waves and the seismic rotational events, RSE, can also be calculated according to the scheme shown in Fig. 2.13 (Jaroszewicz et al. 2005).

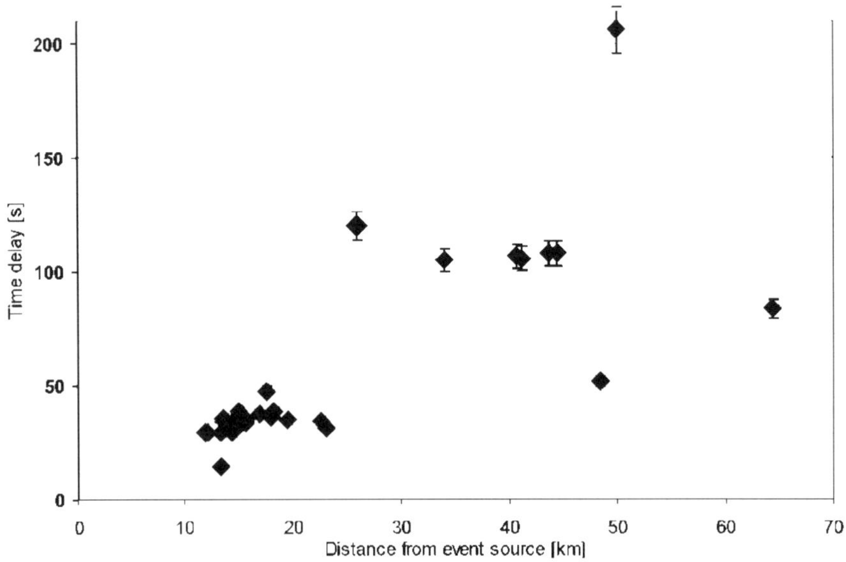

Fig. 2.14 Time delays versus the distance from the seismic event epicentre of the rotational events for the data recorded in Ojców Observatory

The results of the above estimation for the SRE in the seismic events recorded in the Ojców Observatory are summarized in a graphical form in Fig. 2.14. It should be noticed that all events should be treated as near-source rotational ground motions.

2.5 Conclusions

The presented review of methods for short-period weak rotation signals measurement shows the necessity for developing new instrumentation whose principle of operation would eliminate the sensitivity to linear motions. For this reason, seismometers should be used with special care as concerns their positioning, selecting examples with the same response, and calibration. They should be used as an array that compromises between the resolution and frequency of rotation signals. The practically expected sensitivity of less than nrad/s gives preference to systems operating on the basis of the optical Sagnac effect, such as laser or fibre-optic systems or magnetohydrodynamic sensors. The main advantage of such systems is a possibility to detect the absolute rotation, which is impossible to attain in other way. It seems that the data obtained in this manner are clear for identification. A higher sensitivity can be now achieved by a laser system, but it is a stationary equipment. For a portable system, the other two are preferred.

The results obtained in the Ojców Observatory prove that the real seismic rotational events are delayed in time with regard to the classical seismic wave existing during earthquakes. Moreover, the recorded amplitude of these events, connected with the quarry situated near to FORS location, have been identified in the range of $1.5\cdot10^{-6}$ rad/s to $2\cdot10^{-7}$ rad/s, which is less than 5-7 percent of the seismic event amplitude. Besides, it was clearly shown that the TAPS system also detected some events with the time and amplitude correlated with the data recorded by FORS-II.

Acknowledgements. The financial support of the Ministry of Science and Higher Education contract No 2166/B/T02/2007/33 according Grant No N525 2166 33 realization in 2007 year is gratefully acknowledged.

References

Aki K, Richards PG (1980) Quantitative seismology: Theory and methods. WH Freeman and Co, San Francisco

Aronowitz F (1971) The laser gyro. In: Ross M (ed) Laser applications, vol 1. Academic Press, New York, pp 133-200

Bodin P, Gomberg J, Sing SK, Santoyo M (1997) Dynamic deformation of shallow sediments in the valley of Mexico. Part I. Three-dimensional strains and rotations recorded on a seismic array. Bull Seism Soc Am **87**: 528-539

Bradner H, Reichle M (1973) Some methods for determining acceleration and tilt by use of pendulums and accelerometers. Bull Seism Soc Am **63**: 1-7

Bouchon M, Aki K (1982) Strain, tilt, and rotation associated with strong ground motion in the vicinity of earthquake faults. Bull Seism Soc Am **72:** 1717-1738

Droste Z, Teisseyre R (1976) Rotational and displacement components of ground motion as deduced from data of the azimuth system of seismographs. Publs Inst Geophys Pol Acad Sci **97**: 157-167

Duncan CA (1986) Strainmeters and tiltmeters. Rev Geophys **24**: 624-679

Eubank RL (2000) Spline regression in smoothing and regression: approaches, computation, and application. John Wiley & Sons Inc, New York

Farrell WE (1969) A gyroscopic seismometer: Measurements during the Borrego earthquake. Bull Seism Soc Am **59**: 1239-1245

Ferrari G (2006) Note on the historical rotation seismographs. In: Teisseyre R, Takeo M, Majewski E (eds) Earthquake source asymmetry, structural media and rotation effects. Springer-Verlag Berlin, pp 367-376

Flannery BP, Press WH, Teukolsky SA, Vetterling WT 1998, Numerical recipes, The art of scientific computing. 2nd Ed., INTERNET, www.nrcom.

Graizer V (2006) Tilts in strong ground motion, Bull Seism Soc Am **96:** 2090-2102

Gomberg J, Pavlis G, Bodin P (1999) The strain in the array is mainly in the plane (waves below 1 Hz). Bull Seism Soc Am **89**: 1428-1438

Huang BS (2003) Ground rotational motions of the 1999 Chi-Chi, Taiwan earthquake as inferred from dense array observations. Geophys Res Letters **30**: 6, Art. No 1307, 40-1, 40-4

Huang YT (1963) Analytical study of a new seismic sensor, gyro-seismometer. Bull Seism Soc Am **53**: 821-833

Igiel H, Schreiber U, Flaws A, Schuberth B, Velikoseltsev A, Cochard A (2005) Rotational motions induced by the M8.1 Tokachi-oki earthquake. Geophys Res Lett **32:** L08309

Jaroszewicz LR, Krajewski Z (2002) Possibility of fibre-optic rotational seismometer design. Proc SPIE **4900**: 416-423

Jaroszewicz LR, Krajewski Z, Solarz L, Marć P, Kostrzyński T (2003) A new area of the fiber-optic Sagnac interferometer application, Intern. Microwave and Optoelectronics Conference IMOC-2003, 20-23.09.2003 Iguazu Falls: 661-666.

Jaroszewicz LR, Krajewski Z, Solarz L, Teisseyer R (2005) Application of the FORS-II for investigation of the seismic rotation waves. Proc SPIE **5776**: 385-393

Jaroszewicz LR, Krajewski Z, Solarz L, Teisseyre R (2006) Application of the fibre-optic Sagnac interferometer in the investigation of seismic rotational waves. Meas Sci Technol **17:** 4, 1186-1193

Ostrzyżek A (1989) Analyses of rotation velocity measurement accuracy in FOG, [in Polish], doctoral thesis, Military University of Technology, Warsaw

Killpatrick JE (1966) The laser gyro. IEEE Spectrum **67**: 44-55

Kojdecki MA (2002) Private communication, Warsaw

Krajewski Z, Jaroszewicz LR, Solarz L (2005) Optimization of fiber-optic Sagnac interferometer for detection of rotational seismic events. Proc. of SPIE **5952**: 240-246

Macek WM, Davis Jr DTM (1963) Rotation rate sensing with travelling wave ring laser. Appl Phys Lett **2**: 67-71

Moriya T, Marumo R (1998) Design for rotation seismometers and their calibration. Geophys Bull Hokkaido Univ **61**: 99-106

Moriya T, Teisseyre R (1999) Discussion on the recording of seismic rotation waves. Acta Geophys Pol **47**: 351-362

Moriya T, Teisseyre R (2006) Design of rotational seismometer and non-linear behaviour of rotation components of earthquakes. In: Teisseyre R, Takeo M, Majewski E (eds) Earthquake source asymmetry, structural media and rotation effects. Springer-Verlag Berlin Heidelberg, Chap. 32: 439-450

Muramatsu I, Sasatani T, Yokoi I (2001) Velocity-type strong-motion seismometer using a coupled pendulum: design and performance. Bull Seism Soc Am **91**: 604–616

Nigbor RL (1994) Six-degree-of-freedom ground-motion measurement. Bull Seism Soc Am **84**: 1665-1669

Nigbor RL, Evans JR, Hutt CR (2007) Laboratory and field testing of commercial rotational seismometer. Rotational seismology and engineering applications – Online proceedings for the 1st Intern Workshop Menlo Park, CA, USA 18-19.09.2007

Nowożyński K, Teisseyre KP (2003) Time-domain filtering of seismic rotation waves. Acta Geophys Pol **51**: 51-61

Pancha A, Webb TH, Stedman GE, McLeod DP, Schreiber KU (2000) Ring laser detection of rotations from teleseismic waves. Geophys Res Lett **27:** 3553-3556

Post EJ (1967) Sagnac effect. Rev Modern Physics **39**: 475-494

Riedesel MA, Moore RD, Orcutt JA (1990) Limits of sensitivity of inertial seismometers and velocity transducer and electronic amplifiers. Bull Seism Soc Am **80:** 1725-1752

Rosenthal AH (1962) Regenerative circulatory multiple-beam interferometry for the study of light propagation effect. J Opt Soc Am **52**: 1143-1148

Saito M (1968) Synthesis of rotational and dilatational seismograms. J Phys Earth **16**: 53-62

Sagnac G (1913) L'ether lumineux demontre par l'effet du vent relative d'Etherdanus un interferometer en rotation uniforme. Compterendus a l'Academie des Sciences **95**: 708-710

Schreiber U, Schneider M, Rowe CH, Stedman GE, Schlüter W (2001) Aspects of ring lasers as local earth rotation sensors. Surveys in Geophysics **22**: 5-6, 603-611

Schreiber U, Stedman GE, Igel H, Flaws A (2006) Ring laser gyroscopes as rotation sensors for seismic wave studies. In: Teisseyre R, Takeo M, Majewski E (eds) Earthquake source asymmetry, structural media and rotation effects, Springer-Verlag Berlin Heidelberg, Chap. 29: 377-390

Smith S (1966) An array process for SH wave. Trans Am Geophys Un **47:** 171

Smith SW, Kasahara K (1969) Wave and mode separation with strain seismographs. Bull Earthq Res Inst **47**: 831-848

Solarz L, Krajewski Z, Jaroszewicz LR (2004) Analysis of seismic rotations detected by two antiparallel seismometers: Spline function approximation of rotation and displacement velocities. Acta Geophys Pol **52**: 198-217

Streckeisen G, Pfungen AG (1995) Portable Very-Broad-Band Tri-Axial Seismometer STS-2 Manual

Suchcicki J, Skrzyński A, Hościłowicz M, Wiszniowski J (2001) Seismometer calibration method, especially designed for detection and measurements turn vibration (in Polish), Patent application No P-350272

Suryanto W, Igel H, Wassermann J, Cochard A, Schuberth B, Vollmer D, Scherbaun F, Schreiber U, Velikoseltsev (2006) First comparison of array-derived rotational ground motions with direct ring laser measurements. Bull Seism Soc Am **96**: 2059-2071

Takeo M (1998) Ground rotational motions recorded in near-source region. Geophys Rev Lett **25**: 789-792

Takeo M (2006) Ground rotational motions recorded in near-source region of earthquakes. In: Teisseyre R, Takeo M, Majewski E (eds) Earthquake source asymmetry, structural media and rotation effects. Springer-Verlag Berlin Heidelberg, Chap. 12, 157-167.

Takeo M, Ueda H, Matzuzawa T (2002) Development of high-gain rotational-motion seismograph. Research grant 11354004, Earthquake Research Institute, University of Tokyo: 5-29

Teisseyre R, Nagahama H (1999) Micro-inertia continuum: rotations and semi-waves. Acta Geophys Pol **47**: 259-272

Teisseyre R, Majewski E (2001) Earthquake Thermodynamics and Phase Transformations in the Earth's Interior. Academic Press, New York

Teisseyre R, Suchcicki J, Teisseyre KP (2003) Recording the seismic rotation waves: reliability analysis. Acta Geophys. Pol **51**: 37-50

Teisseyre R, Suchcicki J, Teisseyre KP, Wiszniowski J, Palangio P (2003a) Seismic rotational waves: basic elements of theory and recording. Ann Geophys **46**: 671-685

Trifunac MD, Todorovska MI (2001) Evolution of accelerographs, data processing, strong motion arrays and amplitude and spatial resolution in recording strong earthquake motion. Soil Dyn Eartq Eng **21**: 275-286

Wiszniowski J, Skrzyński A, Suchcicki J (2003) Recording rotations with a pendulum seismometer: a sensor with reduced sensitivity to linear motions. Acta Geophys Pol **51**: 433-446

Wiszniowski J (2006) Rotation and twist motion recording – couple pendulum and rigidi seismometers system. In: Teisseyre R, Takeo M, Majewski E (eds) Earthquake source asymmetry, structural media and rotation effects, Springer-Verlag Berlin Heidelberg, Chap. 33: 451-470

Vali V, Shorthill RW (1976) Fiber ring interferometer. Appl Optics **15**: 1099-1100

Zadro M, Braitenberg C (1999) Measurements and interpretations of tilt–strain gauges in seismically active areas. Earth-Science Rev **47**: 151–187

Zembaty Z (2006) Deriving seismic surface rotations for engineering purposes, in: Teisseyre R, Takeo M, Majewski E (eds) Earthquake Source Asymmetry, Structural Media and Rotation Effects, Springer-Verlag Berlin Heidelberg, Chap. 38: 549-568

3 Buildings as Sources of Rotational Waves

Mihailo D. Trifunac

Department of Civil Engineering, University of Southern California
Los Angeles, California, USA, 90089-2531
e-mail: trifunac@usc.edu

3.1 Introduction

Most man-made structures are built above the ground. Single-family residential homes can be several tens of meters high, while modern steel skyscrapers can reach heights of several hundred meters. Supported asymmetrically at their base, with their center of gravity near mid-height, these structures undergo rocking motions when excited by earthquakes, strong winds, and man-made transient and steady excitations. Through the rocking compliance, the soil-structure interaction then acts as a mechanism for conversion of the incident wave energy into rotational motions of the foundation, which then radiates this wave energy back into the soil. During earthquake and ambient noise excitations, the incident waves are scattered and diffracted by the foundation-soil interface, and together with the waves generated by soil-structure interaction radiate rotational motions back into the soil. During wind and man-made excitation, a part of the wave energy in the building is converted into rotational excitation of the soil. The early work on the waves created by soil structure-interaction dates back to the 1930s (Sezawa and Kanai 1935, 1936) and 1940s (Biot 2006). Full-scale experiments of soil-structure interaction have provided data to measure and quantify the nature of the motions at the interface between the soil and the building foundations (Luco et al. 1986, Todorovska 2002, Trifunac and Todorovska 2001). The emphasis in these full-scale tests, so far, has been on the response of structures, and on how this response is affected by soil-structure interaction. Some experiments, however, did investigate the nature of the near-field deformation of soil surrounding the building foundation (Luco et al. 1975, 1988, Foutch et al. 1975, Wong et al. 1977a). It has been found that for stiff foundation-structure systems, the soil-foundation interaction can be approximated by a rigid foundation model having only six degrees of freedom. For flexible foundations (Trifunac et al. 1999) and multiple foundations, the soil deformation is far more complex, and the translational and rotational waves in the near field, radiated by the motion of the foundations, require complex three-dimensional analyses. In densely populated metropolitan areas where separation of adjacent buildings is small or negligible and there are building-soil-building interactions, and where long bridges have multiple supports resting on soil, detailed two- and three-dimensional analyses are required (Werner et al. 1979, Wong and Trifunac 1975). Analytical studies of two-dimensional soil-structure interaction (of long buildings on rigid foundations) have shown how the interference of the incident waves, and of the scattered waves from the foundation, can lead to nearly standing wave motions on the ground surface, which, at the nodes, result in strong torsional motions (Trifunac 1972, Trifunac et al. 2001c, Todorovska et al. 1988). Analytical studies of the response of three-dimensional models show amplification of the torsional response of building-foundation-soil systems and the radiation of torsional scattered waves for near-horizontal incidence of SH waves (Lee 1979). Studies of the wave passage effects around rigid embedded foundations have explained amplification of the rocking foundation motions and the more energetic radiation of rotational waves when half wave-lengths of the incident waves are comparable to the foundation width (Todorovska and Trifunac 1990, 1991, 1993). Observational and analytical studies of buildings in an urban setting have examined the site-city interaction (Boutin and Roussillon 2004, Gueguen et al. 2000, 2002, Kham et al. 2006, Tsogka and Wirgin 2003), and have interpreted the prolonged duration of strong ground motion in urban settings (Wirgin and Bard 1996) in terms of the waves delayed by prolonged paths up and down the buildings (Gicev 2005).

Experiments using forced vibration of full-scale structures have been used to investigate the wave motion in the far field radiated due to soil-structure interaction (Luco et al. 1975, Favela 2004). The radiated waves in the far field have been used as monochromatic sources of waves to investigate the relative significance of irregular topography and irregular geometry of sedimentary layers on amplification of surface displacements (Wong et al. 1977b).

In the following, the linear theory of soil-structure interaction will be illustrated, emphasizing the rotational aspects of motion, which result from (1) the presence of an inclusion (foundation) in the half space (Lee and Trifunac 1982), and (2) from the interaction with a structure (Lee 1979). A discussion of the non-linear aspects of this class of problems is beyond the scope of this chapter, but the reader may find introductory examples of analyses and observations in Gicev (2005) and Trifunac et al. (2001a,b).

3.2 Soil-Foundation Interaction – Near Field

The dynamic response of a rigid foundation embedded in an elastic medium to seismic waves can be separated into two parts. The first part corresponds to the determination of the restraining forces due to the rigid body motion of the inclusion. The second part deals with the evaluation of the driving forces due to the scattering of incident waves by the inclusion, which is presumed to be immobile.

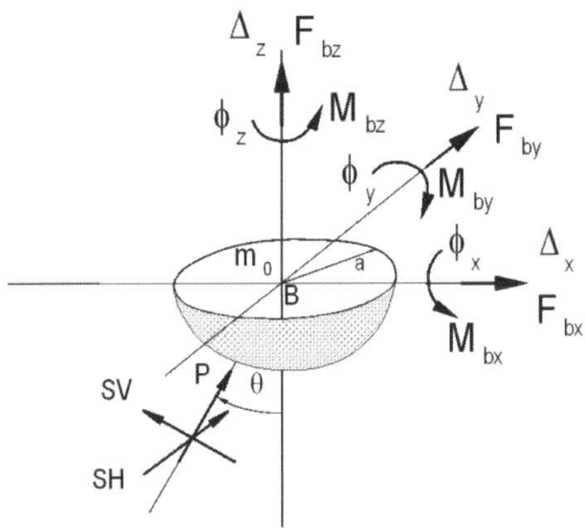

Fig. 3.1 Forces F_{bx}, F_{by}, F_{bz}, M_{bx}, M_{by}, M_{bz} acting on the foundation, and its displacements Δ_x, Δ_y, Δ_z, ϕ_x, ϕ_y, ϕ_z

Consider a foundation embedded in an elastic medium and supporting an elastic superstructure. The steady-state harmonic motion of the foundation having frequency ω can be described by a vector $\{\Delta_x, \Delta_y, \Delta_z, \phi_x, \phi_y, \phi_z\}^T$ (Fig. 3.1), where Δ_x and Δ_y are horizontal translations, Δ_z is vertical translation, ϕ_x and ϕ_y are rotations about horizontal axes, and ϕ_z is torsion about the vertical axis. Using superposition, displacement of the foundation is the sum of two displacements

$$\{U\} = \{U^*\} + \{U_0\}, \tag{3.1}$$

where $\{U^*\}$ is the foundation input motion corresponding to the displacement of the foundation under the action of the incident waves in the absence of external forces, and $\{U_0\}$ is the relative displacement corresponding to the displacement of the foundation under the action of the external forces in the absence of incident wave excitation.

The interaction force $\{F_s(\omega)\}\exp\{-i\omega t\}$ generates the relative displacement $\{U_0\}\exp\{\ i\omega t\}$. It corresponds to the force that the foundation exerts on the soil, and it is related to $\{U_0\}$ by $\{F_s\} = [K_s(\omega)]\{U_0\}$, where $[K_s(\omega)]$

is the 6×6 complex stiffness matrix of the embedded foundation. It depends upon the material properties of the soil medium, the characteristics and shape of the foundation, and the frequency of the harmonic motion. It describes the force-displacement relationship between the rigid foundation and the soil medium.

The driving force of the incident waves is equal to

$$\{F_s^*\} = [K_s]\{U^*\}, \tag{3.2}$$

where the input motion $\{U^*\}$ is measured relative to an inertial frame. The "driving force" is the force that the ground exerts on the foundation when the rigid foundation is kept fixed under the action of incident waves. It depends upon the properties of the foundation and the soil and on the nature of excitation.

The displacement $\{U\}$ is then related to the interaction and driving forces via

$$[K_s]\{U\} = \{F_s\} + \{F_s^*\}.$$

For a rigid foundation having a mass matrix $[M_0]$ and subjected to external force, $\{F_{ext}\}\exp\{-i\omega t\}$, the dynamic equilibrium equation is

$$-\omega^2[M_0]\{U\} = -\{F_s\} + \{F_{ext}\}. \tag{3.3}$$

$\{F_{ext}\} = \{F_{bx}, F_{by}, F_{bz}, M_{bx}, M_{by}, M_{bz}\}$ is the force the structure exerts on the foundation (Fig. 3.1). Then Eq. (3.3) becomes

$$\left(-\omega^2[M_0] + [K_s]\right)\{U\} = \{F_s^*\} + \{F_{ext}\}. \tag{3.4}$$

The solution of $\{U\}$ requires the determination of the mass matrix, the impedance matrix, the driving forces and the external forces.

Foundation Response. The stiffness and damping coefficients associated with the real and imaginary parts of $[K_S]$ are the functions of dimensionless frequency η (Lee 1979). $\eta = \omega a/\pi\beta$ is the ratio of the diameter of the hemispherical foundation to the wavelength of the transverse (S) waves, λ, in the half space. η is also a dimensionless wave number $k_\beta a/\pi$. The range of η considered in the examples here will be from 0 to 1.

Incident SV wave (Fig. 3.1) excites vertical motion, horizontal translation, and rocking, $\{\Delta_x/a_0,\ 0,\ \Delta_z/a_0,\ a\phi_y/a_0,\ 0\}^T$. For the Poisson's ratio, $\nu = 0.25$, the critical angle of incidence is $\theta_{cr} = 35°16'$. The incidence angles $\theta = 0°$ and 30° are thus below the critical angle, while the angles of incidence $\theta = 60°$ and 85° are beyond the critical angle. The case of grazing incidence ($\theta = 90°$) will result in zero motion in the free field, and thus it is not considered here. In Fig. 3.2, the normalized amplitudes $|a\phi_y / a_0|$, where a is the radius of the foundation, and a_0 is the amplitude of incident waves, are plotted versus the dimensionless frequency η for different angles of incidence and for $m_0/m_S = 0$, 2, and 4; m_0 is the mass of the hemispherical foundation; and m_s is the mass of soil removed by the foundation. For vertical incidence, only the horizontal translation and rocking are excited, and there is no vertical motion. At low frequencies, the displacement amplitudes approach the limit of the free field displacement amplitudes.

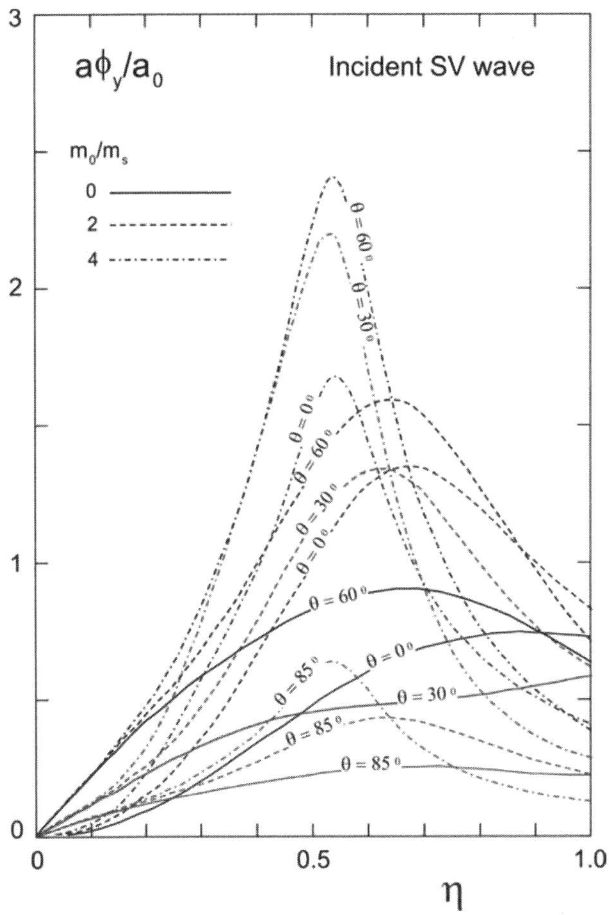

Fig. 3.2 Rocking ϕ_y of the foundation, excited by plane SV waves, with incident angles $\theta = 0°$, 30°, 60° and 85°, and dimensionless frequency η

The amplitudes gradually decrease with increasing η because of the embedment, which introduces coupling and rocking. The increase of the m_0/m_S factor influences the displacement amplitudes in a way that can be viewed through an analogy with a single-degree-of-freedom system (Lee 1979). It is seen that the foundation converts a significant part of the incident wave energy into rotational motions and that this conversion is most efficient for incident wavelengths about twice the diameter of the foundation.

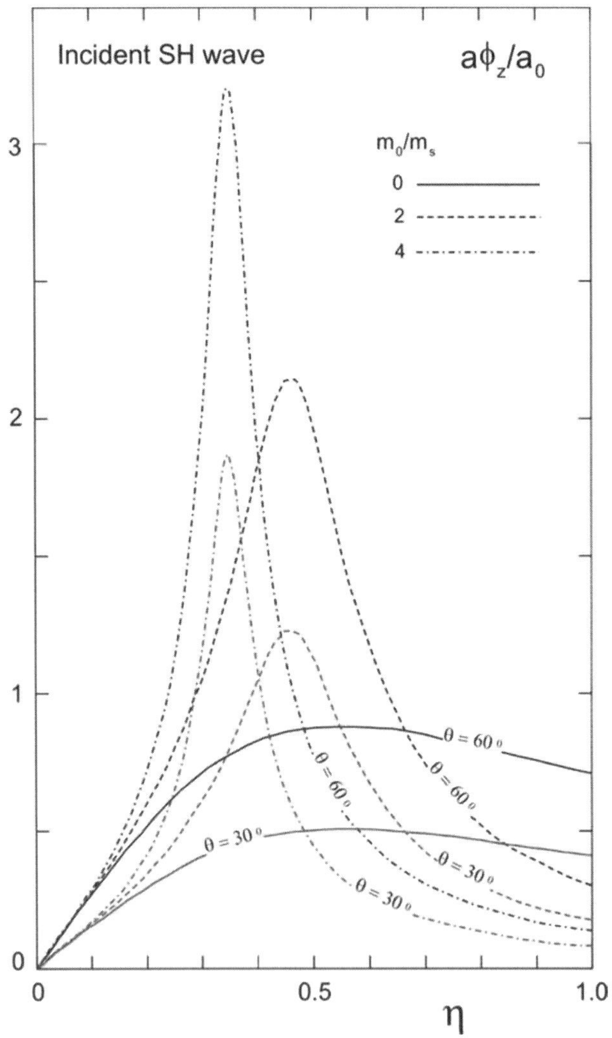

Fig. 3.3 Torsion ϕ_z of the foundation, excited by plane SH waves, with incident angles θ= 30°, 60° and 85°, and dimensionless frequency η

Incident P wave excites vertical motion, horizontal translation, and rocking of the foundation $\{\Delta_X/a_0,\ 0,\ \Delta_Z/a_0,\ 0,\ a\phi_y/a_0,\ 0\}^T$. For vertical incidence, only the vertical motion is excited. The embedment introduces coupling of the horizontal translation and rocking, and, for a general incidence angle, a significant component of rocking is present. The normalized rocking amplitudes $|a\phi_y/a_0|$ are similar to those for excitation by SV waves (Fig. 3.2) but are smaller (Lee 1979).

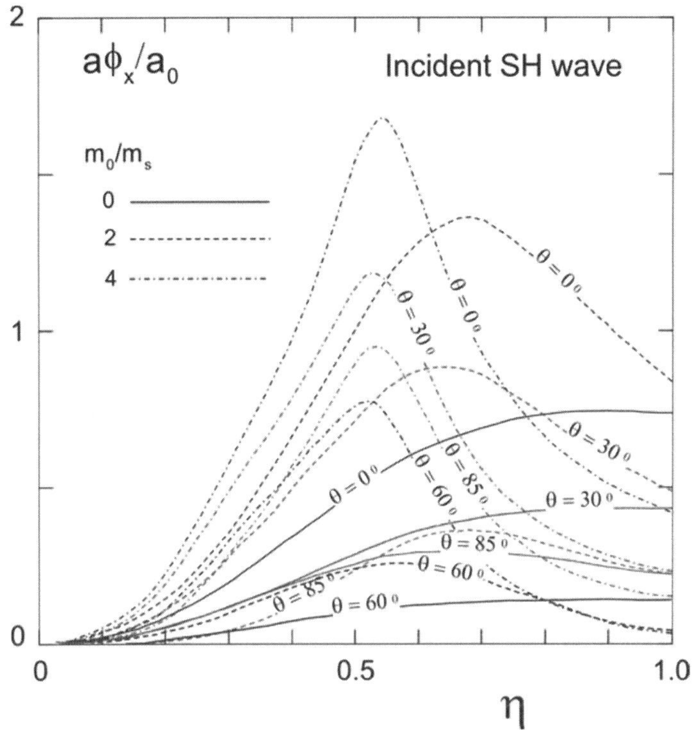

Fig. 3.4 Normalized rocking ϕ_x of the foundation, excited by plane SH waves, with incident angles $\theta = 0°$, $30°$, $60°$ and $85°$, and dimensionless frequency η

Incident SH wave excites torsion, horizontal translation, and rocking of the foundation. Unlike for incident P and SV waves, no vertical motion is excited: $\{0,\ \Delta_y/a_0,\ 0,\ a\phi_x/a_0,\ 0,\ a\phi_z/a_0\}^T$. In the absence of the foundation, the free field surface displacement amplitude for an incident SH wave of unit amplitude is two for the horizontal y-component and zero for the x- and z-components, for all angles of incidence. The presence of the foundation changes this simplicity. In Figs. 3.3 and 3.4, the amplitudes $|a\phi_z/a_0|$ and $|a\phi_x/a_0|$ are plotted versus the dimensionless frequency η for different angles of incidence and for $m_0/m_S = 0$, 2, and 4. For vertical incidence, only the horizontal translation and rocking are excited, and there is no torsion ($a\phi_z a_0$). At low frequency, the y-component displacement amplitudes approach the free field displacement amplitude for all angles of incidence. As for incident P and SV waves, the motions decrease with increasing frequency because of wave scattering by the embedded foundation. Coupling and a significant component of rocking are introduced by the embedment. The increase in the m_0/m_S ratio illustrates again the analogy with single-degree-of-freedom system.

Hill's Equation. From (3.4), the matrix equation of motion of the foundation, with external forces present, is

$$[M_0]\{\ddot{U}\}+[K_s]\{U\}=\{F_s^*\}+\{F_{ext}\}. \tag{3.5}$$

After the mass matrix $[M_0]$, the stiffness matrix $[K_s]$, and the force $\{F_s^*\}$ have all been evaluated, those can be used to determine the foundation displacement $\{U\}$. For soil-structure interaction problems, in the presence of the structure, $\{F_{ext}\}$ is the force that the structure exerts on the foundation.

The first model illustrated here is shown in Fig. 3.5. It considers the cases involving incident P and SV waves. The structure is represented by an equivalent single-degree-of-freedom system, with a concentrated mass m_b at a height h above the foundation. It has a radius of gyration r_b and a moment of inertia $I_b = m_b r_b^2$ about 0. The degree-of-freedom is chosen to correspond to the rocking ψ_r. This rotation is restrained by a spring with rocking stiffness K_r and by a dashpot with rocking damping C_r (both not shown in Fig. 3.5). The gravitational force $m_b g$ is considered. Taking moments about B results in the equation of motion

$$\begin{aligned}\ddot{\phi}_y+\ddot{\psi}_r+2\omega_r\varsigma_r\dot{\psi}_r+\omega_r^2\psi_r=(1/\varepsilon)\big\{&-\big(\ddot{\Delta}_x/a\big)\cos\big(\phi_y+\psi_r\big)\\&+\big(\omega_r^2\varepsilon_g+\ddot{\Delta}_z/a\big)\sin\big(\phi_y+\psi_r\big)\big\},\end{aligned} \tag{3.6}$$

where $\varepsilon = h(1+(r_b/h)^2)/a$, $\omega_r^2 = K_r/[m(h^2+r_b^2)]$; the natural frequency of rocking squared, ζ_r is a fraction of the critical damping in $2\omega_r\zeta_r = C_r/[m(h^2+r_b^2)]$; and $\varepsilon_g = 2/\omega_r^2 a$. Equation (3.6) is a differential equation coupling the rocking of the foundation and the structure with the horizontal and vertical motions of the foundation. It is a nonlinear equation, whose solution will require numerical analysis. In this work we will consider only the case when $\phi_y+\psi_r$ is small. Then

$$\ddot{\psi}_r + 2\omega_r\zeta_r\dot{\psi}_r + \left\{\omega_r^2\left(1-\varepsilon_g/\varepsilon\right) - \ddot{\Delta}_z/\varepsilon a\right\}\psi_r = \\ = -\ddot{\phi}_y + \left(1/\varepsilon\right)\left\{-\ddot{\Delta}_x/a + \left(\omega_r^2\varepsilon_g + \ddot{\Delta}_z/a\right)\phi_y\right\}. \tag{3.7}$$

For steady-state excitation by the incident P and SV waves, with frequency ω, Δ_x, ϕ_y, and Δ_z, and therefore the coefficients of (3.7), will be periodic. Equation (3.7) is then a special form of Hill's equation. Analysis of the stability of this equation can be found in the report by Lee (1979).

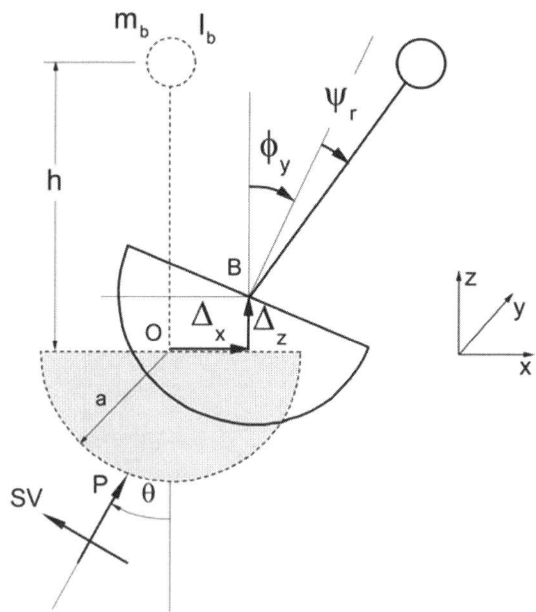

Fig. 3.5 In-plane rocking of foundation-structure model excited by plane P or SV waves

Solution of the Interaction Equation. For steady-state harmonic incident P and SV waves with excitation frequency ω, the vertical component of the foundation displacement, Δ_z, will also be harmonic with the same frequency ω. Considering only the period $T = 2\pi/\omega$, the most general solutions will take the form of Fourier series

$$\Delta_x = \sum \Delta_n e^{-in\omega t}, \quad \phi_y = \sum \phi_n e^{-in\omega t}, \quad \psi_r = \sum \psi_n e^{-in\omega t}, \tag{3.8}$$

where the summation is from $n = -\infty$ to $+\infty$. Δ_n, ϕ_n, and ψ_n are Fourier coefficients to be determined (Lee 1979). For the examples that follow, the model parameters have been chosen to take on the following values:

$m_0/m_s = 1$, $m_b/m_s = 3$, $h/a = 1$, $r_b/h = 0.5$, $\zeta_r = 0.02$, $\varepsilon_g = 0.1$, $a_0/a = 0.05$, and the dimensionless fixed-base frequency of the structure $\eta_r = \omega_r a/\pi\beta = 0.1$, 0.3, 0.5, and 0.8. $m_0/m_s = 1.0$ corresponds to the case of a rigid hemispherical foundation of the same density as the elastic medium of the half space.

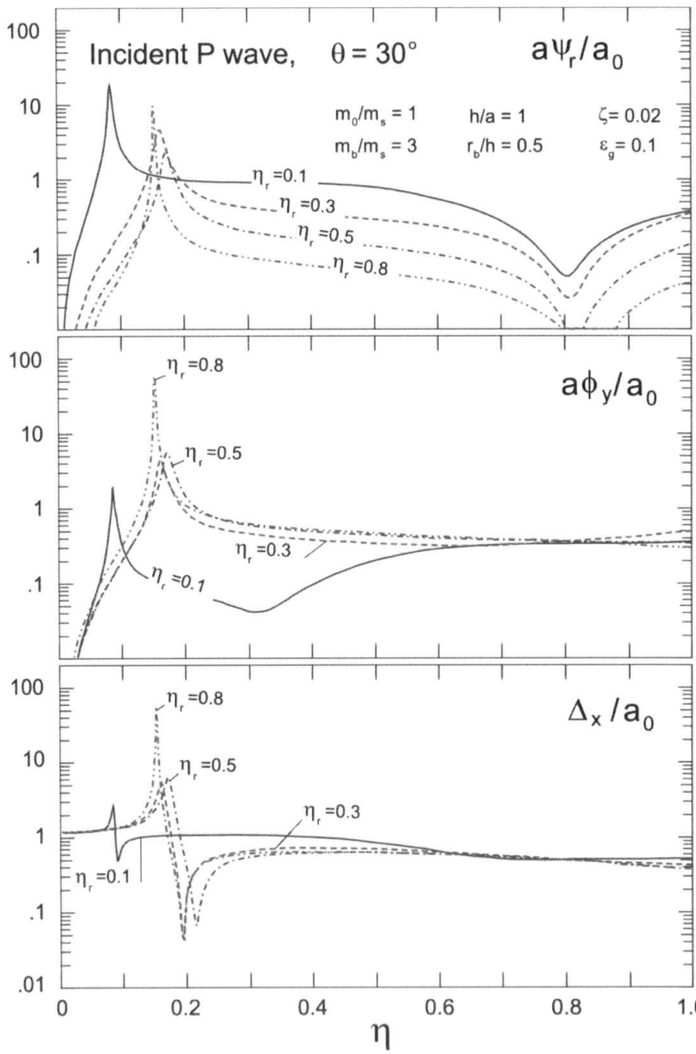

Fig. 3.6 Normalized rocking angles ψ_r and ϕ_y and translation Δ_x versus dimensionless excitation frequency η, for normalized model frequencies $\eta_r = 0.1$, 0.3, 0.5 and 0.8, and for incident P wave with $\theta = 30°$

The amplitudes of the coefficients Δ_1/a_0, $a\phi_1/a_0$, and $a\psi_1/a_0$ of the first harmonic components ($n = 1$) corresponding to the harmonic frequency of excitation, $e^{-i\omega t}$, are the dominating contributors to response. For $n > 1$, the coefficients Δ_n/a_0, $a\phi_n/a_0$, and $a\psi_n/a_0$, among other parameters, are dependent upon the ratio a_0/a, which is chosen to be 0.05. Figure 3.6 illustrates the response amplitudes of the first harmonic components ($n = 1$; the amplitudes for $n = 2$ are much smaller and may be neglected) of the Fourier series of Δ_x/a_0, $a\phi_y/a_0$, and $a\psi_r/a_0$, representing foundation translation, foundation rocking, and relative rocking response.

In the analysis and design of earthquake-resistant structures, it is necessary to estimate the maximum amplitudes of the relative responses at the top of the structure, which are then used to calculate the strain and the maximum stresses in the structure. In the absence of soil-structure interaction, the structure's relative rocking response, ψ_r, would be maximum at the fixed-base natural frequency, ω_r, and the relative response would approach infinity, as the fraction of critical damping, ζ_r, approaches zero. Interaction of the foundation with half space introduces "damping" into the relative response and results in a reduction of the "natural frequency" of the complete system. This reduction is more pronounced for larger values of fixed-base natural frequencies (Todorovska and Trifunac 1991).

At each of the frequencies where the relative responses experience maxima, the foundation rocking has an associated peak. At the fixed-base natural frequencies of long buildings (Trifunac 1972), the corresponding component of the external force will experience a maximum, and the corresponding displacement component has a "node" in the half space. Similar nodes are observed in the foundation rocking component of the three-dimensional models. However, because of the coupling of foundation rocking with the foundation horizontal translation, these nodes do not occur at exactly the fixed-base natural frequencies of the structure.

As the values of m_b/m_s and h/a increase, the rocking components of the foundation and the structure become more prominent–so prominent that the peak relative responses occur at frequencies characteristic of the natural rocking frequencies of the total system. For large m_b/m_s and h/a, the rocking stiffness of the structure, $K_r = m_b\omega_r^2(h^2 + r_b^2)$, becomes large. As the structure becomes stiffer, its relative rocking response becomes small, and the structural response, ψ_r, contributes a negligible amount to $\phi_y + \psi_r$. The total system then behaves like a rigid, partially embedded mass, $m = m_0 + m_b$, vibrating on an elastic half space (Biot 2006, Lee et al. 1982).

The model of the structure for the case of an incident SH wave (Fig. 3.7) is similar to the one for the incident P and SV waves. Its displacement vector

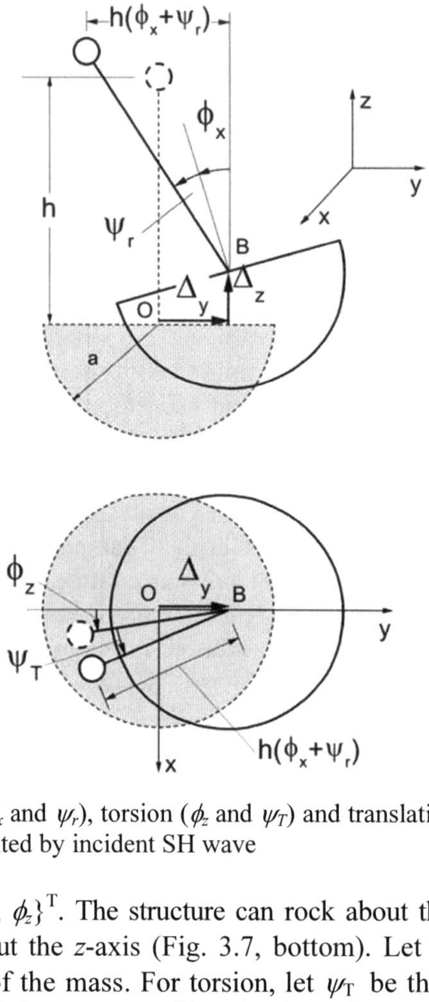

Fig. 3.7 Rocking (ϕ_x and ψ_r), torsion (ϕ_z and ψ_T) and translation Δ_y of foundation-structure model, excited by incident SH wave

is $\{0, \Delta_y, \Delta_z, \phi_x, 0, \phi_z\}^T$. The structure can rock about the x-axis (Fig. 3.7, top) and twist about the z-axis (Fig. 3.7, bottom). Let ψ_r be the relative angle of rocking of the mass. For torsion, let ψ_T be the relative angle of twist of the mass; I_T its moment of inertia about the z axis; K_T the torsional stiffness; C_T the torsional damping; ω_T the torsional natural frequency, $\omega_T^2 = K_T / I_T$; and ζ_T the fraction of critical damping for torsion, where $2\omega_T \varsigma_T = C_T / I_T$. Writing the rocking moment equilibrium equations about B (Fig. 3.7, top), torsional moment equilibrium about a vertical axis through B (Fig. 3.7, bottom), and assuming that $\phi_x + \psi_r$ and $\phi_z + \psi_r$ are small, gives

$$\begin{aligned} &\ddot{\psi}_r + 2\omega_r \varsigma_r \dot{\psi}_r + \left\{\omega_r^2 \left(1 - \varepsilon_g / \varepsilon\right) - \ddot{\Delta} / \varepsilon a\right\} \psi_r = \\ &\qquad = -\ddot{\phi}_x + \left(1/\varepsilon\right)\left\{\ddot{\Delta}_y / a + \left(\omega_r^2 \varepsilon_g + \ddot{\Delta}_z / a\right)\phi_x\right\} \end{aligned} \tag{3.9}$$

and

$$\ddot{\psi}_{\mathrm{T}} + 2\omega_{\mathrm{T}}\varsigma_{\mathrm{T}}\dot{\psi}_{\mathrm{T}} + \omega_{\mathrm{T}}^{2}\psi_{\mathrm{T}} = -\ddot{\phi}_{z}\,. \tag{3.10}$$

It is seen that for small relative response and small rocking, the torsional twist is uncoupled from the horizontal motions. Figure 3.8 illustrates the amplitudes of Δ_y/a_0, $a\phi_x/a_0$, and $a\psi_r/a_0$, for incident SH waves when $\theta = 30°$.

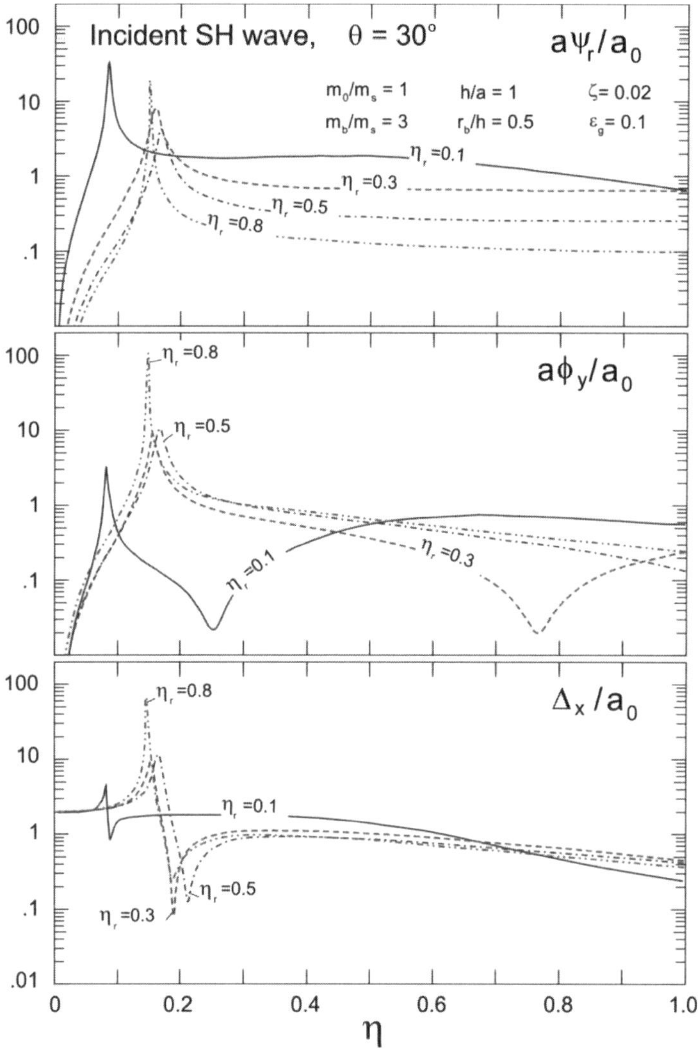

Fig. 3.8 Normalized rocking angles ϕ_x and ψ_r, and out-of-plane foundation motion Δ_y versus dimensionless excitation frequency η, for normalized model frequencies $\eta_r = 0.1$, 0.3, 0.5 and 0.8, and for incident SH wave with $\theta = 30°$

3.3 Soil-Foundation Interaction – Far Field

For distances greater than about $10a$ (Fig. 3.1) the motions of the half space, generated by building vibration, can be evaluated in terms of a point source consisting of two horizontal forces, F_{bx}, F_{by}, two rocking moments, M_{by}, M_{bx}, and a torsional moment, M_{bz}. A building oscillating in the x-z plane, for example, will produce strong Rayleigh waves in the x direction, while Love waves will be generated in the y direction (Bycroft 1956, Cherry 1962, Favela 2004). During forced vibration tests of a building, sufficiently large displacements of the ground can be generated, so that the recording and analysis can be made with standard recording devices at distances approaching 10 km (Luco et al. 1975). With specialized recording and data analyses, surface waves generated by building rocking can be detected at distances approaching 1.000 km (Favela 2004). The recorded motions contain valuable information about the properties of the medium of propagation and can serve as a full-scale laboratory for testing and verification of the wave propagation models of sedimentary and soil layers (Wong et al. 1977b), and of the influence of the surface topography on the amplitudes of surface motions (Wong et al. 1976).

3.4 Summary

The above single-degree-of-freedom system model of buildings erected on flexible soil shows how, through soil-structure interaction, a portion of the energy of incident waves is converted into the rotational motions. The eccentric location of the foundations, below the center of mass of the structures they support, makes the foundations the sources of rotational motions in the soil. Through the soil-structure interaction, the translational energy of the in-plane body P- and the SV-waves is converted into the foundation rocking (in the same plane), while the energy of the incident SH waves becomes the source of out-of-plane rocking and torsion about the vertical axis. It can be shown that the additional rotational excitation associated with the passage of Rayleigh (rocking) and Love (torsion) surface waves further amplifies the rotational motions of the foundations.

The amplitudes of the rotational motions of the foundations depend most upon the natural frequency of the structure above and on the wavelength of the incident waves below. The largest rotations of the foundation occur for relatively stiff buildings (when $\eta_r = \omega_r a/\pi\beta$ is large; e.g., 0.8 in Figs. 3.6 and 3.8) excited by incident waves with wavelengths that are approximately double the characteristic width of the foundation (when η is near one half; see Figs. 3.2, 3.3, and 3.4).

For multiple foundations of long structures, like bridges, for example, all of the above will hold for individual foundations, but additional rotational motions in the soil will also result from differential motions of the supports and from the wave passage effects (Trifunac and Todorovska 1997, Trifunac and Gicev 2006). These additional rotations will tend to be associated with longer wave-lengths, comparable to the inter-foundation distances of the multiple foundation systems, and can be viewed as resulting from a couple, or from a chain of couples, whose component forces lie in a vertical (for in-plane excitation) or horizontal (for out-of-plane excitation) plane.

In an urban setting, a distribution of buildings will act as an extended surface source area, consisting of a large number of closely spaced sources of translational and rotational motions, which will cause the warping of the half space surface in the near field and a seemingly random distribution of strong, high-frequency surface waves in the far field. For a distribution of buildings 1–50 stories high, the waves generated by the movement of their foundations will be in the range 0.1–10 Hz.

In this chapter, the basic source of rotational motions has been illustrated in terms of a spherical rigid foundation supporting a single-degree-of-freedom oscillator as a model of a "simple building". Mutatis mutandis, many of the above-described phenomena can be generalized to interpret the changes in the free-field wave motions resulting from a broad spectrum of other eccentrically supported "oscillators" ranging from individual trees to large and small geological formations like those in Monument Valley in Utah, down to Meteora in Greece or Sigiria in Sri Lanka, for example.

References

Biot MA (2006) Influence of Foundation on Motion of Blocks. Soil Dynamics & Earthquake Eng **26**: 6-7, 486-490

Boutin C, Roussillon P (2004) Assessment of the Urbanization Effect on Seismic Response. Bull Seism Soc Amer **94**: 1, 251-268

Bycroft GN (1956) Forced Vibrations of a Rigid Circular Plate on a Semi-Infinite Elastic Space and on an Elastic Stratum. Philosophical Trans of the Royal Society of London Series A – Math and Physical Sc **248**: 327-368

Cherry JT (1962) The Azimuthal and Polar Radiation Patterns Obtained from a Horizontal Stress Applied at the Surface of an Elastic Half Space. Bull Seism Soc Amer **52**: 1, 27-36

Favela J (2004) Energy Radiation from a Multi-Story Building. PhD Thesis, Calif Inst of Tech, Pasadena, California

Foutch DA, Luco JE, Trifunac MD, Udwadia FE (1975) Full Scale Three-Dimensional Tests of Structural Deformations During Forced Excitation of a Nine-Story Reinforced Concrete Building. Proc. U.S. National Conference on Earthquake Engineering, Ann Arbor, Michigan, 206-215

Gicev V (2005) Investigation of soil-flexible foundation-structure interaction for incident plane SH waves. Ph.D. Dissertation, Dept of Civil Engineering, Univ Southern California, Los Angeles, California

Gueguen P, Bard P-Y, Oliveira CS (2000) Experimental and Numerical Analysis of Soil Motions Caused by Free Vibrations of a Building Model. Bull Seism Soc Amer **90**: 6, 1464-1479

Gueguen P, Bard P-Y, Chavez-Garcia F (2002) Site-City Interaction in Mexico City-Like Environments: An Analytical Study. Bull Seism Soc Amer **92**: 2, 794-811

Kham M, Semblat J-F, Bard P-Y, Dangla P (2006) Seismic Site-City Interaction: Main Governing Phenomena Through Simplified Numerical Models. Bull Seism Soc Amer **96**: 5, 1934-1951

Lee VW (1979) Investigation of Three-Dimensional Soil-Structure Interaction. Department of Civil Engineering, Report CE 79-11, Univ of Southern California, Los Angeles, California

Lee VW, Trifunac MD (1982) Body Wave Excitation of Embedded Hemisphere. ASCE, EMD, **108**: 3, 546-563

Lee VW, Trifunac MD, Feng CC (1982) Effects of Foundation Size on Fourier Spectrum Amplitudes of Earthquake Accelerations Recorded in Buildings. Soil Dynamics and Earthquake Engineering **1**: 2, 52-58

Luco JE, Trifunac MD, Udwadia FE (1975) An Experimental Study of Ground Deformations Caused by Soil-Structure Interaction, Proc. U.S. National Conf. on Earthq. Eng., Ann Arbor, Michigan, 136-145

Luco JE, Wong HL, Trifunac MD (1986) Soil-Structure Interaction Effects on Forced Vibration Tests. Department of Civil Engineering, Report CE 86-05, University of Southern Calif, Los Angeles, California

Luco JE, Trifunac MD, Wong HL (1988) Isolation of Soil-Structure Interaction Effects by Full-Scale Forced Vibration Tests. Earthquake Engineering and Structural Dynamics **16**: 1, 1-21

Sezawa K, Kanai K (1935) Decay in the Seismic Vibration of a Simple or Tall Structure by Dissipation of Their Energy into the Ground. Bull Earth Res Inst **XIII**: 3, 681-697

Sezawa K, Kanai K (1936) Improved Theory of Energy Dissipation in Seismic Vibrations on a Structure. Bull Earth Res Inst **XIV**: 2, 164-168

Todorovska MI (2002) Full-scale Experimental Studies of Soil-Structure Interaction. Indian Society of Earthquake Technology Journal **39**: 3, 139-166

Todorovska MI, Trifunac MD (1989) Antiplane Earthquake Waves in Long Structures. ASCE, EMD **115**: 2, 2687-2708

Todorovska MI, Trifunac MD, Lee VW (1988) Investigation of Earthquake Response of Long Buildings. Department of Civil Engineering, Report CE 88-02, Univ of Southern Calif, Los Angeles, California

Todorovska MI, Trifunac MD (1990) Analytical Model for In-Plane Building-Foundation-Soil Interaction: Incident P-, SV-, and Rayleigh Waves. Department of Civil Eng Report CE 90-01, Univ Southern Calif, Los Angeles, California

Todorovska MI, Trifunac MD (1991) Radiation Damping During Two-Dimensional In-Plane Building-Soil Interaction. Department of Civil Eng Report CE 91-01, Univ Southern California, Los Angeles, California

Todorovska MI, Trifunac MD (1993) The Effects of the Wave Passage on the Response of Base-Isolated Buildings on Rigid Embedded Foundations. Dept of Civil Eng Rep CE 93-10, Univ Southern Calif, Los Angeles, California

Trifunac MD (1972) Interaction of a Shear Wall with the Soil for Incident Plane SH Waves. Bull Seism Soc Amer **62**: 1, 63-83

Trifunac MD, Gicev V (2006) Response Spectra for Differential Motion of Columns, Paper II: Out-of-Plane Response. Soil Dyn & Earthq Eng **26**: 12, 1149-1160

Trifunac MD, Todorovska MI (1997) Response Spectra and Differential Motion of Columns. Earthquake Eng and Struct Dyn **26**: 2, 251-268

Trifunac MD, Todorovska MI (2001) Recording and Interpreting Earthquake Response of Full-Scale Structures. Proc NATO Advanced Research Workshop on Strong-Motion Instrumentation for Civil Eng Structures, June 2-5, 1999, Istanbul, Turkey, Kluwer Academic Publ, Dordrecht, 131-155

Trifunac MD, Ivanovic SS, Todorovska MI (2001a) Apparent Periods of a Building I: Fourier Analysis. J Struct Eng, ASCE, **127**: 5, 517-526

Trifunac MD, Ivanovic SS, Todorovska MI (2001b) Apparent Periods of a Building II: Time-Frequency Analysis. J Struct Eng, ASCE, **127**: 5, 527-537

Trifunac MD, Hao TY, Todorovska MI (2001c) Response of a 14-Story Reinforced Concrete Structure to Nine Earthquakes: 61 Years of Observation in the Hollywood Storage Building. Dept of Civil Eng, Report CE 01-02, Univ of Southern California, Los Angeles, California

Trifunac MD, Ivanovic SS, Todorovska MI, Novikova EI, Gladkov AA (1999) Experimental Evidence for Flexibility of a Building Foundation Supported by Concrete Friction Piles. Soil Dyn & Earthq Eng **18**: 3, 169-187

Tsogka C, Wirgin A (2003) Simulation of Seismic Response in an Idealized City. Soil Dynamics and Earthquake Engineering **23**: 5, 391-402

Werner SD, Lee LC, Wong HL, Trifunac MD (1979) Structural Response to Travelling Seismic Waves, ASCE. J Struct Div **105**: ST12, 2547-2564

Wirgin A, Bard P-Y (1996) Effects of Buildings on the Duration and Amplitude of Ground Motion in Mexico City. Bull Seism Soc Amer **86**: 3, 914-920

Wong HL, Trifunac MD (1975) Two-Dimensional, Antiplane, Building-Soil-Building Interaction for Two or More Buildings and for Incident Plane SH-Waves. Bull Seism Soc Amer **65**: 1863-1885

Wong HL, Trifunac MD, Lo KK (1976) Influence of a Canyon on Soil-Structure Interaction. J Eng Mech Div, ASCE **102**: EM4, 671-684

Wong HL, Luco JE, Trifunac MD (1977a) Contact Stresses and Ground Motion Generated by Soil-Structure Interaction. Earthquake Engineering and Structural Dynamics **5**: 1, 67-79

Wong HL, Trifunac MD, Westermo B (1977b) Effects of Surface and Subsurface Irregularities on the Amplitudes of Monochromatic Waves. Bull Seism Soc Amer **67**: 2, 353-368

4 Two-Pendulum Systems for Measuring Rotations

Vladimir Graizer

California Division of Mines and Geology
Strong Motion Instrumentation Program
801 K Street, MS 13-35, Sacramento, California 95814
graizer@comcast.net

4.1 Introduction

During the last half of the 20th century, a number of attempts were made to measure or estimate rotational component of strong ground motion (Farrell 1969, Kharin and Simonov 1969, Bradner and Reichle 1973, Graizer et al. 1989, Graizer 1991, Nigbor 1994, Takeo 1998, Huang 2003, Zahradnik and Plesinger 2005, Graizer 2006, Schreiber et al. 2006), but still there are no consistent measurements of rotations during earthquake shaking. Technological advances in the recent decade made a number of technologies like a combination of gyroscopes and accelerometers wildly used in inertial navigation much cheaper and compact and available for use in seismic measurements. Considering different possible directions in rotation measurements is beyond the scope of this study. We will discuss classic approach to measuring rotations and translational motion by using a two-pendulum system. The classic way of measuring rotations by using two identical pendulums was apparently first suggested by Golitsyn (1912). It was later implemented by Kharin and Simonov (1969) in an instrument called VBPP (seismograph of large translational motions and rotations). This instrument used two identical pendulums on a same axis and moving in the same plane (Fig. 4.1). In case of purely translational input motion, both pendulums are producing exactly same output. In case of rotation (tilt), outputs of pendulums are opposite due to rotational acceleration. The output of the instrument was either a sum of the two signals, or a difference of them. Actually, both versions of the instruments were made. Summation of the two signals was supposed to result in a "purely" translational signal, and the difference was supposed to result in rotational motion only. In reality, summation of the two signals resulted in reliable translational motions, but the difference in the two signals (of about same amplitude) produced unreliable rotation measurements. The main problem occurred because of difficulty in constructing two identical mechanical systems (pendulums), and it became clear that the difference of the two signals is mainly determined by non-equality of pendulums. We modified Golitsyn's idea by using same configuration of pendulums (two-pendulum system) without requirement of pendulums to be identical. Instead of building two identical pendulums we need to measure precisely natural parameters of each pendulum, and apply post-processing to separate rotational and translational motion. Designed system was tested using a special shake-table, and later successfully applied for recording translational motion and tilt in the near-field of two large underground explosions (Graizer et al. 1989, Graizer 1991).

4.2 Theory

Most of seismological sensors (seismometers and accelerometers) used in conventional seismological instruments are pendulums of the mass-on-rod type. Complete equation of small oscillations (i.e., $\sin\theta \cong \theta$) of the horizontal pendulum of the mass-on-rod type can be expressed as (Graizer 1989, 2005):

$$y_1'' + 2\omega_1 D_1 y_1' + \omega_1^2 y_1 = -x_1'' + g\alpha - l_1\psi'' + x_2''\theta_1 \tag{4.1}$$

where: y_n is the recorded response of the instrument; l_n is the length of pendulum arm; θ_n is the angle of pendulum rotation from the equilibrium; $y_n = l_n\theta_n$ for small angles θ_n; ω_n and D_n are, respectively, the natural circular frequency and fraction of critical damping of the oscillator; x_n'' is the ground motion acceleration in the *n-th* direction; and ψ'' is the angular acceleration.

System of two horizontal pendulums

Let us consider response of the two-pendulum system shown in Fig. 4.1 and recording ground motion in the horizontal direction x_1. Complete equations for sensors 1 and 2 shown in Fig. 4.1 are:

$$\begin{cases} y_1'' + 2\omega_1 D_1 y_1' + \omega_1^2 y_1 = -x_1'' + g\alpha - l_1\psi'' + x_2''\theta_1 \\ y_2'' + 2\omega_2 D_2 y_2' + \omega_2^2 y_2 = -x_1'' + g\alpha + l_2\psi'' + x_2''\theta_2 \end{cases} \tag{4.2}$$

The left sides of Eq. (4.2) represent the response of sensors. The right sides of Eq. (4.2) are complex inputs that include:

- Acceleration along the x_1 axis.
- Tilt α of the base around the x_2 axis.

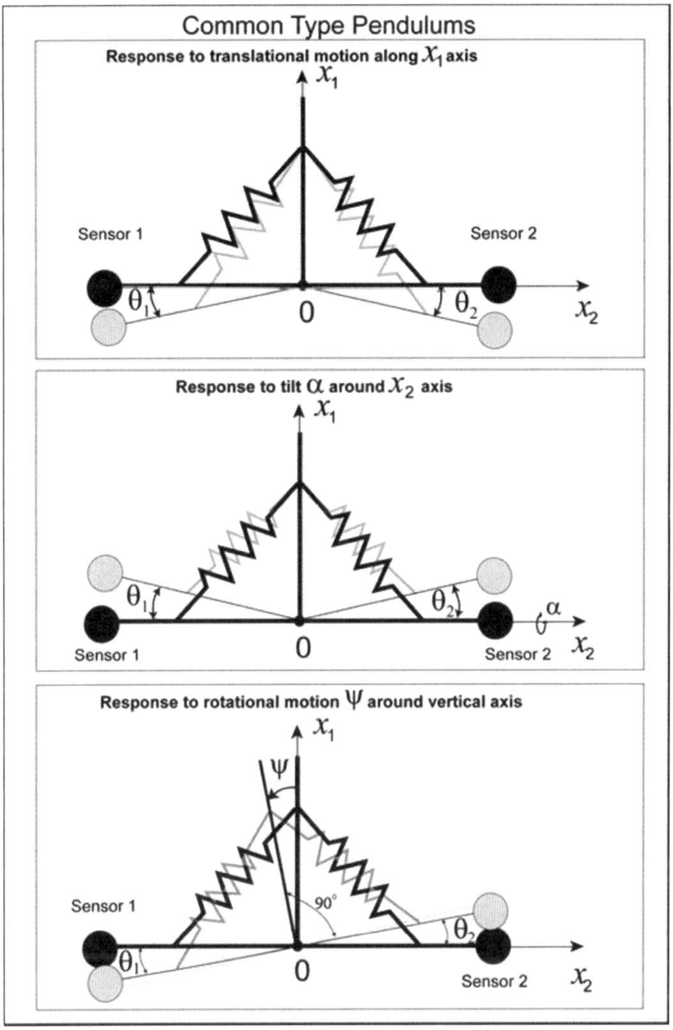

Fig. 4.1 Responses of two identical pendulums on a same axis to horizontal translational motion along the x_1 direction, tilt α and rotation ψ around the vertical axis (torsion)

- Inertial force due to rotation ψ'' around vertical axis.
- Cross-axis sensitivity.

It was previously shown by Graizer (1989, 2006), Todorovska (1998) and Trifunac and Todorovska (2001) that cross-axis sensitivity (terms 4 in the right side of Eq. (4.2)) is relatively low for recent instruments, and can be neglected (most of recent instruments are using force-balance feedback system minimizing the actual motion of a pendulum):

$$\begin{cases} y_1'' + 2\omega_1 D_1 y_1' + \omega_1^2 y_1 = -x_1'' + g\alpha - l_1\psi'' \\ y_2'' + 2\omega_2 D_2 y_2' + \omega_2^2 y_2 = -x_1'' + g\alpha + l_2\psi'' \end{cases} \tag{4.3}$$

The system of two Eq. (4.3) contains three unknown terms (x_1, α, ψ). This system can be resolved for rotations around vertical axis by subtracting Eq. (4.3):

$$(y_2'' - y_1'') + (2\omega_2 D_2 y_2' - 2\omega_1 D_1 y_1') + (\omega_2^2 y_2 - \omega_1^2 y_1) = (l_1 + l_2)\psi'' \tag{4.4}$$

The difference is not sensitive to tilt, being only sensitive to angular acceleration.

Assuming that sensors 1 and 2 are equal:

$$\begin{aligned} &\omega_1 = \omega_2 = \omega \\ &l_1 = l_2 = l \\ &y_R = y_2 - y_1 \\ &y_S = y_2 + y_1 \end{aligned}$$

results in:

$$\begin{cases} y_R'' + 2\omega D y_R' + \omega^2 y_R = 2l\psi'' \\ y_S'' + 2\omega D y_S' + \omega^2 y_S = -2(x_1 - g\alpha) \end{cases} \tag{4.5}$$

When purely translational motion is applied to the system of identical pendulums, both sensors are moving in the same direction, and their outputs are identical (Fig. 4.1a). When purely rotational motion around vertical axis is applied to the same system, sensors are moving in opposite directions (Fig. 4.1b): sensor 1 in positive direction, and sensor 2 in negative direction. Russian instrument VBPP (Kharin and Simonov 1969) was based on this principle by measuring the difference of electrical outputs of the two identical pendulums.

Based on Eqs. (4.3)-(4.5) the following observations can be made:

- Sensitivity to rotations ψ'' around vertical axis is higher for pendulums with long pendulum arm.
- If the signal is low, and sensors are not identical, system is measuring errors instead of rotations.
- Summation of the signals from the two sensors results in acceleration plus tilt (if tilt exists). Tilt sensitivity of both horizontal sensors is the same.

System of two vertical pendulums

Vertical sensors are not sensitive to tilts for small tilt angles. A general case of combination of the two vertical sensors can be written as:

$$\begin{cases} y_1'' + 2\omega_1 D_1 y_1' + \omega_1^2 y_1 = -z'' - l_1\alpha'' \\ y_2'' + 2\omega_2 D_2 y_2' + \omega_2^2 y_2 = -z'' + l_2\alpha'' \end{cases} \tag{4.6}$$

Consequently, the system of Eq. (4.6) can be resolved against both ground motion parameters (vertical acceleration and angular acceleration of tilt):

$$\begin{cases} y_R'' + 2\omega D y_R' + \omega^2 y_R = 2l\alpha'' \\ y_S'' + 2\omega D y_S' + \omega^2 y_S = -2z'' \end{cases} \tag{4.7}$$

In contrast to horizontal sensors, combination of the two identical vertical sensors allows resolution of both vertical and angular acceleration. Unfortunately, difficulty of building two identical mechanical pendulums did not result in any reliable measurements using VBPP.

Learning from our previous experience, we decided to use another approach to measuring rotations (Graizer et al. 1989, Graizer 1991). Since it is easier to measure precisely natural parameters of each sensor (ω_n , D_n , l_n) than to build identical pendulums, we based our two-pendulum instrument on a post-measurement processing. We used same sensor arrangement as proposed by Golytsin (1912), but instead of trying to build two identical mechanical systems, we measured parameters of each sensor and applied post-correction to get rotation and translation.

Integrating both sides of Eq. (4.6) twice and assuming:

$$\begin{aligned} F_1(t) &= y_1 + 2\omega_1 D_1 \int_0^t y_1 d\tau + \omega_1^2 \int_0^t d\tau \int_0^t y_1 d\tau \\ F_2(t) &= y_2 + 2\omega_2 D_2 \int_0^t y_2 d\tau + \omega_2^2 \int_0^t d\tau \int_0^t y_2 d\tau \end{aligned} \tag{4.8}$$

results in:

$$\begin{cases} F_1(t) = -z(t) - l_1\alpha(t) \\ F_2(t) = -z(t) + l_2\alpha(t) \end{cases} \tag{4.9}$$

and

$$z(t) = -\frac{l_2 F_1(t) + l_1 F_2(t)}{l_1 + l_2}$$
$$\alpha(t) = \frac{F_2(t) - F_1(t)}{l_1 + l_2} \tag{4.10}$$

Since the sum of linear and angular displacement functions in the right side of Eq. (4.9) should demonstrate same type of behavior as displacement function alone, same integration scheme as used in strong-motion to get displacements can be used for data processing and baseline correction (e.g., Trifunac 1971, Graizer 1979).

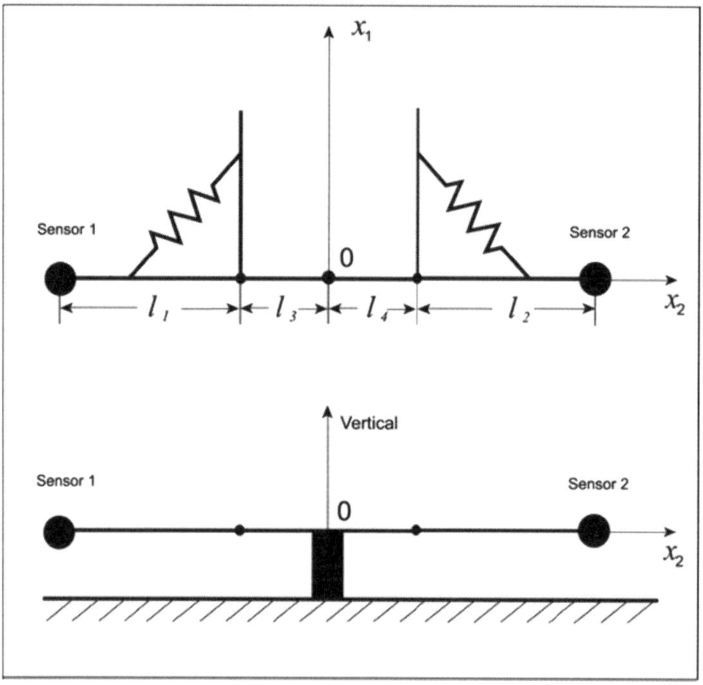

Fig. 4.2 Increasing sensitivity to rotations around the vertical axis

Increasing sensitivity to rotations

To increase the sensitivity of the system to rotations ψ'' around the vertical axis (torsion), we can put sensor 1 at a distance l_3 and sensor 2 at a distance l_4 from the centre of rotation 0 (Fig. 4.2). Assuming the centre of rotation at the same place 0 we will increase the sensitivity and:

$$(y_2'' - y_1'') + (2\omega_2 D_2 y_2' - 2\omega_1 D_1 y_1') + (\omega_2^2 y_2 - \omega_1^2 y_1) \\ = (l_1 + l_3 + l_2 + l_4)\psi'' \tag{4.11}$$

4.3 Testing and Measurements

A registration system similar to that described above consisting of two pairs of accelerometers was built to record large ground motions. One pair of accelerometers was sensitive to vertical and another one to horizontal ground motion. It was based on commercially made Soviet accelerometers ASZ (similar to the American SMA-1). The system was first tested at the Institute of the Physics of the Earth using a specially designed shake-table (Graizer et al. 1989). It was later applied to measuring ground motion in the near-field of the two underground nuclear explosions. Explosions of different power (m_b of 4.5 and 4.4, respectively) were recorded at the same station at the hypocentral distance of less than 1 km (at reduced distances of 14.3 and 18.6 $m/kg^{1/3}$).

The results of separate determination of displacements and tilts are shown in Fig. 3. The maximum amplitude of displacement reached 14 mm at the vertical component, and tilt reached $3.7*10^{-3}$ rad (0.21^0). The maximum tilt during the second less powerful explosion reached $9.2*10^{-4}$ rad (0.053^0). The similarity in the shapes of tilts and displacement pulses for both explosions gives additional confidence in the results. Ratio of tilt motions between the two explosions is ~4, and the displacements differ 2.5 times. The more rapid decay of tilt with distance to the source is consistent with the theory of an explosion. Residual tilts reached $1.5*10^{-3}$ and $2.4*10^{-4}$ rad for the first and second explosions, respectively. Those results are not contradicting the published data. For example, amplitudes of relatively slow tilts measured in lakes at distances of few km from the epicenter of the CANNIKIN underground explosion were of the order of 10^{-5} to 10^{-4} rad (Dickey et al. 1972).

4.4 Discussion and Conclusions

Measuring rotations of the ground and structures during earthquake shaking is not part of common strong motion measurement practice, and there are only few measurements (estimates) of rotations during strong ground shaking. We considered modified classical approach to rotation measurements by using pairs of pendulums. The registration system based on these principles was used to record tilts in the vicinity of two large underground explosions, with maximum tilts reaching $3.7*10^{-3}$ and $9.2*10^{-4}$ rad, respectively.

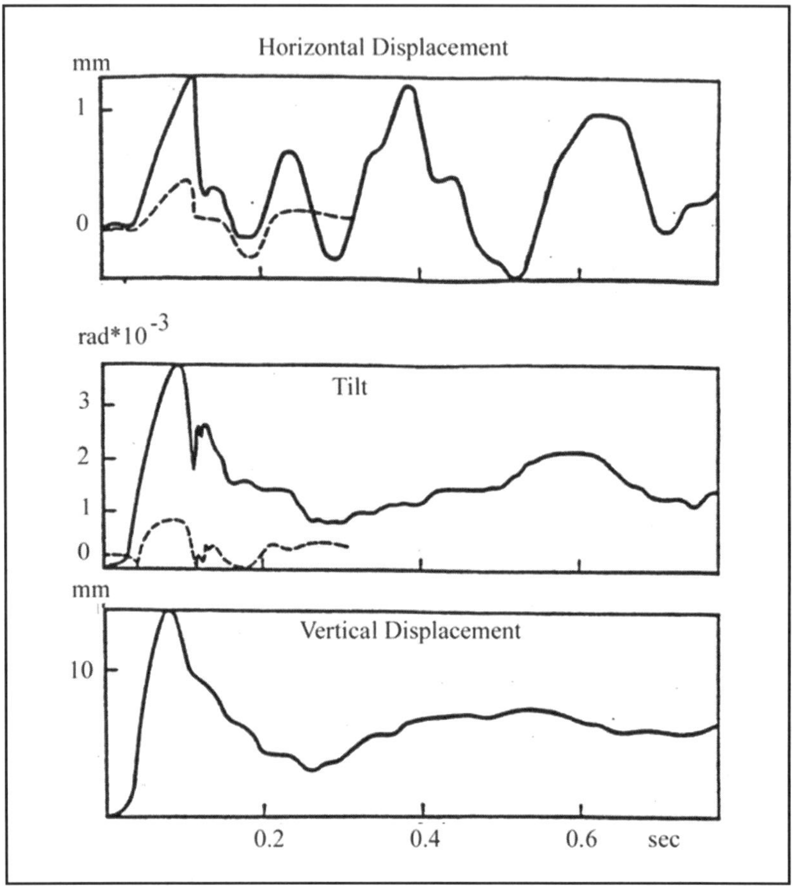

Fig. 4.3 Motions in the near-field of the two underground explosions: horizontal displacement, tilt and vertical displacement (from Graizer et al. 1989)

In recent years, the situation started to change when researchers realized the necessity of recording translational and rotational earthquake motions simultaneously. Recent technological advances provide new opportunities for rotation measurements since some developments previously available only for defense industries become cheaper and widely available. The six-component strong-motion measuring systems that include three translational and three rotational sensors should bring new insights into earthquake engineering studies. Two-pendulum systems similar to that described above can be used for measuring rotations during strong ground shaking.

Acknowledgments. I am grateful to all my colleagues who at different stages participated in studies of rotational effects. This work was first started at the Institute of the Physics of the Earth in the former Soviet Union. Any opinions, findings, and conclusions or recommendations expressed in this material are those of the author solely and do not necessarily reflect the views of the California Geological Survey.

References

Bradner H, Reichle M (1973) Some methods for determining acceleration and tilt by use of pendulums and accelerometers. Bull Seism Soc Am **63**: 1-7

Dickey DD, McKeown FA, Bucknam RC (1972) Preliminary results of ground deformation measurements near the CANNIKIN explosion. Bull Seism Soc Am **62**: 1505-1518

Farrell WE (1969) A gyroscopic seismometer: measurements during the Borrego Earthquake. Bull Seism Soc Am **59**: 1239-1245

Golitsyn BB (1912) Lectures on Seismometry. Russian Acad Sci, St Petersburg (in Russian)

Graizer VM (1979) Determination of the true displacement of the ground from strong-motion recordings. Izv USSR Acad Sci, Physics Solid Earth **15**: 12, 875-885

Graizer VM (1989) Bearing on the problem of inertial seismometry. Izv USSR Acad Sci, Physics Solid Earth **25**: 1, 26-29

Graizer VM, Kuznetsov OP, Nedoshivin NI, Sultanov DD (1989) Bearing on ground tilt measurements near the explosion source. Reports of USSR Acad Sci **305**: 2, 314-318 (in Russian)

Graizer VM (1991) Inertial seismometry methods. Izv USSR Acad Sci, Physics Solid Earth **27**: 1, 51-61

Graizer VM (2005) Effect of tilt on strong motion data processing. Soil Dyn Earthq Eng **25**: 197-204

Graizer V (2006a) Tilts in strong ground motion. Bull Seism Soc Am **96**: 2090-2106

Graizer VM (2006b) Equation of pendulum motion including rotations and its implications to the strong-ground motion. In: Teisseyre R, Takeo M, Majewski E (eds) Earthquake source asymmetry, structural media and rotation effects, Springer, Berlin, pp 471-485

Huang B-S (2003) Ground rotational motions of the 1999 Chi-Chi, Taiwan earthquake as inferred from dense array observations. Geophys Res Letters **30**: 6, 1307-1310

Kharin DA, Simonov LI (1969) VBPP seismometer for separate registration of translational motion and rotations. In: Seismic Instruments **5**: 51-66 (in Russian)

Niazi M (1986) Inferred displacements, velocities and rotations of a long rigid foundation located at El Centro differential array site during the 1979 Imperial Valley, California earthquake. Earthquake Eng Struct Dyn **14**: 531-542

Nigbor RL (1994) Six-degree-of-freedom ground-motion measurements. Bull Seism Soc Am **84**: 1665-1669

Oliveira CS, Bolt BA (1989) Rotational components of surface strong ground motion. Earthquake Eng Struct Dyn **18**: 517-526

Schreiber KU, Stedman GE, Igel H, Flaws A (2006) Ring laser gyroscopes as rotation sensors for seismic wave studies. In: Teisseyre R, Takeo M, Majewski E (eds) Earthquake source asymmetry, structural media and rotation effects, Springer, Berlin, pp 377-390

Takeo M (1998) Ground rotational motions recorded in near-source region of earthquakes. Geophys Res Letters **25**: 6, 789-792

Todorovska MI (1998) Cross-axis sensitivity of accelerographs with pendulum like transducers – mathematical model and the inverse problem. Earthq Eng Struct Dyn **27**: 1031-1051

Trifunac MD (1971) Zero baseline correction of strong-motion accelerograms. Bull Seism Soc Am **61**: 1201-1211

Trifunac MD, Todorovska MI (2001) A note on the usable dynamic range of accelerographs recording translation. Soil Dyn Earthq Eng **21**: 275-286

Zahradnik J, Plesinger A (2005) Long-period pulses in broadband records of near earthquakes. Bull Seism Soc Am **95**: 1928-1939

5 Theory and Observations: Some Remarks on Rotational Motions

Roman Teisseyre

Institute of Geophysics, Polish Academy of Sciences
ul. Księcia Janusza 64, 01-452 Warszawa, Poland
e-mail: rt@igf.edu.pl

5.1 Ten Motions and Deformations

In the first chapter we have discussed the basic system of motions and deformations. To the displacements and rotations (spin), theoretically forming six basic independent motions, we shall add four basic deformations: the axial deformation, E_{kk} (see: Chap. 1), and the twist vector describing the string-string deformation (see: Chap. 1) and related to deviatoric strain tensor, E^D_{sk}, $(E^D_{ss} = 0)$; twist can be defined as a vector perpendicular to the string-string plane (cf: Chap. 7). We may note that these motions form the reference system with ten components.

Further on (see: Chaps. 6 and 7), we will demonstrate that the axial and deviatoric strains can be presented in the invariant form as axial and twist motions.

In our considerations (see: Chap. 1) we have mentioned that the independent displacements, u_k (except the axial part, $\partial u_n/\partial x_n$), and rotations, $\omega_{[k]}$, might be mutually replaced using the appropriate potentials (see: Chap. 1); such equivalence holds for any continuum with an intrinsic length, l, greater than zero (or realistically, greater than the Planck length unit). However, we shall note that as a result we may obtain two displacements, or two rotation fields shifted in phase; e.g., the original displacements, u, and those related to rotations, U (U = curl ω), may be shifted in their phases. Moreover, the displacement and rotations differ essentially when considering their origin and effects produced. However, in a theoretical description we can demonstrate that the spin motions, $\omega_{[k]}$, and twist motions, $\omega_{(s)}$, defined on the basis of the shear motions

$$E^D_{ns} = \frac{1}{2}\left(\frac{\partial u_s}{\partial x_n} + \frac{\partial u_n}{\partial x_s}\right), \qquad s \neq n,$$

form a system governed by the equations of motion quite similar to the Maxwell equtions (see: Chap. 7).

We have already considered numerous defaults of the classical elasticity (see: Chap. 1) and we have also quoted many attempts to improve this theory. A new standard asymmetric continuum theory is presented in Chapter 7; this new theory not only eliminates the above-mentioned defaults but also can be extended for larger deformations, including some aspects of fracture processes. Moreover, this relatively simple theory leads to new relations, like those related to defect distribution, and permits to consider the interaction fields (see: Chap. 16).

We hope that in this way we can approach an adequate description of the modern observation results concerning the seismic rotation waves.

5.2 Recording Spin and Twist

The first seismic rotation observations with the ring laser system (Igel et al. 2005) started in the Wetzell Observatory, Germany. Considering a plane wave propagation, it is possible to directly compare the observed rotation motions with the transverse displacement velocity rotations obtained by the array of seismographs; such very successful comparisons were presented by Suryanto et al. (2006) and Cochard et al. (2006); it was proved that there exists almost exact fit between the rotation rate converted from the array-derived transverse acceleration and rotation rate obtained from the ring laser system.

Of course, the new very precise measurements based on the Sagnac interferometer systems are more adequate to determine rotations and require much less logistic efforts than the derivation of rotations from the array systems, which, moreover, can be sensitive to heterogeneous ground condition in an array region.

The above-mentioned comparisons between the rotation rate records obtained from the ring laser system and the rates derived from the transverse accelerations, based on records at the array of broadband seismographs, confirm the theoretical equivalence between the rotations and the deviatoric strains derived from the displacement velocity motions, as expected from the theory.

Rotational motions recorded on the Izu peninsula, Japan, with the use of the fiber-optics Sagnac sensor provided the important near-source data; Takeo (2006), inspired by these results, has developed an important theoretical consideration on the influence of material defects, including dislocations and disclinations, on a possible magnification of rotational motions in a zone near the earthquake source. This is confirmed by some near-source observations made with the use of the rotation seismometer systems (Moriya and Teisseyre 2006, Teisseyre et al. 2003) in Hokkaido, Japan, and the Appenines, Italy. However, we may mention also an exceptionally small rotational part of motions for the local surface volcanic eruptions in a soft volcanic material as observed at extremely small distances (hundreds of meters).

It is to be noted that the measuring system based on the rotation seismographs records simultaneously both the spin and twist motions (Moriya and Marumo 1998, Teisseyre et al. 2003, Wiszniowski 2006); e.g., to obtain the spin and twist motions around the vertical axis we need the seismograph system composed of two parallel horizontal pendulums of opposite orientations. Spin is achieved from a mean of the recorded values, while twist from their difference; however, the twist field thus obtained presents only the angular variations of the off-diagonal shear axes.

Only in the special coordinate system related to the off-diagonal shear axes, the twist measured with the use of rotation seismographs represents the exact shear field. With the help of a complete system of strainmeters, we can achieve reliable data on the shear-twist variations. However, in Chapter 7 we present the invariant definition of the twist motion, therefore, we shall explain, in advance, how we can compare such invariantly defined twist motion when we have a possibility to record the full shear oscillations: to achieve the required comparison we shall transform, at each time moment, the measured shear values into the off-diagonal system:

$$\{E_{11},\ E_{22},\ E_{33},\ E_{23},\ E_{31},\ E_{12}\} \rightarrow \{E_{23}^{D},\ E_{31}^{D},\ E_{12}^{D}\} = \{\omega_{(s)}\}.$$

On a free ground surface, $z = 0$, we have $E_{31} = E_{32} = 0$ and we need to measure only E_{11}, E_{22}, and E_{12}.

5.3 Rotation Motions in the Universe

We have considered the independent motion and deformation fields; we will continue this approach.

A significant role of rotation motions in fracture processes is discussed further on (see: Chap. 8); however, we shall present already here some further remarks. Fracture process under confining pressure load can lead to the synchronized significant spin motions related to fragmentation centers, while under a shear load the spin motions could be partly compensated by the opposite sense of the friction-related rotations along the main fracture plane and the other ones along the auxiliary perpendicular fracture planes.

Finally, we quote here a particular observation: the squeezing process related to the depression region in the lower part of the Pasterze glacier, Grossglockner, Austria. The circular fracture zones (Fig. 5.1) present an interesting example of the co-action of different deformation motions.

Fig. 5.1 Depression in the Pasterze glacier, Grossglockner, Austria

We must emphasize here a very significant role of rotations observed in ground tilts (Kaczorowski 2006) and in rocking and torsion of the elevated structures on a ground surface (Zembaty 2006, Trifunac 2006).

Numerous observations, in the whole spectrum of ranks, kinds and dimensions, show the universal role of rotational motions. The rotational structures and motions appear in astronomical observations, are observed in geological and tectonical structures and the continent evolution, become of recent interest in seismology and fracture mechanics and enter the micro-domains and their physics. In the related context, we may mention two books devoted to this point of view: "Vortex-Related Events of the Geological Processes" (Vikulin 2004) and "Rotational Processes in Geology and Physics" (Milanovsky 2007).

Starting with a point of view based on Asymmetric Continuum Theory (see Chap. 7) we present some remarks concerning the Universe origin. Such implications, when do not interfere or deny the essential results of the quantum and relativity theories, may deliver a simpler and intuitional approach to the metric tensor perturbations in relativity and hypotheses related to the inflation phase effects in cosmology.

In Chapter 1 we discussed the basic motions and deformations in asymmetric continuum: the axial motion, displacements, rotations and twist (oscillations of the off-diagonal shear axes); these fields being function of time. With initial explosion (or quantum virtual process) appearing at zero moment (determined with Planck precision $\sim 10^{-44}$s) and spreading axially in a symmetric system at the "inflation" velocity, $C \gg c$, the initial "condensate" (string and quarks) could be subjected to motion similar to axial motion in elastic continuum.

The end of this phase of rapid expansion could relate to phase transformations ($\sim 10^{-35}$s) leading to the physical state of our Universe with the light velocity limit; the initial "condensate" transforms to our Riemannian 4D space and related vacuum formed by the net of primordial elements. This phase transition would mean transition from symmetric phase to asymmetric one: the resulting World might correspond to the fluid phase with the asymmetric properties (see Chap. 14)

At this phase, an enormous expansion energy should be transformed to other forms. Here, we may put forward another hypothesis: both the continuation of expansion at $V < c$, and the formation of initial rotations can lead to spiral structures, where we might also introduce a notion of spin black hole.

References

Cochard A, Igel H, Schuberth B, Suryanto W, Velikoseltsev A, Schreiber U, Wassermann J, Scherbaum F, Vollmer D (2006) Rotational motions in seismology: theory, observation, simulation. In: Teisseyre R, Takeo M, Majewski E (eds) Earthquake source asymmetry, structural media and rotation effects. Springer, Berlin, pp 391-411

Igel H, Schreiber U, Flaws A, Schuberth B, Velikoseltsev A, Cochard A (2005) Rotational motions induced by the M8.1 Tokachi-oki earthquake, September 25, 2003. Geophys Res Lett **32**: 8, L08309, doi: 10.1029/2004GL022336

Kaczorowski M (2006) High-resolution wide-range tiltmeter: observations of Earth free oscillations excited by the 26 December 2004 Sumatra-Andaman earthquake. In: Teisseyre R, Takeo M, Majewski E (eds) Earthquake source asymmetry, structural media and rotation effects, Springer, Berlin, pp 493-520

Milanovsky EE (ed) (2007) Rotational processes in geology and physics. Izd KomKniga, Moscow (in Russian), 628 pp

Moriya T, Marumo T (1998) Design for rotation seismometers and their calibration. Geophys Bull Hokkaido Univ **61**: 99-106

Moriya T, Teisseyre R (2006) Design of rotation seismometer and non-linear behaviour of rotation components of earthquakes. In: Teisseyre R, Takeo M, Majewski E (eds) Earthquake source asymmetry, structural media and rotation effects. Springer, Berlin, pp 439-450

Suryanto W, Igel H, Wassermann J, Cochard A, Schuberth B, Vollmer D, Scherbaum F, Schreiber U, Velikoseltsev A (2006) First comparison of array-derived rotational ground motions with direct ring laser measurements. Bull Seismol Soc Am **69**: 6, doi: 10.1785/0120060004

Teisseyre R, Suchcicki J, Teisseyre KP, Wiszniowski J, Palangio P (2003) Seismic rotation waves: basic elements of theory and recording. Ann Geophys **46**: 4, 671-685

Trifunac MD (2006) Effects of torsional and rocking excitations on the response of structures. In: Teisseyre R, Takeo M, Majewski E (eds) Earthquake source asymmetry, structural media and rotation effects. Springer, Berlin, pp 569-582

Vikulin AV (ed) (2004) Vortex–related events of the geological processes. KSPU, Petropavlovsk-Kamchatsky (in Russian), 297 pp

Wiszniowski J (2006) Rotation and twist motion recording – couple pendulum and rigid seismometers system. In: Teisseyre R, Takeo M, Majewski E (eds) Earthquake source asymmetry, structural media and rotation effects. Springer, Berlin, pp 451-470

Zembaty Z (2006) Deriving seismic surface rotations for engineering purposes. In: Teisseyre R, Takeo M, Majewski E (eds) Earthquake source asymmetry, structural media and rotation effects. Springer, Berlin, pp 549-568

PART II

CONTINUUM WITH DEFECT DENSITIES AND ASYMMETRY OF FIELDS

6 Field Invariant Representation: Dirac Tensors

Jan Wiszniowski and Roman Teisseyre

Institute of Geophysics, Polish Academy of Sciences,
ul. Księcia Janusza 64, 01-452 Warszawa, Poland
e-mail: rt@igf.edu.pl

6.1 Introduction

Using the Dirac matrices (Dirac 1982, Good 1955) we can give invariant presentation of any symmetric and antisymmetric tensor forms. Splitting a symmetric tensor into the axial and deviatoric parts we come, with the help of appropriate Dirac matrices, to their invariant representation in any system. Here we present the diagonal and off-diagonal tensor forms of the deviatoric tensors.

6.2 Axial and Deviatoric Parts of any Symmetric Tensor

In Chapter 1 we have discussed the basic motions in an asymmetric continuum: the physical fields are represented by tensors. Any antisymmetric tensor preserves its general form; however, the symmetric tensors may have different forms in different systems: diagonal form, off-diagonal or mixed one. Therefore, it is useful to split a symmetric tensor into its axial, E_{ik}^{A}, and deviatoric, E_{ik}^{D}, parts:

$$E_{ik} = E_{ik}^{A} + E_{ik}^{D} = \frac{1}{3} E_{ss} \delta_{ik} + \left(E_{ik} - \frac{1}{3} E_{ss} \delta_{ik} \right). \tag{6.1}$$

Of course, a tensor with vanishing trace is equal to its deviatoric part:

$$\omega_{(ik)} = \omega_{(ik)}^{D}, \quad \text{at} \quad \omega_{(ss)} = 0. \tag{6.2}$$

In a specially chosen coordinate system we can describe a deviatoric tensor (6 values in general) in its diagonal or off-diagonal forms (3 values only):

$$\omega_{(ik)}^{D} = \begin{bmatrix} \omega_{(11)}^{D} & 0 & 0 \\ 0 & \omega_{(22)}^{D} & 0 \\ 0 & 0 & \omega_{(33)}^{D} \end{bmatrix}, \quad \text{or} \quad \omega_{(ik)}^{D} = \begin{bmatrix} 0 & \omega_{(3)} & -\omega_{(2)} \\ \omega_{(3)} & 0 & \omega_{(1)} \\ -\omega_{(2)} & \omega_{(1)} & 0 \end{bmatrix} \tag{6.3}$$

where we have put $\omega^D_{(12)} = \omega_{(3)}$, $\omega^D_{(13)} = -\omega_{(2)}$, and $\omega^D_{(23)} = \omega_{(1)}$.

When comparing the observational/recording data related to a symmetric tensor with the theoretical results presented in such a special system, we shall be aware that we must transform the measured tensor values to the appropriate system (diagonal or off-diagonal) for any time moment. However, for a theoretical consideration, there exists a method to preserve such forms invariant with the help of Dirac tensors, maintaining these forms (diagonal or off-diagonal) in any 4D system. We will consider the off-diagonal case.

6.3 Dirac Tensors

We introduce the following system of the Dirac ε^α tensors:

$$\varepsilon^1 = \begin{bmatrix} 0 & 0 & 0 & -1 \\ 0 & 0 & 1 & 0 \\ 0 & 1 & 0 & 0 \\ -1 & 0 & 0 & 0 \end{bmatrix}, \quad \varepsilon^2 = \begin{bmatrix} 0 & 0 & -1 & 0 \\ 0 & 0 & 0 & -1 \\ -1 & 0 & 0 & 0 \\ 0 & -1 & 0 & 0 \end{bmatrix},$$
$$\varepsilon^3 = \mathrm{i}\begin{bmatrix} 0 & 0 & 0 & 1 \\ 0 & 0 & 1 & 0 \\ 0 & -1 & 0 & 0 \\ -1 & 0 & 0 & 0 \end{bmatrix}, \quad \varepsilon^4 = \mathrm{i}\begin{bmatrix} 1 & 0 & 0 & 0 \\ 0 & 1 & 0 & 0 \\ 0 & 0 & -1 & 0 \\ 0 & 0 & 0 & -1 \end{bmatrix}. \tag{6.4}$$

For the system $\{x_s, \mathrm{i}ct\}$, $\eta^{\alpha\beta} = \begin{bmatrix} 1 & 0 & 0 & 0 \\ 0 & 1 & 0 & 0 \\ 0 & 0 & 1 & 0 \\ 0 & 0 & 0 & -1 \end{bmatrix}$, these tensors fulfil the following condition:

$$\frac{1}{2}\left(\varepsilon^\alpha \varepsilon^\beta + \varepsilon^\beta \varepsilon^\alpha\right) = \eta^{\alpha\beta}. \tag{6.5}$$

Some other Dirac tensors can be obtained as their products, e.g.:

$$\varepsilon^1\varepsilon^3 = \mathrm{i}\begin{bmatrix} 1 & 0 & 0 & 0 \\ 0 & -1 & 0 & 0 \\ 0 & 0 & 1 & 0 \\ 0 & 0 & 0 & -1 \end{bmatrix}, \quad \varepsilon^2\varepsilon^3 = \mathrm{i}\begin{bmatrix} 0 & 1 & 0 & 0 \\ 1 & 0 & 0 & 0 \\ 0 & 0 & 0 & -1 \\ 0 & 0 & -1 & 0 \end{bmatrix},$$

$$\varepsilon^1\varepsilon^2 = \begin{bmatrix} 0 & 1 & 0 & 0 \\ -1 & 0 & 0 & 0 \\ 0 & 0 & 0 & -1 \\ 0 & 0 & 1 & 0 \end{bmatrix}, \qquad \varepsilon^1\varepsilon^4 = \mathrm{i}\begin{bmatrix} 0 & 0 & 0 & 1 \\ 0 & 0 & -1 & 0 \\ 0 & 1 & 0 & 0 \\ -1 & 0 & 0 & 0 \end{bmatrix},$$

$$\varepsilon^2\varepsilon^4 = \mathrm{i}\begin{bmatrix} 0 & 0 & 1 & 0 \\ 0 & 0 & 0 & 1 \\ -1 & 0 & 0 & 0 \\ 0 & -1 & 0 & 0 \end{bmatrix}, \qquad \varepsilon^3\varepsilon^4 = -\begin{bmatrix} 0 & 0 & 0 & 1 \\ 0 & 0 & 1 & 0 \\ 0 & 1 & 0 & 0 \\ 1 & 0 & 0 & 0 \end{bmatrix}, \tag{6.6}$$

$$\varepsilon^4\varepsilon^2\varepsilon^3 = \begin{bmatrix} 0 & -1 & 0 & 0 \\ -1 & 0 & 0 & 0 \\ 0 & 0 & 0 & -1 \\ 0 & 0 & -1 & 0 \end{bmatrix}.$$

Now, for the deviatoric tensor, e.g., (Eq. 6.3), we can define its 4D equivalent:

$$\omega^D_{(\lambda\kappa)} = \begin{bmatrix} 0 & -\omega_{(3)} & -\omega_{(2)} & -\omega_{(1)} \\ -\omega_{(3)} & 0 & \omega_{(1)} & -\omega_{(2)} \\ -\omega_{(2)} & \omega_{(1)} & 0 & -\omega_{(3)} \\ -\omega_{(1)} & -\omega_{(2)} & -\omega_{(3)} & 0 \end{bmatrix} \tag{6.7}$$

and we arrive at its invariant representation in any system:

$$\omega^D_{(\lambda\kappa)} = \omega_{(3)}\varepsilon^1 + \omega_{(2)}\varepsilon^2 + \omega_{(1)}\varepsilon^4\varepsilon^2\varepsilon^3 \tag{6.8}$$

where $\omega_{(3)}$, $\omega_{(2)}$, $\omega_{(1)}$ play a role of scalar fields (there remains only a change of orientation of the Dirac tensors).

Further on (see the next chapter), we will show that the form (Eq. 6.8) is suitable to write the wave equation, e.g.:

$$\mu\frac{\partial^2\omega_{(\lambda\kappa)}}{\partial x_k\partial x_k} - \rho\frac{\partial^2\omega_{(\lambda\kappa)}}{\partial t^2} = 0. \tag{6.9}$$

This relation describes the motion of the twist field, $\omega_{(s)} = \{\omega_{(1)}, \omega_{(2)}, \omega_{(3)}\}$, forming a kind of strange vector; we consider this point of view further on.

Again we stress that a comparison between these theoretical fields and the observational/recording data requires to transform the measured tensor values to their off-diagonal form (for any time moment).

Finally, we may add that any 4D antisymmetric tensor can be presented by means of the Dirac tensors (see: Eq. 6.6) as well:

$$F_{[\alpha\beta]} = \mathrm{i}\varepsilon^1\varepsilon^4 F_1 + \mathrm{i}\varepsilon^2\varepsilon^4 F_2 + \varepsilon^1\varepsilon^2 F_3 \tag{6.10}$$

where we have put

$$F_{[\alpha\beta]} = \begin{bmatrix} 0 & F_3 & -F_2 & -F_1 \\ -F_3 & 0 & F_1 & -F_2 \\ F_2 & -F_1 & 0 & -F_3 \\ F_1 & F_2 & F_3 & 0 \end{bmatrix}.$$

This form is especially useful for presenting the balance relation for the antisymmetric tensor fields, e.g., the Maxwell equations for the electromagnetic tensor (here, without a current term):

$$\frac{\partial}{\partial x_\beta} F_{[\alpha\beta]} = 0. \tag{6.11}$$

6.4 Motion Equations: Classical Elasticity

The classical form of motion equation can be presented as follows:

– for the central motion:

$$\frac{\partial^2}{\partial x_s \partial x_s} S_{kk} - \rho \frac{\partial^2}{\partial t^2} E_{ss} = (3\lambda + 2\mu) \frac{\partial^2}{\partial x_s \partial x_s} E_{kk} - \rho \frac{\partial^2}{\partial t^2} E_{ss} = \frac{\partial}{\partial x_s} F_s \tag{6.12}$$

– for the shear motion as described by the deviatoric tensor parts:

$$\frac{\partial^2}{\partial x_i \partial x_s} S^D_{sk} + \frac{\partial^2}{\partial x_k \partial x_s} S^D_{si} - 2\rho \frac{\partial^2}{\partial t^2} E^D_{ik} = \frac{\partial}{\partial x_i} F_k + \frac{\partial}{\partial x_k} F_i - \frac{1}{3} \frac{\partial}{\partial x_s} F_s \delta_{ik}$$

– and finally

$$\mu \frac{\partial^2}{\partial x_s \partial x_s} E^D_{ik} - \rho \frac{\partial^2}{\partial t^2} E^D_{ik} = \frac{1}{2} \left(\frac{\partial}{\partial x_i} F_k + \frac{\partial}{\partial x_k} F_i - \frac{1}{3} \frac{\partial}{\partial x_s} F_s \delta_{ik} \right). \tag{6.13}$$

6.5 Diagonal and Off-Diagonal Symmetric Tensor Representation

Two dimensions

Any symmetric deviatoric tensor, $E_{KK} = 0$, can be transformed to the diagonal form to be further easily transformed into the off-diagonal form:

$$E_{KS} = \begin{bmatrix} m & n \\ n & -m \end{bmatrix} \rightarrow E_{KS} = \begin{bmatrix} e & 0 \\ 0 & -e \end{bmatrix} \rightarrow E_{KS} = \begin{bmatrix} 0 & e \\ e & 0 \end{bmatrix}. \tag{6.14}$$

Three dimensions

A three dimensional case is more complicated; we may consider the following sequence:

$$E_{KS} = \begin{bmatrix} E_{11} & E_{12} & E_{13} \\ E_{12} & E_{22} & E_{23} \\ E_{13} & E_{23} & E_{33} \end{bmatrix} \rightarrow E_{KS} = \begin{bmatrix} \lambda_1 & 0 & 0 \\ 0 & \lambda_2 & 0 \\ 0 & 0 & \lambda_3 \end{bmatrix} \rightarrow E_{KS} = \begin{bmatrix} 0 & e_3 & e_2 \\ e_3 & 0 & e_1 \\ e_2 & e_1 & 0 \end{bmatrix} \tag{6.15}$$

where $\lambda_1 + \lambda_2 + \lambda_3 + = 0$.

The eigenvalues appearing in the diagonal form are determined from the characteristic equation:

$$\lambda^3 - 2\left(e_1^2 + e_2^2 + e_3^2\right)\lambda - 2e_1e_2e_3 = 0. \tag{6.16}$$

However, the transition from the diagonal form to the off-diagonal one is not unique; we get a multitude of representations. Substituting λ with λ_1, λ_2 and λ_3 we achieve a set of three equations but with only two independent variables:

$$\Pi = e_1e_2e_3 \quad \text{and} \quad \Sigma = e_1^2 + e_2^2 + e_3^2. \tag{6.17}$$

The shear energy in a pure shear system is $E = \dfrac{1}{2} S_{ks} E_{ks} = 2\mu E_{12} E_{12}$. In the off-diagonal presentation of strain deviatoric tensor we obtain

$$E_{kk} = 0, \quad E = \frac{1}{2} S_{ks} E_{ks} = 2\mu\left(E_{12}E_{12} + E_{13}E_{13} + E_{32}E_{32}\right);$$

$$e_1^2 + e_2^2 + e_3^2 = \frac{1}{2\mu} E, \quad \text{and} \quad \Sigma = \frac{1}{2\mu} E.$$

Since the tensor is deviatoric, the three equations are consistent and we can express these variables as follows

$$\Pi = -\frac{1}{2}\left(\lambda_K \lambda_L^2 + \lambda_L \lambda_K^2\right); \qquad \Sigma = \lambda_K^2 + \lambda_K \lambda_L + \lambda_L^2 \tag{6.18}$$

where indexes K and L are any two different numbers of the set {1, 2, 3}, e.g. $K = 1$, $L = 2$.

Using the quantities Π and Σ, we can describe the deviatoric tensor; we can find parametric solution for the unknown quantities e_1, e_2 and e_3 in (Eq. 6.15). To this end we can treat one of these quantities as parameter s, e.g., $e_3 = s$. For this case we obtain:

$$\begin{aligned} e_1 &= \frac{\sqrt{s\Sigma - s^3 + 2\Pi} + \sqrt{s\Sigma - s^3 - 2\Pi}}{2\sqrt{s}}, \\ e_2 &= \frac{\sqrt{s\Sigma - s^3 + 2\Pi} - \sqrt{s\Sigma - s^3 - 2\Pi}}{2\sqrt{s}}, \\ e_3 &= s. \end{aligned} \tag{6.19}$$

In order to achieve the real values of e_1, e_2, we demand that

$$s^2 + \left|\frac{2\Pi}{s}\right| \le \Sigma. \tag{6.20}$$

From the three values, λ_1, λ_2, λ_3, we can find two with the same sign; expressing Π and Σ by these two values (Eq. 6.18) we can prove the existence of s fulfilling the condition (Eq. 6.20).

We have demonstrated that in 3D a transition from the diagonal form to the off-diagonal one results in a multidude of forms related to any real parameter s fulfilling the condition (Eq. 6.20).

6.6 Particular Cases

Further on, we will consider some particular forms of the off-diagonal tensor.

First, we shall note that the off-diagonal tensor is related to the three values forming a kind of strange vector $\omega_{(s)} = \{\omega_{(1)}, \omega_{(2)}, \omega_{(3)}\}$. We have already shown that these values are not uniquely determined from the deviatoric diagonal tensor; we may have a multitude of representations related to the 2D different orientations around the twist vector. However, for the twist direction perpendicular to a free surface, $x_3 = 0$, under conditions for strains

$E_{31} = E_{32} = 0$, we will obtain a semi two dimensional case $E = 2\mu E_{12}E_{12}$; the problem reduces to the 2D case already discussed.

In 3D we can introduce an extra condition to obtain a unique representation of shears in the off-diagonal system.

We may demand, e.g., that two parameters of $\{e_1, e_2, e_3\}$ be equal

$$e_1 = e_2 \to s\Sigma' - s^3 = 2\Pi. \tag{6.21}$$

This equation is similar to the characteristic equation and has the solutions

$$e_2 = s = \begin{cases} -\lambda_1 \\ -\lambda_2 \\ -\lambda_3 \end{cases}. \tag{6.22}$$

In that case, the off-diagonal values will be

$$e_2 = -\lambda_K, \qquad e_1 = e_3 = \sqrt{\frac{\Pi}{-\lambda_K}}. \tag{6.23}$$

To have a real solution, we must choose the values λ_K for which

$$\operatorname{sign}\lambda_K = -\operatorname{sign}\Pi. \tag{6.24}$$

Further, we present some applications for the seismic waves.

1. A simple squeeze (the P wave):

$$E_{KS} = \begin{bmatrix} 2e & 0 & 0 \\ 0 & -e & 0 \\ 0 & 0 & -e \end{bmatrix} \to E_{KS} = \begin{bmatrix} 0 & e & e \\ 0 & 0 & e \\ e & e & 0 \end{bmatrix}. \tag{6.25}$$

For a simple squeeze this is unique off-diagonal representation of deviatoric tensor because the condition (Eq. 6.20) is true only for $s = e$.

2. Simple shear (the S wave)

The diagonal form

$$E_{KS} = \begin{bmatrix} e & 0 & 0 \\ 0 & -e & 0 \\ 0 & 0 & 0 \end{bmatrix} \tag{6.26}$$

may have a parametric off-diagonal representation in the form

$$E_{KS} = \begin{bmatrix} 0 & s & \sqrt{e^2 - s^2} \\ s & 0 & 0 \\ \sqrt{e^2 - s^2} & 0 & 0 \end{bmatrix} \tag{6.27}$$

or other equivalent representations appropriately rotated.

We can demonstrate the other off-diagonal representations that fulfill the condition (Eq. 6.21):

$$E_{KS} = \begin{bmatrix} 0 & 0 & \pm e \\ 0 & 0 & 0 \\ \pm e & 0 & 0 \end{bmatrix} \tag{6.28}$$

and

$$E_{KS} = \begin{bmatrix} 0 & \frac{e}{\sqrt{2}} & 0 \\ \frac{e}{\sqrt{2}} & 0 & \frac{e}{\sqrt{2}} \\ 0 & \frac{e}{\sqrt{2}} & 0 \end{bmatrix} \tag{6.29}$$

or other equivalent representations appropriately rotated. The simple shear is an exception because $\Pi = 0$.

The deviatoric tensor with minus e_1 and minus e_3 is also correct in all the cases. Now we can write relation (Eq. 6.15) in the form

$$E_{KS} = \begin{bmatrix} E_{13} & E_{12} & E_{13} \\ E_{12} & E_{22} & E_{23} \\ E_{13} & E_{23} & E_{33} \end{bmatrix} \rightarrow E_{KS} = \begin{bmatrix} \lambda_1 & 0 & 0 \\ 0 & \lambda_2 & 0 \\ 0 & 0 & \lambda_2 \end{bmatrix} \rightarrow$$

$$E_{KS} = \begin{bmatrix} 0 & \pm\sqrt{\frac{\Pi}{-\lambda_1}} & -\lambda_1 \\ \pm\sqrt{\frac{\Pi}{-\lambda_1}} & 0 & \pm\sqrt{\frac{\Pi}{-\lambda_1}} \\ -\lambda_1 & \pm\sqrt{\frac{\Pi}{-\lambda_1}} & 0 \end{bmatrix} \tag{6.30}$$

under the assumption that λ_1 fulfills (Eq. 6.24). The sequence of components in the diagonal tensor can be chosen freely.

6.7 Conclusion

We have considered the different invariant representations of the symmetric and antisymmetric tensors; for the symmetric tensor we have separately discussed its axial and deviatoric parts.

Confining ourselves to the classical approach, we have discussed the equation of motions for the symmetric and antisymmetric tensors.

References

Dirac PAM (1982) Principles of quantum mechanics. 4th edn Oxford Univ Press, Oxford

Good RH Jr (1955) Properties of Dirac matrices. Rev Mod Phys **27**: 187-211

7 Asymmetric Continuum: Standard Theory

Roman Teisseyre

Institute of Geophysics, Polish Academy of Sciences
ul. Księcia Janusza 64, 01-452 Warszawa, Poland
e-mail: rt@igf.edu.pl

7.1 Introduction

Classical treatment of the continuum mechanics found its basis in considerations on deformations caused by displacement field, including the moment of momentum and angular deformation counterpart. However, when treating the displacements and rotations appearing in the field relations as the equally and similarly treated independent fields, we will arrive at the theory of asymmetric continuum; our present work relates to our former studies (Teisseyre 2004, 2005) as well as to the recent monograph (Teisseyre and Boratyński 2006).

Thus, we shall recall that in the micropolar and micromorphic theories with the infinite small nuclei (see: Eringen 1999) or in the other advanced continuum treatments, the spin motions or, say, angular fields, appear as independent variables; however, in our present treatment we deal with the ideal elastic continuum, or elastic continuum with the defect fields (dislocation and disclination densities), in which there appear, besides the constitutive laws for the symmetric stresses and strains, also the bonds joining the antisymmetric stresses and the rotation motions with a related rigidity modulus.

In the asymmetric continuum, defined as that including both the symmetric stresses and the antisymmetric stresses, the rotational deformations split into the pure rotation and twist motions, the latter relating to the shear deformations of point-nuclei. The twist motions are defined as those related to oscillations of the shear off-diagonal axes. This notion appeared when analyzing the seismic rotation fields. The rotation motions, spin and twist, can be measured by means of the ring laser or fiber optics interferometers, based on the Sagnac principle or by strainmeteres and rotation seismographs; the experimental evidence for the spin and twist motions is based on the records of seismic rotation fields obtained both with a help of the ring laser or fiber optics interferometers, based on the Sagnac principle (Schreiber et al. 2006, Takeo 2006, Jaroszewicz et al. 2006) or by the rotation seismographs (Moriya and Teisseyre 2006, Wiszniowski 2006). For some events of explosive nature (explosions and some volcanic events) the above-mentioned "true" spin and twist motions are not detectable. The twist deformations present the grain deformations caused by elastic strain; however, when considering the grains as the "rigid" points of continuum, such a bend-twist deformation converts to a kind of 3D space curvature (Teisseyre et al. 2005, Teisseyre et al. 2006).

In the present study we confine ourselves to a new concise theoretical approach to the asymmetric continuum.

The other important remark is that in a homogeneous continuum the influence of rotational processes generated in earthquake sources remains theoretically not attenuated, as it is for the classical ideal elasticity, and the related waves can propagate to distant sites. First let us note that in the classical elastic continuum with the symmetric stresses, the rotations are introduced with the help of the stress moment tensor (related to some arm length). Instead, we assume, in our theoretical approach, the existence of the asymmetric stresses, strains and rotation fields. The antisymmetric elastic stresses form the system equivalent to stress moments, while the symmetric rotations describe the twist motions being equivalent in classic theory to shear oscillations of the point-grains of a continuum (or the equivalent notions appearing in the micromorphic theories) – compare papers by Teisseyre (2002), Teisseyre and Boratyński (2003, 2006) and Boratyński and Teisseyre (2004).

The elastic theory with the asymmetric stress field is presented here; in this theory the antisymmetric part of stresses plays a role equivalent to the stress moments.

In the Appendix to this chapter we recall the Kröner approach, with its self fields and interaction nuclei, we define the elastic fields as given by the difference between the total and self fields, where the total fields – stresses, strains and rotations – mean the fields defined as the respective derivatives of the displacement motion. In this way we arrive at the deviations from the classical elasticity and the independent rotation motions – spin and twist; the first being a counterpart to rotation of displacements, while the other to oscillations of shear strain. Usually the deviations from ideal elasticity are described by the appriopriate elasto-plastic constitutive laws; the other approach follows the Kröner method, in which we maintain the ideal elastic relation for the stresses and strains, supplemented by constitutive law joining the antisymmetric stresses with rotations, and we introduce the self/inner stresses, strains and rotations, as related to the internal nuclei or defects: $\mathbf{S}^S$, $\mathbf{E}^S$, ω^S. Therefore, we distinguish there between the total stresses, strains and rotations $\mathbf{S}^T$, $\mathbf{E}^T$, $\omega^T = 0$, and the asymmetric elastic stresses, strains and rotations $\mathbf{S}$, $\mathbf{E}$, ω. This method is shortly presented in the Appendix (Kröner 1981, Teisseyre 2005, Teisseyre and Boratyński 2003).

Here, we introduce a new approach: the standard asymmetric continuum theory.

7.2 Standard Asymmetric Theory: Basic Assumptions

As mentioned above, in opposition to other approaches, we may construct the asymmetric standard theory entirely related to displacement field. Such a theory shall be based on both the symmetric and asymmetric stresses and on the related constitutive laws and motion cquations. The asymmetric deformations contain the symmetric strain and antisymmetric rotation. Thus, our theory is based on two groups of relations, for the symmetric and antisymmetric fields:

$$S_{kl} = S_{(kl)} + S_{[kl]}, \qquad E_{kl} = E_{(kl)}, \qquad \omega_{kl} = \omega_{[kl]}. \tag{7.1}$$

However, when introducing the new material parameters (structure indexes: e^0, χ^0), we may connect these deformation fields, in an independent way, with some displacement motion:

$$\begin{aligned} E_{kl} &= e^0 E_{kl}^0 = e^0 \frac{1}{2}\left(\frac{\partial u_l}{\partial x_k} + \frac{\partial u_k}{\partial x_l}\right), \\ \omega_{kl} &= \chi^0 \omega_{kl}^0 = \chi^0 \frac{1}{2}\left(\frac{\partial u_l}{\partial x_k} - \frac{\partial u_k}{\partial x_l}\right). \end{aligned} \tag{7.2}$$

For the internal energy stored in such a medium we obtain

$$E = S_{(ks)} E_{ks} + S_{[ks]}\omega_{ks}. \tag{7.3}$$

For the particular values of these indexes, e^0 and χ^0, we have:

- the classic elasticity for $\chi^0 = 0$;
- for $e^0 = 0$ we obtain a granular/crushed medium filled with rigid spheres with a friction interaction; when applying a torque load on its surface boundary, e.g., a cylindrical one, we obtain only some angular deformation, and torque energy stored would be given as $E = S_{[ks]}\omega_{ks}$;
- for solids we put $e^0 = 1$, $\left|\chi^0\right| = 1$; hence

$$E_{kl} = E_{kl}^0, \qquad \omega_{kl} = \chi^0 \omega_{kl}^0, \tag{7.4}$$

where χ^0 represents the phase shift between the motions E_{kl}^0 and ω_{kl}^0;

– for granulated or partly crashed material the structural indexes may take different values.

Further on, we will consider a more general continuum with the constitutive laws including also the time rates processes; for such cases we might discuss in a similar way the different particular cases of the material structure indexes including the dynamic objects.

For the symmetric part of stresses we can assume the classical constitutive relation:

$$S_{(kl)} = \lambda \delta_{kl} E_{ss} + 2\mu E_{kl} .$$

But there is no problem to include in it the appropriate linear deviations related to visco-plastic effects.

To construct the asymmetric theory, we assume, after Shimbo (1975, 1995), the appropriate constitutive law for the antisymmetric part of stresses. It joins the friction/fracture rotations with the antisymmetric stresses:

$$S_{[kl]} = 2\mu \omega_{kl} .$$

Using the motion equation for the symmetric part of stresses

$$\frac{\partial}{\partial x_k} S_{(kl)} = \rho \frac{\partial^2}{\partial t^2} u_l + F_l$$

and the scalar and vector potentials

$$u_l = l^2 \frac{\partial}{\partial x_l} \varphi + l^2 \varepsilon_{lps} \frac{\partial}{\partial x_p} \psi_s , \qquad F_l = l^2 \frac{\partial}{\partial x_l} \Phi + l^2 \varepsilon_{lps} \frac{\partial}{\partial x_p} \Psi_s ,$$

we can arrive at the equations for strain tensor

$$\begin{aligned} E_{lq} &= \frac{1}{3} \delta_{lq} E_{kk} + E_{lq}^D = \\ &= l^2 \frac{\partial^2}{\partial x_l \partial x_q} \varphi + \frac{1}{2} l^2 \varepsilon_{lps} \frac{\partial^2}{\partial x_p \partial x_q} \psi_s + \frac{1}{2} l^2 \varepsilon_{qps} \frac{\partial^2}{\partial x_p \partial x_l} \psi_s , \end{aligned} \tag{7.5}$$

where we have introduced the intrinsic length unit l.

We write the expected equations separately for the axial and deviatoric parts $E_{lq} = \frac{1}{3} \delta_{lq} E_{kk} + E_{lq}^D = \frac{1}{3} \delta_{lq} E_{kk} + \left(E_{lq} - \frac{1}{3} \delta_{lq} E_{kk} \right)$, obtaining:

$$\mu \Delta E_{ss} - \rho \frac{\partial^2 E_{ss}}{\partial t^2} = l^2 \Delta \Phi , \tag{7.6}$$

$$\mu \Delta E_{lq}^{D} - \rho \frac{\partial^2 E_{lq}^{D}}{\partial t^2} = \\ = l^2 \left(\left(\frac{\partial^2}{\partial x_l \partial x_q} - \frac{\delta_{lq}}{3} \Delta \right) \Phi + \frac{\partial}{2 \partial x_p} \left(\varepsilon_{lps} \frac{\partial}{\partial x_q} + \varepsilon_{qps} \frac{\partial}{\partial x_l} \right) \Psi_s \right), \tag{7.7}$$

where

$$E_{kk} = l^2 \frac{\partial^2 \varphi}{\partial x_s \partial x_s},$$

$$E_{lq}^{D} = l^2 \left(\frac{\partial^2 \varphi}{\partial x_l \partial x_q} - \frac{\delta_{lq}}{3} \frac{\partial^2 \varphi}{\partial x_s \partial x_s} + \frac{1}{2} \frac{\partial}{\partial x_p} \left(\varepsilon_{lps} \frac{\partial}{\partial x_q} + \varepsilon_{qps} \frac{\partial}{\partial x_l} \right) \psi_s \right).$$

The motion equation for the antisymmetric stresses $S_{[ni]}$ shall replace the balance law for the stress moments. To this end, we take the divergence of rotation force moment acting on a body element due to the antisymmetric stresses (rotational moment of forces per infinitesimal arm length corresponding to stress moments), and, on the other hand, the balancing term – the acceleration related to angular momentum (Teisseyre and Boratyński 2003):

$$\varepsilon_{lki} \frac{\partial^2}{\partial x_k \partial x_n} S_{[ni]} = \rho \varepsilon_{lki} \frac{1}{\xi} \frac{\partial^2}{\partial t^2} \omega_{ki} + \varepsilon_{lki} \rho K_{[ki]} = \\ = \chi^0 \rho \varepsilon_{lki} \frac{1}{2} \frac{\partial^2}{\partial t^2} \left(\frac{\partial u_l}{\partial x_k} - \frac{\partial u_k}{\partial x_l} \right) + \varepsilon_{lki} \rho K_{[ki]}. \tag{7.8}$$

With the compatibility condition, introduced in a similar way as for the symmetric strains:

$$I_{[ij]} = \varepsilon_{imk} \varepsilon_{jns} \frac{\partial^2}{\partial x_m \partial x_n} \omega_{ks} = 0$$

we obtain from (7.8):

$$\frac{\partial^2 S_{[ki]}}{\partial x_s \partial x_s} = 2\rho \frac{\partial^2 \omega_{[ki]}}{\partial t^2} + 2\rho K_{[ki]} \quad \text{or} \quad \mu \frac{\partial^2 \omega_{[ki]}}{\partial x_s \partial x_s} - \rho \frac{\partial^2 \omega_{[ki]}}{\partial t^2} = \rho K_{[ki]}, \tag{7.9}$$

where we have introduced the body couples $K_{[ki]}$ or body moment $K_{[l]} = \varepsilon_{lki} \rho K_{[ki]}$.

Otherwise, we can write:

$$\mu \frac{\partial^2}{\partial x_k \partial x_k} \omega_{[l]} - \rho \frac{\partial^2}{\partial t^2} \omega_{[l]} = \rho K_{[l]}, \tag{7.10}$$

where the left-hand side of these forms presents the basic expression for the resulting stress moment divergence in the continuum with asymmetric rotation nuclei.

These relations are equivalent to the following ones:

$$\frac{1}{l^2} \frac{\partial}{\partial x_k} M_{lk} = \varepsilon_{lki} \frac{\partial^2}{\partial x_k \partial x_n} S_{[ni]} = \varepsilon_{lki} \frac{\partial}{\partial x_n} \frac{\partial}{\partial x_n} S_{[ki]},$$

$$\frac{1}{l^2} M_{lk} = \varepsilon_{lki} \frac{\partial}{\partial x_n} S_{[ni]}$$

or defining the angular moment Ξ_i, we obtain:

$$\frac{\partial}{\partial x_k} M_{ik} = 2\mu \Xi_i, \qquad \Xi_i = l^2 \varepsilon_{iks} \frac{\partial}{\partial x_n} \frac{\partial}{\partial x_s} \omega_{[kn]}.$$

7.3 Spin and Twist Motions

The spin motion is governed by the Eq. (7.9), or equivalently by its vector form (7.10). In the system related to main shear axes or in that with the off-diagonal components, the motion equation for the deviatoric strains (7.7) can be presented in the form related to the other rotation vector – the twist, $\omega_{(s)}$ (see: the former chapter):

$$\{\omega_{(s)}\} = \{E_{23}^D,\ E_{31}^D,\ E_{12}^D\}. \tag{7.11}$$

The defined twist motion, $\omega_{(s)}$, means the rotational oscillation of the off-diagonal shear axes of the deviatric tensor (corresponding to oscillation of the main shear axes), E_{lq}^D, accompanied with the changes of the shear magnitude; such perturbation of the shear load may be caused by the internal fracturing processes.

Once having defined the twist vector field we can maintain its form due to the invariant properties of the Dirac tensors applied for the symmetric off-diagonal tensor $\omega_{(ik)}$ in its 4D form (see: the former chapter):

$$\omega_{(\lambda\kappa)} = \omega_{(1)} \varepsilon^1 + \omega_{(2)} \varepsilon^2 + \omega_{(3)} \varepsilon^4 \varepsilon^2 \varepsilon^3 = \tag{7.12}$$

$$= \begin{bmatrix} 0 & -\omega_{(3)} & -\omega_{(2)} & -\omega_{(1)} \\ -\omega_{(3)} & 0 & \omega_{(1)} & -\omega_{(2)} \\ -\omega_{(2)} & \omega_{(1)} & 0 & -\omega_{(3)} \\ -\omega_{(1)} & -\omega_{(2)} & -\omega_{(3)} & 0 \end{bmatrix},$$

where the Dirac matrices, ε^{μ}, $\mu = \{1,4\}$, were in the former chapter:

In a similar way, we may define the external off-diagonal part of the right-side expression of (7.14):

$$Y_{(lq)} = l^2 \left(\frac{\partial^2}{\partial x_l \partial x_q} \Phi + \frac{\partial}{2\partial x_p} \left(\varepsilon_{lps} \frac{\partial}{\partial x_q} + \varepsilon_{qps} \frac{\partial}{\partial x_l} \right) \Psi_s \right).$$

For its 4D form we can write:

$$Y_{(\lambda\kappa)} = Y_{(12)}\varepsilon^1 + Y_{(13)}\varepsilon^2 + Y_{(23)}\varepsilon^4\varepsilon^2\varepsilon^3 =$$

$$= \begin{bmatrix} 0 & -Y_{(12)} & -Y_{(13)} & -Y_{(23)} \\ -Y_{(12)} & 0 & Y_{(23)} & -Y_{(13)} \\ -Y_{(13)} & Y_{(23)} & 0 & -Y_{(12)} \\ -Y_{(23)} & -Y_{(13)} & -Y_{(12)} & 0 \end{bmatrix}.$$

Using these definitions for the off-diagonal form (7.12) we obtain

$$\mu \frac{\partial^2 \omega_{(\lambda\kappa)}}{\partial x_k \partial x_k} - \rho \frac{\partial^2 \omega_{(\lambda\kappa)}}{\partial t^2} = \mathbf{Y}_{(\lambda\kappa)}. \tag{7.13}$$

The defined twist motion, $\omega_{(\lambda\kappa)}$, means the rotational oscillation of the off-diagonal shear axes of the deviatric tensor, E_{lq}^{D}, accompanied with the changes of the shear magnitude; such perturbation of the shear load may be caused by the internal fracturing processes (see Fig. 7.1).

For the elastic continuum with defects the spin and twist motions form the complex rotation field defined as:

$$\omega_s = \omega_{[s]} + \mathrm{i}\omega_{(s)}. \tag{7.14}$$

From the related balance relation (see further on) we will obtain the relations joining the spin and twist motions.

7.4 Defects: Dislocation and Disclination Densities

The classical approach to the dislocation and disclination densities is based on the Kröner description of continuum with the self-fields (Kossecka and

DeWitt 1977, compare: Appendix). In the asymmetric homogeneous continuum the defect density can be introduced using the modified definition of disclosure, B_l, and the following definition of the twist-bend vector (equivalent to Eq. A.11 in Appendix); we define:

$$B_l = \oint \left[E_{kl} - \varepsilon_{lqr} \chi^T_{kq} x_r \right] \mathrm{d}l_k = \oint \left[E_{kl} + \omega_{kl} \right] \mathrm{d}l_k ,$$
$$\Omega_q = \oint \chi^T_{kq} \mathrm{d}l_k = \iint \theta_{kq} \mathrm{d}s_k . \tag{7.15}$$

In our approach (see, e.g. Teisseyre and Boratyński 2003) the tensor χ^T_{mq} is defined differently than by Kossecka and DeWitt (1977) in order to obtain a proper expression for dislocation density; we put

$$\chi^T_{mq} = \chi^0 \varepsilon_{ksq} \frac{\partial}{\partial x_s} \overset{0}{\omega}_{mk} \tag{7.16}$$

and we obtain from (7.15):

$$B_l = \oint \left(E_{kl} + \omega_{kl} \right) \mathrm{d}l_k = \oint \left[\overset{0}{E}_{kl} + \chi^0 \overset{0}{\omega}_{kl} \right] \mathrm{d}l_k = \iint \left(\alpha_{pl} - \frac{1}{2} \delta_{pl} \alpha_{ss} \right) \mathrm{d}s_p , \tag{7.17}$$

where we have introduced the definition of the dislocation and disclination densities: α_{pl} and θ_{pq}.

From (7.15) and (7.16) it follows that the disclination density vanishes, in accordance with the compatibility condition:

$$\theta_{pq} = \varepsilon_{pmk} \frac{\partial \chi^T_{kq}}{\partial x_m} = \varepsilon_{pmk} \varepsilon_{qns} \frac{\partial^2 \overset{0}{\omega}_{ks}}{\partial x_m \partial x_n} = 0 .$$

For the dislocation field we obtain from (7.17):

$$\alpha_{pl} - \frac{1}{2} \delta_{pl} \alpha_{ss} = \varepsilon_{pmk} \frac{\partial}{\partial x_m} \left[E_{kl} + \omega_{kl} \right] = \varepsilon_{pmk} \frac{\partial}{\partial x_m} \left[\overset{0}{E}_{kl} + \chi^0 \overset{0}{\omega}_{kl} \right] . \tag{7.18}$$

Supplementing the constitutive relations with that for the antisymmetric stresses and rotations we arrive at the relation between the dislocation density and asymmetric stresses:

$$\alpha_{pl} - \frac{1}{2} \delta_{pl} \alpha_{ss} = \frac{\varepsilon_{pmk}}{2\mu} \frac{\partial}{\partial x_m} \left(S_{(kl)} - \frac{\nu}{1+\nu} \delta_{kl} S_{ii} + S_{[kl]} \right) . \tag{7.19}$$

We note that in the classic theory with defects, we distinguish also the different definitions for a dislocation field, e.g., the Burgers and Nye dislocations.

For a coincidence of the spin and twist motions, $\chi^0 = 1$, we may obtain an extreme shear deformation; for this particular case the defect vanishes:

$$B_l = \oint \left[E_{kl} + \omega_{kl} \right] \mathrm{d}l_k = \oint \left[E_{kl}^0 + \omega_{kl}^0 \right] \mathrm{d}l_k = \oint \frac{\partial u_l}{\partial x_k} \mathrm{d}l_k = 0, \quad \alpha_{pl} = 0. \tag{7.20}$$

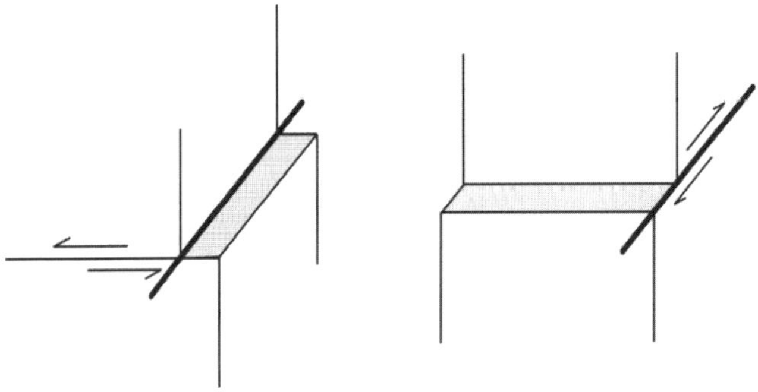

Fig. 7.1 The edge and screw dislocation types

With $\chi^0 = -1$, we obtain for the edge dislocations:

$$\alpha_{pl} = \frac{1}{2\mu} \varepsilon_{pmk} \frac{\partial \left(S_{(kl)} - \frac{\nu}{1+\nu} \delta_{kl} S_{ss} + S_{[kl]} \right)}{\partial x_m}; \qquad p \neq l \tag{7.21}$$

and the screw ones:

$$\alpha_{pl} = \frac{1}{\mu} \varepsilon_{pmk} \frac{\partial (S_{(kl)} + S_{[kl]})}{\partial x_m}; \qquad l = p. \tag{7.22}$$

The both considered cases could relate to the formation of the respective slip-discontinuities, Fig. 7.1.

7.5 Balance Laws for the Rotation Field and the EM Analogy

The complex rotation field (7.14), $\omega_s = \omega_{[s]} + \mathrm{i}\omega_{(s)}$, may be presented in the tensor form:

$$\varepsilon_{kis} \omega_s = \omega_{ki} = \begin{pmatrix} 0 & -\omega_{[3]} - \mathrm{i}\omega_{(3)} & -\omega_{[2]} - \mathrm{i}\omega_{(2)} \\ -\omega_{[3]} - \mathrm{i}\omega_{(3)} & 0 & \omega_{[1]} + \mathrm{i}\omega_{(1)} \\ \omega_{[2]} + \mathrm{i}\omega_{(2)} & -\omega_{[1]} - \mathrm{i}\omega_{(1)} & 0 \end{pmatrix}. \tag{7.23}$$

We can write the balance condition as

$$\iint \varepsilon_{kps} \frac{\partial}{\partial x_p} \omega_s ds_k = \iint \left(\frac{\partial}{\mathrm{i} V \partial t} \omega_k + \frac{4\pi}{\upsilon} J_k \right) \mathrm{d}s_k , \tag{7.24}$$

where we introduced the current field J_k and velocity V.

Hence, we obtain the field equations for the complex rotation motions:

$$\varepsilon_{kps} \frac{\partial}{\partial x_p} \omega_s - \frac{\partial}{\mathrm{i} V \partial t} \omega_k = \frac{4\pi}{\upsilon} J_k \tag{7.25}$$

or the system for spin and twist motions:

$$\varepsilon_{kps} \frac{\partial \omega_{[s]}}{\partial x_p} - \frac{1}{V} \dot{\omega}_{(k)} = \frac{4\pi}{V} J_k , \qquad \varepsilon_{kps} \frac{\partial \omega_{(s)}}{\partial x_p} + \frac{1}{V} \dot{\omega}_{[k]} = 0. \tag{7.26}$$

These equations are quite similar to the EM equations; the velocity V shall obey the relativistic rule for a sum of velocities.

Appendix: Continuum with Internal Nuclei

First let us mention that in the classical elastic continuum with the symmetric stresses, the rotations are introduced with a help of the stress moment tensor (related to some arm length). Instead we may assume after Kröner, 1982, an existence of the asymmetric stresses, strains and rotation fields related to the self fields of the internal nuclei. The antisymmetric elastic stresses form the system equivalent to stress moments, while the symmetric rotations describe the twist motions being equivalent in classic theory to shear oscillations of the point-grains of a continuum (the equivalent notions appearing in the micromorphic theories) – compare papers by Teisseyre (2002), Teisseyre and Boratyński (2003, 2006), Boratyński and Teisseyre (2004).

Usually the deviations from ideal elasticity are described by the appriopriate elasto-plastic constitutive laws; here following the Kröner approach, we maintain the ideal elastic relation for the stresses and strains, supplemented by constitutive law joining the antisymmetric stresses with rotations, and we introduce the self/inner stresses, strains and rotations as related to the internal nuclei or defects: $\mathbf{S}^S$, $\mathbf{E}^S$, ω^S.

Therefore we distinguish between the total stresses strains and rotations $\mathbf{S}^T$, $\mathbf{E}^T$, $\omega^T = 0$, as defined by its direct relations to the displacement field, from the asymmetric elastic stresses, strains and rotations $\mathbf{S}$, $\mathbf{E}$, ω. These elastic fields shall obey the constitutive laws given in the ideal elasticity.

However, the asymmetric continuum includes the displacements and also rotations; the related balance equations split into the parts related to symmetric and antisymmetric stresses (Teisseyre 2004); in such a continuum the elastic strains, rotations and stresses can be expressed as differences between total and self-fields (Kröner 1981):

$$\mathbf{E} = \mathbf{E}^T - \mathbf{E}^S, \quad \omega = \omega^T - \omega^S, \quad \beta = \beta^T - \beta^S, \quad \mathbf{S} = \mathbf{S}^T - \mathbf{S}^S. \tag{A.7.1}$$

The elastic and self deformations, rotations and stresses are, in general, asymmetric; the antisymmetric parts for stresses and strains are mutually compensated:

$$E_{ki}^T = u_{(i,k)} = \frac{1}{2}\left(\frac{\partial u_i}{\partial x_k} + \frac{\partial u_k}{\partial x_i}\right), \qquad E_{[ki]} + E_{[ki]}^S = 0 \tag{A.7.2}$$

and similarly the symmetric parts for elastic and self rotations are compensated:

$$\omega_{ki}^T = u_{[i,k]} = \frac{1}{2}\left[\frac{\partial u_i}{\partial x_k} - \frac{\partial u_k}{\partial x_i}\right]; \qquad \omega_{(ik)} + \omega_{(ik)}^S = 0. \tag{A.7.3}$$

However, referring to our earlier paper we assume that the self-parts of the antisymmetric part of strain and rotation are put as equal to each other:

$$E_{[ik]}^S = \omega_{[ik]}^S. \tag{A.7.4}$$

The elastic fields **S**, **E**, ω represent the physical fields, while the total fields $\mathbf{S}^T$, $\mathbf{E}^T$, ω^T relate, according to the compatibility condition, to the displacement motions u_i, while the self-fields relate to the internal nuclei, defect densities and continuum structure.

First, we refer to the definitions of the twist-bend tensor; first after Kossecka and DeWitt (1977) related to gradient of the rotation vector (vanishing of both the Frank vector and the disclination density):

$$\overline{\chi}_{mq} = \frac{\partial \omega_{[q]}}{\partial x_m} = \frac{1}{2}\varepsilon_{nsq}\frac{\partial \omega_{[ns]}}{\partial x_m} \rightarrow \overline{\chi} = \text{grad}\ \omega_{[..]} = \nabla \omega_{[..]} \tag{A.7.5}$$

and then with rotation of the transposed rotation tensor (Teisseyre 2001, Teisseyre and Boratyński 2002):

$$\chi_{mq}^T = \varepsilon_{ksq}\frac{\partial \omega_{mk}^T}{\partial x_s} \rightarrow \chi^T = -\text{curl}\ \omega^T. \tag{A.7.6}$$

These two definitions can describe different types of rotation nuclei; Kossecka and DeWitt's definition directly leads to the pure rotation nuclei, while the other definition describes nuclei which can undergo the rotation and twist vibrations and includes the Frank vector and the disclination density. The total fields related to these two definitions coincide (apart of the sign) $\overline{\chi}^T_{mq} = -\chi^T_{mq}$ and we could combine them (Teisseyre and Boratyński 2002), however we will relate further on to the definition (A.7.6).

For the incompatibility tensors with their symmetric and antisymmetric parts we obtain:

$$I_{(ij)} = -\varepsilon_{ikm}\varepsilon_{jtn}\frac{\partial^2 E_{(mn)}}{\partial x_k \partial x_t} = \varepsilon_{ikm}\varepsilon_{jtn}\frac{\partial^2 E^S_{(mn)}}{\partial x_k \partial x_t} = -\varepsilon_{ikm}\varepsilon_{jtn}\frac{\partial^2 \omega_{(mn)}}{\partial x_k \partial x_t} = \varepsilon_{ikm}\varepsilon_{jtn}\frac{\partial^2 \omega^S_{(mn)}}{\partial x_k \partial x_t},$$

$$I_{[ij]} = -\varepsilon_{ikm}\varepsilon_{jtn}\frac{\partial^2 E_{[mn]}}{\partial x_k \partial x_t} = \varepsilon_{ikm}\varepsilon_{jtn}\frac{\partial^2 E^S_{[mn]}}{\partial x_k \partial x_t} = -\varepsilon_{ikm}\varepsilon_{jtn}\frac{\partial^2 \omega_{[mn]}}{\partial x_k \partial x_t} = \varepsilon_{ikm}\varepsilon_{jtn}\frac{\partial^2 \omega^S_{[mn]}}{\partial x_k \partial x_t} \tag{A.7.7}$$

and we can arrive to (Teisseyre 2005):

$$I_{pq} = -\left(\varepsilon_{pmk}\frac{\partial(\alpha_{qk} - \chi^S_{kq})}{\partial x_m}\right)_{(pq)} - \left[\varepsilon_{pmk}\frac{\partial \chi^S_{kq})}{\partial x_m}\right]_{[pq]}. \tag{A.7.8}$$

The symmetric tensor $I_{(pq)}$ relates to the dislocation and disclination densities, α and θ (Kossecka and DeWitt 1977):

$$I_{(pq)} = -\left(\varepsilon_{pmk}\frac{\partial \alpha_{qk}}{\partial x_m} + \theta_{pq}\right)_{(pq)}. \tag{A.7.9}$$

We introduce the Shimbo (1975 and 1995) constitutive law for antisymmetric stresses and rotation

$$\omega^S_{[mn]} = E^S_{[mn]} = -E^S_{[mn]} = -\frac{1}{2\mu^*}S_{[ik]}, \tag{A.7.10}$$

where the constant μ^* represents the rotation rigidity modulus: rotation related bonds (inner friction).

Antisymmetric stresses relate to an internal rotation motion; these stresses become important in zones with higher dislocation densities or under high stresses or in zones where microfracturings nucleate; in such zones we can expect the presence of rotation nuclei.

These considerations, related to defects – dislocation and disclination densities and to rotation nuclei, contain some partial results presented by Kossecka and DeWitt (1977), Teisseyre (2002), Teisseyre and Boratyński (2002, 2003); however, some necessary modifications are introduced.

Following Kossecka and De Witt (1977), we write, with a help of the twist-bend tensor and with Eq. A.7.6), for a total disclosure and twist along a closed circuit (the Burgers vector and the Frank vector):

$$B_l = -\oint \left[E^S_{(kl)} - \varepsilon_{lqr} \chi^S_{kq} x_r \right] \mathrm{d}l_k , \qquad \Omega_q = \oint \chi^S_{kq} \mathrm{d}l_k = \iint \theta_{pq} \mathrm{d}s_p \qquad \text{(A.7.11)}$$

and the dislocation and disclination densities, α and θ, become:

$$\alpha_{pl} = -\varepsilon_{pmk} \left(\frac{\partial E^S_{(kl)}}{\partial x_m} + \varepsilon_{klq} \chi^S_{mq} \right), \qquad \theta_{pq} = \varepsilon_{pmk} \frac{\partial \chi^S_{kq}}{\partial x_m} . \qquad \text{(A.7.12)}$$

Further we obtain

$$\alpha_{pl} = -\varepsilon_{pmk} \frac{\partial E^S_{(kl)}}{\partial x_m} - \delta_{pl} \chi^S_{qq} + \chi^S_{lp} . \qquad \text{(A.7.13)}$$

From the definitions of the twist-bend tensor (A.7.6) follows that $\chi^S_{mq} = \varepsilon_{nsq} \dfrac{\partial \omega^S_{mn}}{\partial x_s}$ and we obtain

$$\alpha_{pl} = -\varepsilon_{pmk} \frac{\partial E^S_{(kl)}}{\partial x_m} - \delta_{pl} \varepsilon_{smk} \frac{\partial \omega^S_{[ks]}}{\partial x_m} - \varepsilon_{pmk} \frac{\partial \omega^S_{lk}}{\partial x_m} ,$$

$$\theta_{pq} = \varepsilon_{pmk} \varepsilon_{qns} \frac{\partial^2 \omega^S_{ks}}{\partial x_m \partial x_n} .$$

In the former papers by Kossecka and DeWitt (1977) and Teisseyre (2001), the dislocation current has been defined as the related deviation between the total plastic distortion and total plastic flow, here more convenient appears the modified definition (Teisseyre 2002):

– for the dislocation current (Teisseyre 2002)

$$J_{kl} = -\left(\frac{\partial \upsilon^S_l}{\partial x_k} \right)_{(kl)} + \dot{E}^S_{(kl)} + \dot{\omega}^S_{(kl)} , \qquad \text{(A.7.14)}$$

– for the disclination current

$$Y_{kq} = -\varepsilon_{qns} \frac{\partial \dot{\omega}^{S}_{ks}}{\partial x_n} + \varepsilon_{qns} \frac{\partial \dot{\omega}^{S}_{(sk)}}{\partial x_n} = -\varepsilon_{qns} \frac{\partial \dot{\omega}^{S}_{[ks]}}{\partial x_n}. \tag{A.7.15}$$

On the other side, a dislocation current follows according to Teodosiu (1970) as

$$J_{pk} = -\varepsilon_{qmk} \alpha_{qp} \upsilon_m \tag{A.7.16}$$

where υ is a dislocation velocity.

Constitutive law joining dislocation velocity with stresses was given by Mataga et al. (1987):

$$V_m = \varepsilon_{snq} \frac{\alpha_{kl}}{|\alpha_{kl}|} \frac{S_{ql} - R_{ql}}{\sqrt{(s-R) + R^2}}, \tag{A.7.17}$$

where S are the stresses; R is the resistance stress; υ_j and $V_j = \upsilon_j / \upsilon_0$ are, respectively, the dislocation velocity and relative dislocation velocity with respect to the shear wave velocity υ_0.

References

Boratyński W, Teisseyre R (2004) Generalized continuum with defects and asymmetric stresses. Acta Geopyhs Pol **52**: 2, 185-195

Eringen AC (1999) Microcontinuum field theories. Springer, Berlin

Jaroszewicz LR, Krajewski Z, Solarz L (2006) Absolute rotation measurement based on the Sagnac effect. In: Teisseyre R, Takeo M, Majewski E (eds) Earthquake source asymmetry, structural media and rotation effects. Springer, Berlin, pp 413-438

Kossecka, E, DeWitt R (1977) Disclination kinematic. Arch Mech 29: 633-651

Mataga PA, Freund LB, Hutchinson J (1987) Crack tip plasticity in dynamic fracture. J Phys Chem Solid **48**: 985-1005

Moriya T, Teisseyre R (2006) Design of rotation seismometer and non-linear behaviour of rotation components of earthquakes. In: Teisseyre R, Takeo M, Majewski E (eds) Earthquake source asymmetry, structural media and rotation effects. Springer, Berlin, pp 439-450

Schreiber KU, Stedman GE, Igel H, Flaws A (2006) Ring laser gyroscopes as rotation sensors for seismic wave studies. In: Teisseyre R, Takeo M, Majewski E (eds) Earthquake source asymmetry, structural media and rotation effects. Springer, Berlin, pp 377-390

Shimbo M (1975) A geometrical formulation of asymmetric features in plasticity, Bull Fac Eng, Hokkaido Univ. 77: 155-159.

Shimbo M (1995) Non-Riemannian geometrical approach to deformation and friction. In: Teisseyre R (ed) Theory of earthquake premonitory and fracture processes. PWN, Warszawa, pp 520-528

Takeo M (2006) Ground rotational motions recorded in near-source region of earthquakes. In: Teisseyre R, Takeo M, Majewski E (eds) Earthquake source asymmetry, structural media and rotation effects. Springer, Berlin, pp 157-167

Teisseyre R (2001) Evolution, propagation and diffusion of dislocation fields. In: Teisseyre R, Majewski E (eds) Earthquake thermodynamics and phase transformations in the Earth's interior. (International Geophysics Series vol 76). Academic Press, San Diego, pp 167-198

Teisseyre R (2002) Continuum with defect and self rotation fields. Acta Geophys Pol **50**: 1, 51-68

Teisseyre R (2004) Spin and twist motions in a homogeneous elastic continuum and cross-band geometry of fracturing. Acta Geophys Pol **52**: 2, 173-183

Teisseyre R (2005) Asymmetric continuum mechanics: deviations from elasticity and symmetry. Acta Geophys Pol **53**: 2, 115-126

Teisseyre R (2006) Asymetric continuum and anisotropy. Acta Geophys **54**: 3, 225-238

Teisseyre R, Boratyński W (2002) Continua with self-rotation nuclei: Evolution of defect fields and equations of motion. Acta Geophys Pol **50**: 2, 179-206

Teisseyre R, Boratyński W (2003) Continua with self-rotation nuclei: evolution of asymmetric fields. Mech Res Commun **30**: 235-240

Teisseyre R, Boratyński W (2006) Deviations from symmetry and elasticity: asymmetric continuum Mechanics. In: Teisseyre R, Takeo M, Majewski E (eds) Earthquake source asymmetry, structural media and rotation effects. Springer, Berlin, pp 32-4

Teisseyre R, Białecki M, Górski M (2005) Degenerated mechanics in homogenous continuum: Potentials for spin and twist. Acta Geophys Pol **53**: 3, 219-231

Teisseyre R, Górski M, Teisseyre KP (2006) Fracture-band geometry and rotation energy release. In: Teisseyre R, Takeo M, Majewski E (eds) Earthquake source asymmetry, structural media and rotation effects. Springer, Berlin, pp 169-184

Teodosiu C (1970) A dynamic theory of dislocations and its applications to the theory of the elastic plastic continuum. In: Simmons JA, De Witt R, Bullough R (eds) Fundamental aspects of dislocation theory. Nat Bur Stand Spec Publ **317**, II: 837-876

Wiszniowski J (2006) Rotation and twist motion recording – couple pendulum and rigid seismometers system. In: Teisseyre R, Takeo M, Majewski E (eds) Earthquake source asymmetry, structural media and rotation effects. Springer, Berlin, pp 451-470

8 Fracture Processes: Spin and Twist-Shear Coincidence

Roman Teisseyre, Marek Górski, and Krzysztof P. Teisseyre

Institute of Geophysics, Polish Academy of Sciences
ul. Księcia Janusza 64, 01-452 Warszawa, Poland
e-mail: rt@igf.edu.pl

8.1 Introduction

We consider rotation processes in a material under load leading to fracture process: in the preseismic stage as well as in the main rupture phase and in the post-earthquake stage. The rotation processes of different scale help to understand these extremely complicated phenomena in which dynamic processes proceed simultaneously with changes of material parameters. We recall a special role of rotations in the energy release effectiveness under different load conditions, and further on we discuss the rotation impact on the granulation processes accompanying the material crushing processes.

As fracturing phenomena proceed, the constitutive laws undergo considerable changes, from description of elastic to plastic and, further, to mylonite-type material. In the narrow zone adjacent to the zone of fracturing, the shear stresses break the molecular bonds and in the crashed rock the stresses immediately drop down, the increase of stress rates and strain rates occurs, and material becomes partly granulated. Later, in that zone the stresses and strains become gradually less important and progressively replaced by the time-rates of stresses and strains.

Finally in the thin zone adjacent to fracture plane, there appears material somewhat similar to fluid with the transport properties described by Navier-Stokes equations. The bond breaking and granulation processes force co-action of the twist-shear motion with spin of grains in the mylonite zone.

In the earthquake precursory processes, the energy micro-releases relate to a coalescence of dislocation arrays of opposite signs. We may consider two different cases: in the first, we consider a rock body under the confining pressure; in the other – under the external shear load. Under confining pressure and under external shears, the role of micro-fracturing in the bond breaking process is similar; however, we observe here essential differences in larger-scale rotations. Confining condition leads to formation of the induced different arrays of dislocations, resulting in fragmentation processes and related macro-rotations. On the other hand, shear load leads to more concentrated fracturing along some planes. We can underline that macro-rotations and related energy release are more effective for the fracturing under confining pressure.

For the first case, we have reexamined Dieterich's (1978) compression experiments which led us to conclusion that the precursory induced shear stresses appear due to confining load and presence of defects. Rocks are not homogeneous and we have to assume that inside a body there appear regions with the induced shear stresses of different signs. This case is symbolized in Fig. 8.1; the rotation energy release is postponed until the moment the induced stresses are so high that the material breaks locally and fragmentation starts. The main rupture is delayed and when it finally occurs, a considerable part of rotation energy is released in fragmentations. The total shear stress drop for compression conditions could be quite small, while rebound phenomena (preseismic, coseismic and in the rebound stage) bring an important release of rotation energy. The rotation processes in fragmentation and fracturing play in this case an essential role.

In the second case – the shear load, the development of fracture planes is not chaotic and the sense of shear motions is preserved. The external shearing force promotes the microcracks coalescence in some zones, which gradually merge to form the main rupture. In such circumstances, rearrangement and merging of a main fracture plane can occur. Rotations (in every scale) are then of lesser importance. Therefore, most of the earthquake energy is that of shear stress release.

Asperities and other macro-scale inhomogeneities can locally redistribute the stresses in various modes, changing therefore the preseismic conditions in parts of rock volume.

Further on, we discuss the importance of rotations in mesoscale, between these related to the micro-scale bond breaking process and the macro-rotations at material fragmentation and transport processes.

In any fracturing process, especially if revealed in a more spectacular form (long fractures) under shearing load, we shall consider the processes leading to granulation of material adjacent to shear fracturing. Such processes become intense and lead to the formation of narrow, long mylonite zones adjacent to fracture zones. The related rotations – the mesoscale rotations – are related to bond breaking and friction processes leading to material granulation.

Coincidence and counter-action of the spin and twist-shear motions in bond breaking, granulation and formation of mylonite material help us to understand the fracture mechanism; the simultaneous mylonite formation appears to be caused by the common action of these rotations and fracture transport phenomena.

When applying such an approach to the progressively crushing material we can use the Navier-Stokes equations to describe the fracture propagation.

When in a narrow zone the huge shear stresses would break the molecular bonds, crushing and granulation of the rock material start. Immediately, stress magnitudes drop down to low values, and simultaneously the stress and strain time rates increase to such a degree that the stresses and strains may be neglected in the constitutive relations. In result, we may treat the material in this zone as more similar to a liquid than to a solid – simultaneously to granulation process (in an intact material or in zone already previously crushed), the nucleation progresses and fractures propagate. In such conditions, we may apply the Navier-Stokes transport equation. Thus, we arrive at a new description of the fracturing transport process, in which the shear rates force the phase coincidence of the twist-shear motion and the related spin motion of grains subjected to the bond breaking and granulation processes.

The constitutive law of a material under crushing condition undergoes substantial changes; first we may have the intact rigid rock material with the basic constitutive law given by the elastic properties with the total stress, strain and rotation fields related to displacements; further, in the intermediate stage both these fields as well as the appropriate rates should be taken into account, and finally in the sand-like mylonite material we might neglect the primordial stress, strain and rotation fields – there remain (in consideration) only the rates of stress, strain and rotations.

Searching for the fault slip solutions we use the classical elasticity tools with the additional friction constitutive law basing on the experimental data. When instead we consider the asymmetric elastic continuum, we are able to consider the defect interaction and we can derive the elastodynamic fault solution describing slip propagation with fracturing process and related seismic radiation.

For the fracture zone, the constitutive relation for the asymmetric elastic continuum (Chap. 7) shall be modified to include the field rates.

As we have already mentioned, in the compression case with no initial shear field, we may still assume, owing to the presence of internal defects and the lower value of shear resistance, that there appear some internal regions with induced shear stresses of opposite signs under the condition that the resulting field is neutral. The inner shear stress accumulation relates to defect densities, while the interaction between dislocations gives rise to stress concentration at the first blocking dislocations of the dislocation arrays formed. The fracture processes and the energy releases will relate to a coalescence of dislocation arrays of opposite signs and a related rotation rebound motion. However, the further penetrating fracture processes might consecutively lead to a completely crushed/granulated material up to the molecular scale; in the final stage, an energy release will relate to the bond breaking processes and the rebound rotations of the released molecules.

Of course, we shall consider a fracturing process as a chain of events; let us consider the centers formed by two perpendicular plane fragments; the induced shear stresses are opposite on those plane fragments but have a common orientation of spin motion (Fig. 8.1).

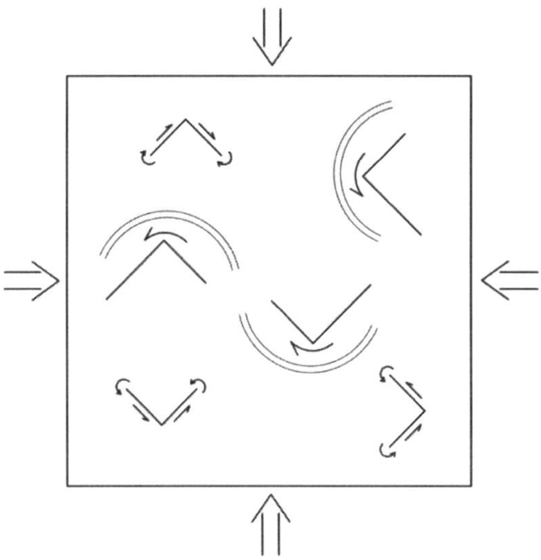

Fig. 8.1 Compression load: induced shear centers and formation of fragments

The induced shears become almost compensated, while the spin fields may support the rotation processes.

However, this spin field, $\omega_{[s]}$, propagates and influences the processes in the adjacent regions; we believe that this propagation synchronizes the spin motions in the adjacent centers in such a way that the sense of spin motion becomes the same over the whole fracture region. This means, the spin propagation assures a synchronization of fracture processes, especially under compression load where the energy release relates to the fragmentation revealed by rotation and granulation processes.

Under the shear load, a fracture process runs differently; as a common shear deformation, E^D_{kn} (or expressed as twist $\omega_{(s)}$), progresses, the spins on the main fracture differ from those on the adjacent perpendicular fragments and attenuate the fracture progress on those fragments, while these motions support formation of long linear fractures (Figs. 8.2 and 8.3).

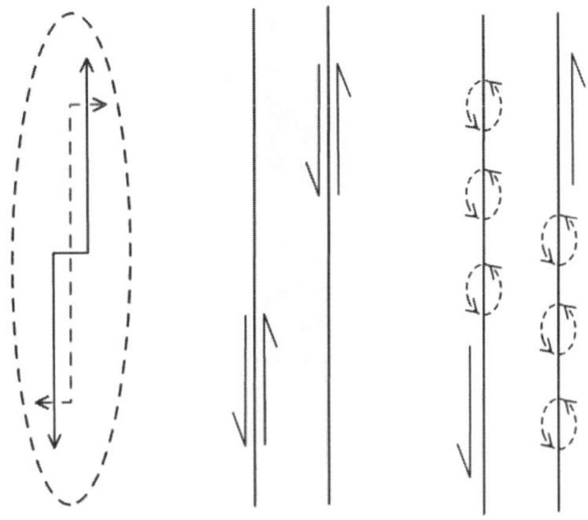

Fig. 8.2 From the extreme shear deformation to formation of granulation and fracturing mylonite zone

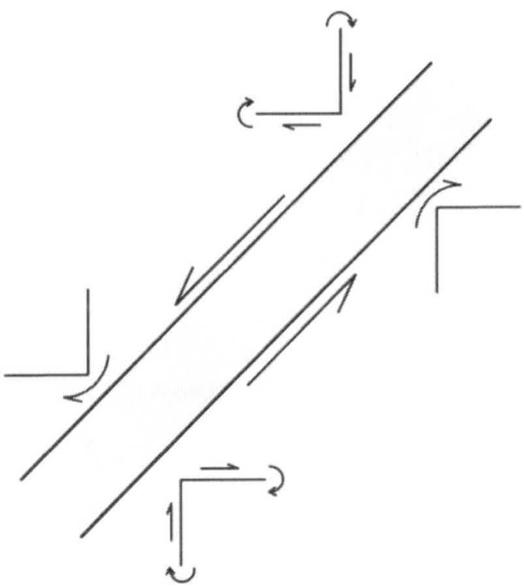

Fig. 8.3 Mylonite zone and adjacent deformations

Accordingly, we can believe that at the compression load, the total shear stress drop will be relatively small, while the rebound rotations will release an important amount of rotation energy, whereas at the shear load, the release of shear stresses will prevail.

Concluding, the rotation processes play an essential role in fragmentation and fracturing under compression load. Under prevailing shear load the rebound process releases shear load with the regional stress drop, while the rotation processes play a minor role.

Further on, we will discuss the importance of granulation processes related to rotations in mesoscale, which we can place between the bond breaking processes in the micro-scale and material fragmentation in the macro-scale.

8.2 Approaching Fracture: Constitutive Laws for Mylonite Zones

The fracturing process, especially under the action of shearing load, is accompanied by material granulation adjacent to shear fracture planes; this becomes spectacular at the formation of narrow, long mylonite zones. In this process we shall take into account a special role of rotations – the mesoscale rotations of different scales; these rotations are related to bond breaking and friction processes.

Co-action of the spin and twist-shear motions in bond breaking, granulation and formation of mylonite material can effectively help us to explain the fracture process; the simultaneous formation of the adjacent mylonite zone appears due to such a co-action of spin and shears with the fracture transport phenomena.

Basing on the standard asymmetric continuum theory, as presented in the former chapter, we will consider the material undergoing a progressive crashing process. We may even arrive at the conditions more similar to fluid material, and thus, finally, shall enter into our consideration the Navier-Stokes transport equations.

Starting with the description of the rock continuum following from the standard asymmetric theory of continuum ($S_{ik} = S_{(ik)} + S_{[ik]}$, $E_{ik} = E_{(ik)}$, $\omega_{ik} = \omega_{[ik]}$), we approach the final stage of crushing/granulation process in zones adjacent to fracture planes. In these zones, simultaneously with dynamic processes, there occur changes of material properties from hard rocks to mylonite granulated material.

Near to the final stage, the stresses, strains and rotations become gradually neglected and progressively replaced by the constitutive relations for time-rates of stresses and strains; further we consider the deviatoric part of the symmetric fields:

$$S^D_{(ks)} + \tau \dot{S}^D_{(ks)} = 2\mu E^D_{ks} + 2\eta \dot{E}^D_{ks}, \qquad S_{[ks]} + \tau \dot{S}_{[ks]} = 2\mu\omega_{ks} + 2\eta\dot{\omega}_{ks}. \tag{8.1}$$

The introduced material constants may depend on the values of slip, **u**, and slip rate, υ.

When in a narrow zone the huge shear stresses break the molecular bonds, the stresses crashing rock material will immediately drop down to the low values and in the crashed mylonite material we observe the immediate increase of the stress and strain rates to such a degree that the stresses and strains may be neglected in the respective constitutive relations for this narrow zone. Finally, these changes will lead to the constitutive laws for the melt and granulated parts of mylonite material in which, practically, there only will remain the field time rates:

$$\tau \dot{S}^D_{(ks)} = 2\eta \dot{E}^D_{ks}, \qquad \tau \dot{S}_{[ks]} = 2\eta \dot{\omega}_{ks} = 2\eta \chi^0 \dot{\omega}^0_{ks}. \tag{8.2}$$

The direct observation of the gauge zone of the 1995 Kobe, Japan, earthquake at the Avaji island suggests that the size of an inner completely melted part of the mylonite zone ranges around a couple of centimeters (private comm. W. Dębski).

Further on, we will assume, for the sake of simplicity, that the mylonite material remains incompressible during the fracturing. In such a way, the nucleation progresses and fracture propagates simultaneously with granulation process in the intact material (or in the compact zone previously crashed).

Now we introduce the structural indexes (see: Chap. 7):

$$\begin{aligned} E_{kl} &= \overline{e}^0 E^0_{kl} = \overline{e}^0 \frac{1}{2}\left(\frac{\partial u_l}{\partial x_k} + \frac{\partial u_k}{\partial x_l}\right), \\ \omega_{kl} &= \overline{\chi}^0 \omega^0_{kl} = \overline{\chi}^0 \frac{1}{2}\left(\frac{\partial u_l}{\partial x_k} - \frac{\partial u_k}{\partial x_l}\right), \end{aligned} \tag{8.3}$$

where the indexes $\overline{e}^0$ and $\overline{\chi}^0$ change along with changing material properties.

The presented relation is kept in the same form for the field rates; moreover, to preserve a concise theoretical approach, we shall treat them as homogenized indexes over the part of material under consideration.

In this new description, the shear rates create the dynamic angular deformations, lead to the bond breaking processes, and then, finally, to the fracturing transport process.

We pass to the final stage: the crashed incompressible mylonite or sand, similar to incompressible fluids. Our relations define the ideal quasi-viscous mylonite for an incompressible crashed material.

For our narrow mylonite zone, existing already near the pre-slip planes or just simultaneously formed, we may, further on, apply the Navier-Stokes transport equation. Such motions, especially in an earthquake source zone, are due to the friction processes.

We may note that when including these complex rotational motions in the theory, we replace the friction constitutive laws, as based on the experimental data, by the constitutive law joining the asymmetric stresses with spin and shear field oscillations or otherwise with spin and twist. Thus, instead the fault slip solutions with the additional friction constitutive law we consider the asymmetric elastic continuum with the defect interaction and with rotational motions; we can derive the elastodynamic fault solution describing slip propagation with fracturing process and related seismic radiation.

Especially, in fracture zone we shall use the constitutive law (8.1); the asymmetric stresses may prevail and the constants introduced there, τ, μ and η, depend on slip and slip rate velocity. As mentioned above, the coincidence of the spin and twist motions leads to destruction of bonds and formation of a granulated material in an intermediate mylonite zone adjacent to slip.

In the former chapter we have considered an action of the spin and twist motions leading to formation of dislocation fields; here, we show another way to present a coincidence resulting in an extreme slip when sum of these motions approaches to the derivative of a displacement velocity (see: Chap. 7, Eq. 7.20):

$$\iint \varepsilon_{sik}\left(\dot{E}^0_{ik} - \dot{\omega}^0_{ik}\right)\mathrm{d}s_s = \iint \varepsilon_{sik}\frac{\partial}{\partial x_i}\upsilon_k\mathrm{d}s_s = \oint \frac{\partial}{\partial x_s}\upsilon_k\mathrm{d}x_k \to \Delta\upsilon .$$

Further, we will try to approach these problems considering the special theoretical solution for the spin and twist fields.

8.3 Slip Propagation and Spin Release Hypothesis

While searching for the fault slip solutions we use the classical elasticity tools with an additional friction constitutive law basing on experimental data. When, instead of it, we consider the asymmetric elastic continuum, we are able to include the defect interaction and we can derive the elastodynamic fault solution describing slip propagation with fracturing process and related seismic radiation.

The angular deformations preceding the bond breaking process lead to the efficient rise of the angular moments around material grains.

In the narrow mylonite zone, we arrive at the equivalence between this expression and the laws introduced in the considerations on the friction resistance and slip.

The co-action of rotation motions, curl $\mathbf{\upsilon}$ (or spin $\dot{\omega}_{[.]}$), and shear, $\mathrm{E}^{\mathrm{D}}_{(..)}$ (or twist, $\dot{\omega}_{(.)}$), can lead further to the slip fracturing motion. We assume that the bond breaking process and granulation of material precede the slip movement: just after the bond breaking micro-process, we would have the released rebound spin motion retarded in phase.

This hypothesis is supported by the following solution of the homogeneous wave equations for the twist and spin vectors, $\omega_{(.)}$, $\omega_{[.]}$:

$$\begin{gathered} \omega_{(s)} = \mathrm{i}\omega_{[s]}, \qquad \omega_{[s]} = \omega^0_{[s]} \exp\left[\mathrm{i}(\mathbf{kx} - \varpi t)\right], \\ \omega_{(s)} = \omega^0_{(s)} \exp\left[\mathrm{i}(\mathbf{kx} - \varpi t)\right], \end{gathered} \tag{8.4}$$

where with the complex constants, $\omega^0_{(s)} = \mathrm{i}\omega^0_{[s]}$, we can fulfill the motion equations (see the former chapter):

$$\operatorname{curl} \omega_{(.)} - \frac{1}{c}\dot{\omega}_{[.]} = 0 \quad \text{and} \quad \operatorname{curl} \omega_{[.]} + \frac{1}{c}\dot{\omega}_{(.)} = 0. \tag{8.5}$$

Here, an important role may be played by the turbulence related solution in the system $\{r, \phi, z\}$:

$$\omega_{(\varphi)}(r) = \mathrm{i}\omega_{[\varphi]}(r) \tag{8.6}$$

With the introduced waves, $\omega_{(s)} = \mathrm{i}\omega_{[s]}$, we arrive at the possibility of explaining the synchronization of the micro-fracturing processes as being due to the influence of propagating waves. For the fracture processes under compression such a synchronization will assure the common sense of the induced twist and spin motions, while for shear load it will lead to the formation of a long shearing fracturing. In the last case, the spin waves related to a given slip on the main fracture plane attenuate those with the opposite spins generated at the perpendicular fragments and, due to the conjugate solution (8.4), reduce the slip motions on those fragments.

The presented conjugate solution suggests that the spin rebound motion is retarded in phase by $\pi/2$ (as we have: $\mathrm{i}\exp\left[\mathrm{i}(\mathbf{kx} - \varpi t)\right] = \exp\left[\mathrm{i}\mathbf{kx} - (\varpi t - \pi/2)\right]$); when slip would start due to the breaking of bonds, the micro-spin motions become released. Following this assumption we expect that such a correlation between the recorded twist motions and spin motions shifted by $\pi/2$ in phase can exist in some wavelets (Fig. 8.4).

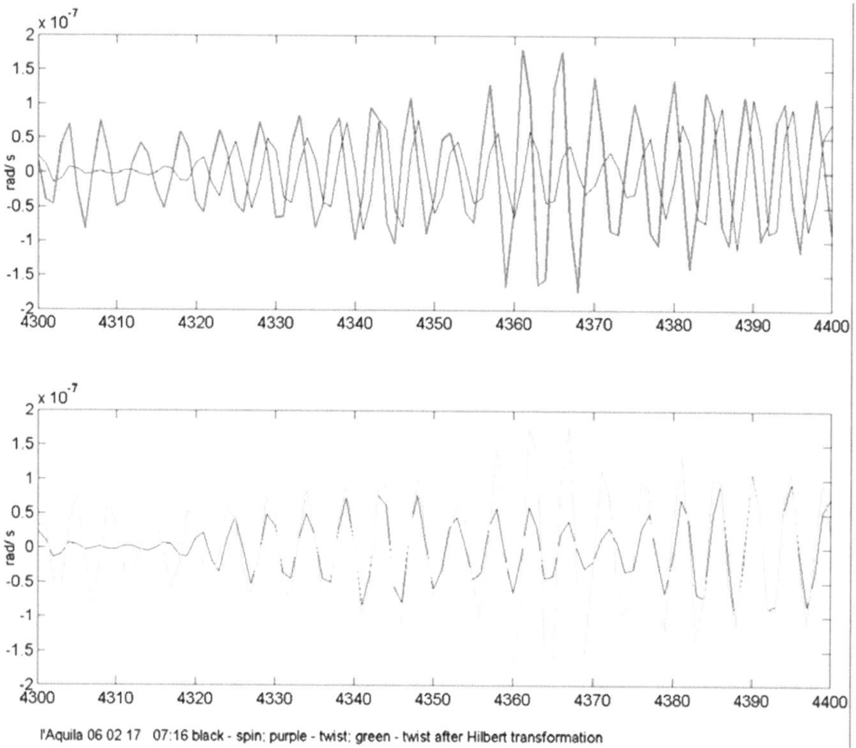

Fig. 8.4 An example of the coincidence of spin and twist motions after the Hilbert transformation shifting the twist record ahead in phase by $\pi/2$; upper panel: the original records, lower panel: the twist record transformed

Now we can propose the following description of the fracture process:

- first, according to external load conditions, the asymmetric stresses rise; following them, the strain and rotation lead to the extreme deformation; the case is illustrated by Fig. 8.5;
- next, approaching the fracture process we may observe the "accumulation" phase with the co-action of the twist and spin;
- finally, the fracturing process starts and we enter into the time rate domain; there will appear the dynamic disclosures and v-dislocation density under the conditions formed by the presented solution.

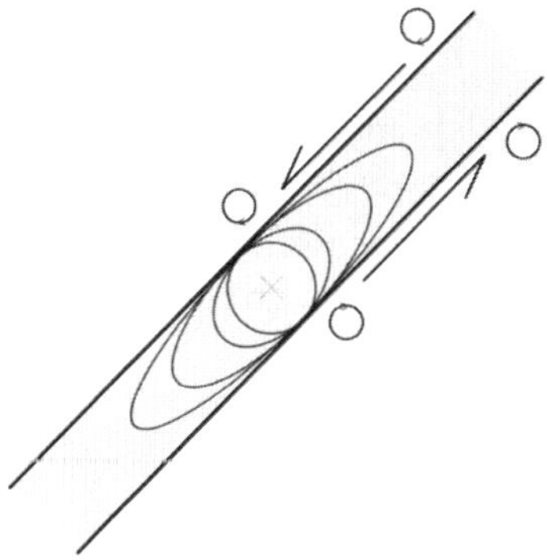

Fig. 8.5 The extreme shear deformation

In the final stage described by the constitutive laws (8.2), we can expect an appearance of the dynamic dislocation objects (v-dislocations); for the v-dislocation field we obtain (see the former chapter):

$$\dot{B}_l = \oint\left(\dot{E}_{kl} + \dot{\omega}_{kl}\right) \mathrm{d}l_k = \oint\left[\overline{e}^0 \dot{E}_{kl}^0 + \overline{\chi}^0 \dot{\omega}_{kl}^0\right] \mathrm{d}l_k = \iint \dot{\alpha}_{pl} \mathrm{d}s_p ,$$
$$\dot{\alpha}_{pl} = \frac{\varepsilon_{pmk}}{2\eta} \frac{\partial}{\partial x_m}\left(\dot{S}_{(kl)} + \dot{S}_{[kl]}\right). \tag{8.7}$$

This case presents the formation of dynamic discontinuities and the related dynamic processes in which the slip and bond breaking leads to the rebound spin motions delayed in phase by $\pi/2$.

The co-action of the spin and twist motions leads to the "accumulation" phase, while the conjugate solution (8.4) presents a fracture process – "release" phase.

We might suppose that the fracture process could proceed with the consecutive accumulation and release micro-processes; in such a situation the related twist and spin motions will appear consecutively as pairs of in phase (or anti-phase) wavelets, and those differing in phase by $\pi/2$.

Such a theory, due to its simplicity, could be very useful for some problems, among others those in which macro-rotation plays an important role in the asymmetric fluid dynamics. The solution proposed can be called the fracture synchronization waves.

Finally, we shall notice that similar solution might exist for the electric and magnetic induction vectors:

$$D_s = \mathrm{i}B_s \leftrightarrow D_s^0 = \mathrm{i}B_{[s]}^0 ,$$

under the condition that it will be assured by the appropriate material constants.

8.4 Conclusions

Rotational deformations play an essential role in fracture processes, especially for the case of compression load.

Fracture processes can be synchronized by the spin and twist waves; the properties of the mylonite zone give grounds for the special solution which leads to fracture synchronization.

References

Dieterich JH (1978) Preseismic fault slip and earthquake prediction. J Geophys Res **83**: B8, 3940-3954

9 Inplane and Antiplane Fracturing in a Multimode Random Sequence

Wojciech Boratyński [1,2]

[1] Institute of Geophysics, Polish Academy of Sciences
ul. Księcia Janusza 64, 01-452 Warszawa, Poland

[2] Faculty of Mathematics and Information Science
Warsaw University of Technology
Plac Politechniki 1, 00-661 Warszawa, Poland
e-mail: wbora@mini.pw.edu.pl

9.1 Introduction

The problem of slip formation can be considered from the point of view of an evolution of dislocation density fields; the slip planes can be formed due to dynamic motion of dense dislocation arrays. To this end, we will consider, after Kossecka and DeWitt (1977), plastic deformations related to dislocation motions; however, the classic Kröner approach with the internal nuclei (represented by the self fields, see: Boratyński and Teisseyre 2006) is replaced by the standard asymmetric theory of continuum. On this basis, some numerical examples presenting dislocation flow patterns are given.

9.2 Standard Asymmetric Theory of Continuum

We recall some elements of the standard asymmetric theory of continuum (see: Chap. 7) as based on both the symmetric and asymmetric stresses and the related constitutive laws and motion equations. The asymmetric deformations contain the symmetric strain and antisymmetric rotation. Thus, our theory is based on two groups of relations, for the symmetric and antisymmetric fields:

$$S_{kl} = S_{(kl)} + S_{[kl]}, \qquad E_{kl} = E_{(kl)}, \qquad \omega_{kl} = \omega_{[kl]} \tag{9.1}$$

We join these deformation fields, in an independent way, with some displacement motion:

$$E_{kl} = e^0 E^0_{kl} = e^0 \frac{1}{2}\left(\frac{\partial u_l}{\partial x_k} + \frac{\partial u_k}{\partial x_l}\right),$$
$$\omega_{kl} = \chi^0 \omega^0_{kl} = \chi^0 \frac{1}{2}\left(\frac{\partial u_l}{\partial x_k} - \frac{\partial u_k}{\partial x_l}\right). \tag{9.2}$$

We should remember that with the structure indexes, e^0 and χ^0, we have different types of media: from the ideal elastic, $\chi^0 = 0$, to fully granulated/crushed material, $e^0 = 0$; here for solids, according to the standard asymmetric continuum theory, we have put: $e^0 = 1$, $|\chi^0| = 1$ (see: Chap. 7).

For the symmetric part of stresses and for the antisymmetric part of stresses, we assume the appropriate constitutive laws:

$$S_{(kl)} = \lambda \delta_{kl} E_{ss} + 2\mu E_{kl}, \qquad S_{[kl]} = 2\mu \omega_{kl} = \chi^0 2\mu \omega^0_{kl}. \tag{9.3}$$

The motion equation for the antisymmetric stresses $S_{[ni]}$ replaces the balance law for the stress moments. Following equations derived in Chapter 7 we write for the antisymmetric stresses:

$$\frac{\partial^2 S_{[ki]}}{\partial x_s \partial x_s} = 2\rho \frac{\partial^2 \omega_{ki}}{\partial t^2} + N_{[ki]} \quad \text{or} \quad \mu \frac{\partial^2 \omega_s}{\partial x_s \partial x_s} - \rho \frac{\partial^2 \omega_s}{\partial t^2} = N_{[s]}, \tag{9.4}$$

where we have introduced also the body couples $K_{[ki]} = \varepsilon_{ski} K_{[s]}$.

Using the scalar and vector potentials (we introduce here the intrinsic length unit l) the motion equation for the symmetric part of stresses

$$\frac{\partial}{\partial x_k} S_{(kl)} = \rho \frac{\partial^2}{\partial t^2} u_l + F_l,$$

$$u_l = l^2 \frac{\partial}{\partial x_l}\varphi + l^2 \varepsilon_{lps} \frac{\partial}{\partial x_p}\psi_s, \qquad F_l = l^2 \frac{\partial}{\partial x_l}\Phi + l^2 \varepsilon_{lps} \frac{\partial}{\partial x_p}\Psi_s$$

takes for the axial and deviatoric strains ($E^D_{lq} = E_{lq} - 1/3\,\delta_{lq} E_{ss}$) the following form:

$$\mu \Delta E_{ss} - \rho \frac{\partial^2 E_{ss}}{\partial t^2} = l^2 \Delta \Phi,$$
$$\mu \Delta E^D_{lq} - \rho \frac{\partial^2 E^D_{lq}}{\partial t^2} = \tag{9.5}$$

$$
= l^2 \left(\left(\frac{\partial^2}{\partial x_l \partial x_q} - \frac{\delta_{lq}}{3} \Delta \right) \Phi + \frac{\partial}{2\partial x_p} \left(\varepsilon_{lps} \frac{\partial}{\partial x_q} + \varepsilon_{qps} \frac{\partial}{\partial x_l} \right) \Psi_s \right),
$$

where

$$
E_{lq} = \frac{1}{3} \delta_{lq} E_{kk} + E_{lq}^D = \frac{l^2}{e^0} \frac{\partial^2}{\partial x_l \partial x_q} \varphi
$$

$$
+ \frac{1}{2} \frac{l^2}{e^0} \varepsilon_{lps} \frac{\partial^2}{\partial x_p \partial x_q} \psi_s + \frac{1}{2} \frac{l^2}{e^0} \varepsilon_{qps} \frac{\partial^2}{\partial x_p \partial x_l} \psi_s .
$$

The second equation of (9.5) can be written in the off-diagonal system as

$$
\mu \Delta E_{lq}^D - \rho \frac{\partial^2 E_{lq}^D}{\partial t^2} = Y_{(lq)} =
$$

$$
= l^2 \left(\Delta \Phi + \frac{\partial}{2\partial x_p} \left(\varepsilon_{lps} \frac{\partial}{\partial x_q} + \varepsilon_{qps} \frac{\partial}{\partial x_l} \right) \Psi_s \right)
$$

and its invariant form in 4D can be achieved with the help of the Dirac tensors (see: Chap. 7):

$$
\mu \frac{\partial^2 \omega_{(\lambda\kappa)}}{\partial x_k \partial x_k} - \rho \frac{\partial^2 \omega_{(\lambda\kappa)}}{\partial t^2} = \mathbf{Y}_{(\lambda\kappa)}, \tag{9.6}
$$

where

$$
\omega_{(\lambda\kappa)} = \omega_{(1)} \varepsilon^1 + \omega_{(2)} \varepsilon^2 + \omega_{(3)} \varepsilon^4 \varepsilon^2 \varepsilon^3 =
$$

$$
= \begin{bmatrix} 0 & -\omega_{(3)} & -\omega_{(2)} & -\omega_{(1)} \\ -\omega_{(3)} & 0 & \omega_{(1)} & -\omega_{(2)} \\ -\omega_{(2)} & \omega_{(1)} & 0 & -\omega_{(3)} \\ -\omega_{(1)} & -\omega_{(2)} & -\omega_{(3)} & 0 \end{bmatrix};
$$

$$
\{\omega_{(s)}\} = \{E_{23}^D, \ E_{31}^D, \ E_{12}^D\}
$$

and

$$
Y_{(\lambda\kappa)} = Y_{(12)} \varepsilon^1 + Y_{(13)} \varepsilon^2 + Y_{(23)} \varepsilon^4 \varepsilon^2 \varepsilon^3 =
$$

$$
= \begin{bmatrix} 0 & -Y_{(12)} & -Y_{(13)} & -Y_{(23)} \\ -Y_{(12)} & 0 & Y_{(23)} & -Y_{(13)} \\ -Y_{(13)} & Y_{(23)} & 0 & -Y_{(12)} \\ -Y_{(23)} & -Y_{(13)} & -Y_{(12)} & 0 \end{bmatrix}.
$$

9.3 Dislocation Flow on Slip Plane

For our continuum with dislocations (see: Chap. 7) and for $\chi^0 = -1$ we can define the deformation tensor:

$$D_{lq} = E_{lq} + \omega_{lq} = E_{lq}^0 - \omega_{lq}^0 = \frac{\partial u_l}{\partial x_q}. \tag{9.7}$$

We recall here the stress-dislocation relation (see: Chap. 7)

$$\alpha_{pl} = -\frac{1}{2\mu}\varepsilon_{pmk}\frac{\partial(S_{(kl)} + S_{[kl]} - \frac{\nu}{1+\nu}\delta_{kl}S_{ii})}{\partial x_m}$$

leading now to

$$\alpha_{pl} = -\varepsilon_{pmk}\frac{\partial(D_{kl} - \frac{\nu}{1+\nu}\delta_{kl}D_{ii})}{\partial x_m} = -\varepsilon_{pmk}\frac{\partial^2 u_k}{\partial x_m \partial x_l} + \frac{\nu}{1+\nu}\varepsilon_{pml}\frac{\partial^2 u_s}{\partial x_m \partial x_s}.$$

After Teisseyre (2001) we write the continuity relation for a dislocation flow including a rate of dislocation generation in some sources P (source/sink term):

$$\frac{\partial \alpha_{lp}}{\partial t} + V^S(1-P)\frac{\partial}{\partial x_j}\alpha_{lp}V_j = 0 \tag{9.8}$$

where **V** is the dislocation flow velocity and V^S is the shear wave velocity; we assume that the dislocation velocity vanishes in the dislocation line direction.

We can estimate a dislocation flow velocity in the following way:

- First, we recall the Koehler expression for a force acting on a dislocation:

$$F_j = \varepsilon_{njk}\, b_i\, S_{in}\zeta_k, \tag{9.9}$$

where ζ is a dislocation line element and the right screw convention relates to vectors **n**, **F**, ξ (**n** is a normal to dislocation plane).

- Next, we establish a formula for dislocation density tensor

$$\alpha_{ki} = \frac{\zeta_k b_i}{\Delta s} = \frac{\zeta_k nb_i}{\Delta s}, \tag{9.10}$$

where Δs is a surface element, and b_i or nb_i are the sums of the Burgers vectors of dislocation lines crossing that surface.

– Finally, we modify the formula for dislocation velocity (Mataga et al. 1987) introducing to it a dislocation density:

$$V_j = V^S \varepsilon_{jnk} \frac{\alpha_{ki}}{|\alpha|} \frac{S_{in} - R_{in}}{\sqrt{(S-R)^2 + R^2}} \approx V^S \varepsilon_{jnk} \operatorname{sgn}(\alpha_{ki}) \frac{S_{in} - R_{in}}{R} \tag{9.11}$$

where **S** are stresses; **R** is the resistance stress; $cB = bR$; B is the drag coefficient; b_i is the Burgers vector of dislocation; V_j is the dislocation velocity; and V^S is the shear wave velocity; the inertia term is neglected here.

The flow Eq. (9.8) becomes

$$\frac{\partial \alpha_{lp}}{\partial t} + c\frac{(1-P)}{R} \operatorname{sgn}(\alpha_{ki}) \frac{\partial}{\partial x_j}\left(\alpha_{lp}\varepsilon_{jnk}(S_{in} - R_{in})\right) = 0. \tag{9.12}$$

An influence of plastic/dislocation flow in the basic Eqs. (9.4) and (9.6) requires that we put there, instead of body forces and moments, the terms related to time rate of plastic velocity and plastic spin; in the first approximation, with the plastic-like dislocation velocity, **V**, we can put for the equations corresponding to (9.4) and (9.6):

$$\mu \Delta \omega_{ks} - \rho \frac{\partial^2 \omega_{ks}}{\partial t^2} = \frac{\partial}{\partial t}\left(\frac{\partial}{\partial x_k} V_s - \frac{\partial}{\partial x_s} V_k\right),$$

$$\mu \Delta \omega_{(ks)} - \rho \frac{\partial^2 \omega_{(ks)}}{\partial t^2} = \frac{\partial}{\partial t}\left(\frac{\partial}{\partial x_k} V_s + \frac{\partial}{\partial x_s} V_k\right)$$

or using (9.7) we arrive at the unique equation:

$$\mu \Delta D_{ls} - \frac{\rho}{\xi} \frac{\partial^2 D_{ls}}{\partial t^2} = \frac{\partial}{\partial t} \frac{\partial}{\partial x_s} V_l . \tag{9.13}$$

We see that the motion equations for strain and rotation fields become more complicated when taking into account a massive dislocation flow on slip/fracture plane. We may remember the approach applied by Teisseyre and Yamashita (1999), in which the motion equations outside a slip plane and just along it were considered separately; it is reasonable to assume that outside the slip plane we may use the homogeneous wave Eqs. (9.4) and (9.6), while along the slip plane we can consider only dislocation flow neglecting the wave part.

However, in the present approach we used a more exact procedure; using (9.11) we can express the last relation as follows:

$$\mu \frac{\partial^2 D_{ls}}{\partial x_k \partial x_k} - \rho \frac{\partial^2 D_{ls}}{\partial t^2} = \frac{V^S}{R} \frac{\partial}{\partial t} \frac{\partial}{\partial x_l} \varepsilon_{snk} \operatorname{sgn}(\alpha_{ki})(S_{in} - R_{in}) . \tag{9.14}$$

This will be our basic equation forming a reference to some numerical examples presenting dislocation flow pattern.

9.4 Numerical Simulation of Dislocation Flow Pattern

The counterparts of the inplane and antiplane motions are mutually related on a common plane – (see Fig. 9.1): for Δs_2 and b_3 we get $F_j = \varepsilon_{jk2} b_3 S_{32} \zeta_k \rightarrow \{F_3 = b_3 S_{32} \zeta_1,\ F_1 = -b_3 S_{32} \zeta_3\}$.

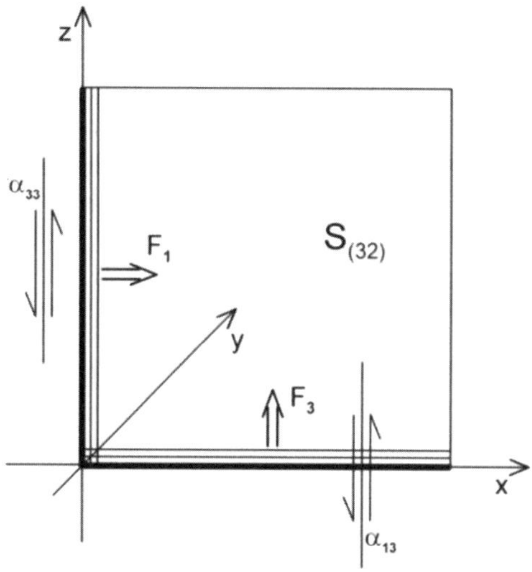

Fig. 9.1 Both motions can occur simultaneously – see Fig. 9.2

Assume that an inplane motion occurs first, forming a fractured segment; its sides parallel to motion direction become formed with the related screw dislocations; we may expect that an antiplane motion might start in a neighbouring segment – see Fig. 9.3 (such a sequence can start reversely: first an antiplane motion and then an inplane one).

According to (9.14) we consider the following equations:

– inplane

$$\frac{\partial^2}{\partial x^2} X - \frac{1}{c_0^2} \frac{\partial^2}{\partial t^2} X + B \frac{\partial}{\partial t} X \frac{\partial}{\partial x}(X + G) + AX = F, \qquad (9.15)$$

where $X = X(x,t)$, $F = F(x,t) = F_0 \sin\theta(x - Vt)$,
$G = G(x,t) = G_0 \sin\theta(x - Vt)$;

– antiplane

$$-\frac{1}{c_0^2}\frac{\partial^2}{\partial t^2}Z+2B\frac{\partial}{\partial t}Z\frac{\partial}{\partial z}(Z+H)+AZ=K, \tag{9.16}$$

where $Z=Z(z,t),\ \ K=K(z,t)=K_0\sin\vartheta(z-Wt),$
$H=H(z,t)=H_0\sin\vartheta(z-Wt).$

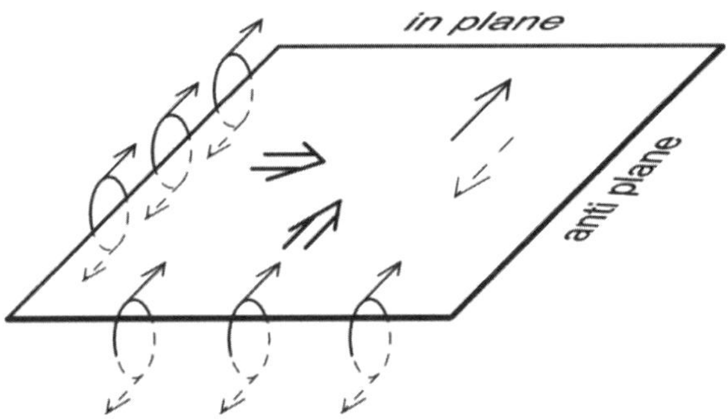

Fig. 9.2 Fault plane: simultaneous inplane and antiplane motions

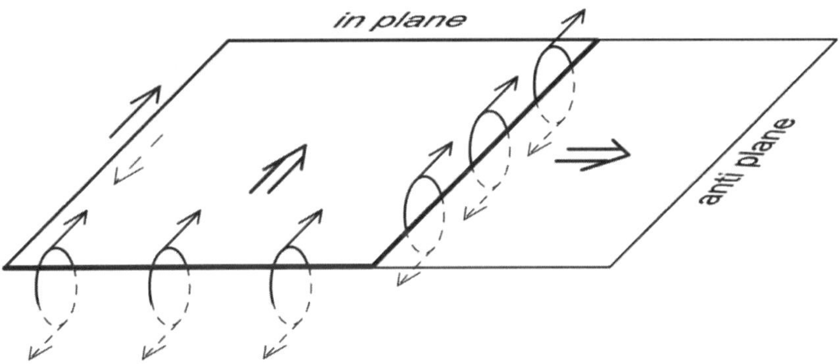

Fig. 9.3 Inplane motion and induced antiplane motion on the neighbouring fault

These equations are solved numerically under the assumption that the final result of numerical solution of the equation will serve as initial conditions for the equation solved in the next step and so on.

Solutions were obtained with the help of *Mathematica Version 5.1 Program*.

We consider two cases, as shown in Figs. (9.3)-(9.10).

The first case

We will consider a sequence of three consecutive solutions as shown in Fig. 9.4.

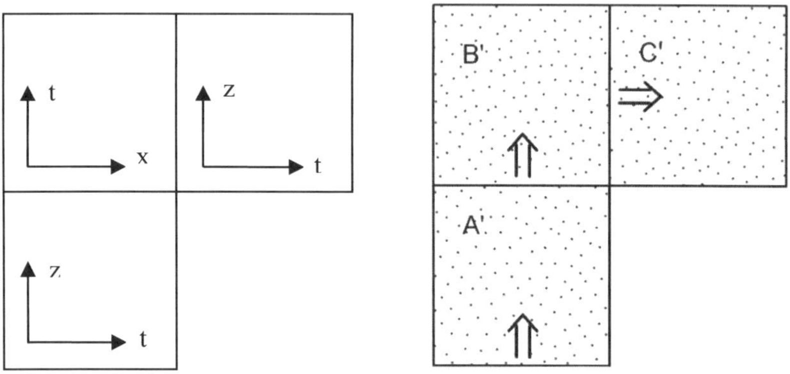

Fig. 9.4 The sequence of three consecutive solutions and explanation of the consecutive axes in the sequence

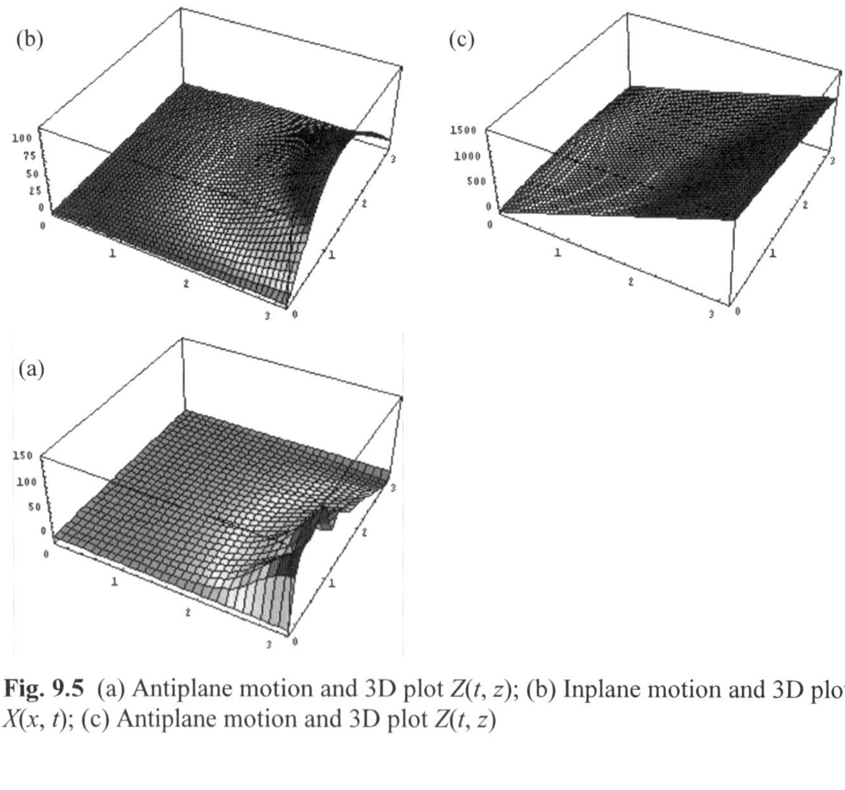

Fig. 9.5 (a) Antiplane motion and 3D plot $Z(t, z)$; (b) Inplane motion and 3D plot $X(x, t)$; (c) Antiplane motion and 3D plot $Z(t, z)$

The first step

Equation (9.16) can be solved numerically with the assumed initial and boundary conditions:

$$Z(z,0)=\sin z,\quad \frac{\partial}{\partial t}Z(z,0)=0,\quad Z(0,t)=0,\quad Z(\pi,t)=t.$$

An example of the obtained solution is presented in Fig. 9.5a in the rectangular domain $(t, z) \in [0, \pi] \times [0, \pi]$. (Time derivatives in Fig. 9.6a).

The second step

Equation (9.15) can be solved numerically with the assumed initial and boundary conditions:

$$X(x,0)=Z(\pi,x),\quad \frac{\partial}{\partial t}X(x,0)=\frac{\partial}{\partial z}Z(z,x)|_{z=\pi},\quad X(0,t)=t.$$

An example of the obtained solution is presented in Fig. 9.5b in the rectangular domain $(x, t) \in [0, \pi] \times [0, \pi]$. (Time derivatives in Fig. 9.6b).

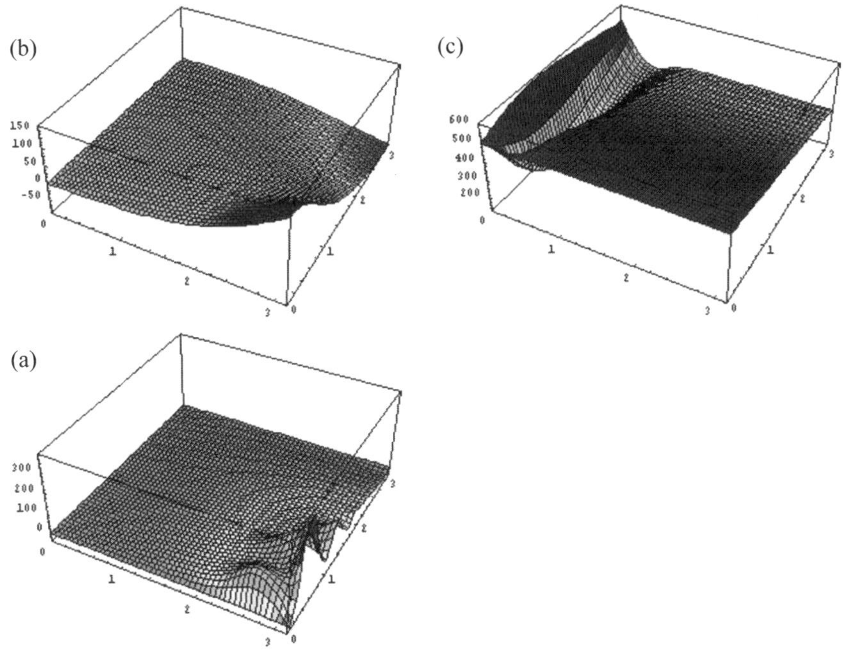

Fig. 9.6 (a) 3D plot $\frac{\partial}{\partial t}Z(t,z)$; (b) 3D plot $\frac{\partial}{\partial t}X(x,t)$; (c) 3D plot $\frac{\partial}{\partial t}Z(t,z)$

The third step

Equation (9.16) can be solved numerically with the assumed initial and boundary conditions:

$$Z(z,0) = X(\pi,z),\ \frac{\partial}{\partial t}Z(z,0) = \frac{\partial}{\partial x}X(x,z)\Big|_{x=\pi},\ Z(0,t) = 500t.$$

An example of the obtained solution is presented in Fig. 9.5c in the rectangular domain $(t, z) \in [0, \pi] \times [0, \pi]$. (Time derivatives in Fig. 9.6c).

The second case

We will consider a sequence of three consecutive solutions as shown in Fig. 9.7.

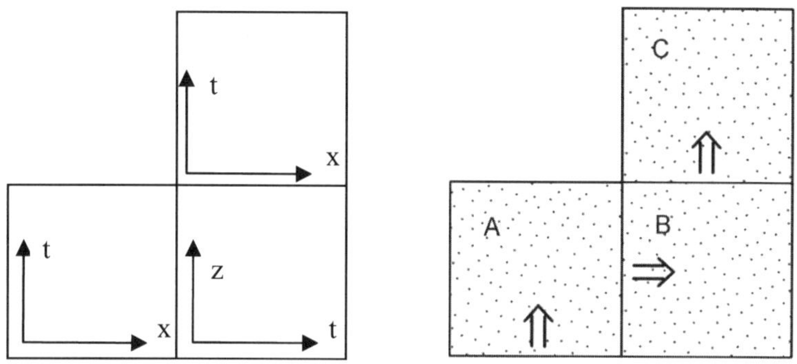

Fig. 9.7 The sequence of three consecutive solutions

The first step

Equation (9.15) can be solved numerically with the assumed initial and boundary conditions:

$$X(x,0) = \sin x,\quad \frac{\partial}{\partial t}X(x,0) = 0,\quad X(0,t) = 0,\quad X(\pi,t) = 0.$$

An example of the obtained solution is presented in Fig. 9.8a in the rectangular domain $(x,t) \in [0,\pi] \times [0,\pi]$.

The second step

Equation (9.16) can be solved numerically with the assumed initial and boundary conditions:

$$Z(z,0) = X(\pi,z),\quad \frac{\partial}{\partial t}Z(z,0) = \frac{\partial}{\partial x}X(x,z)\,|_{x=\pi},\quad Z(0,t) = t^2.$$

An example of the obtained solution is presented in Fig. 9.8b in the rectangular domain $(t,z) \in [0,\pi] \times [0,\pi]$.

The third step

Equation (9.15) can be solved numerically with the assumed initial and boundary conditions:

$$X(x,0) = wp(x,\pi), \quad \frac{\partial}{\partial t} X(x,0) = wpoch(x,\pi), \quad X(0,t) = 0,$$

where $wp(x,\pi)$ denotes the approximation of function $Z(\pi,x)$ by polynomials and $wpoch(x,\pi)$ denotes the approximation of function $\frac{\partial}{\partial z} Z(z,x)\big|_{z=\pi}$ by polynomials. An example of the obtained solution is presented in Fig. 9.8c in the domain $(x,t) \in [0,\pi] \times [0,1.2016]$.

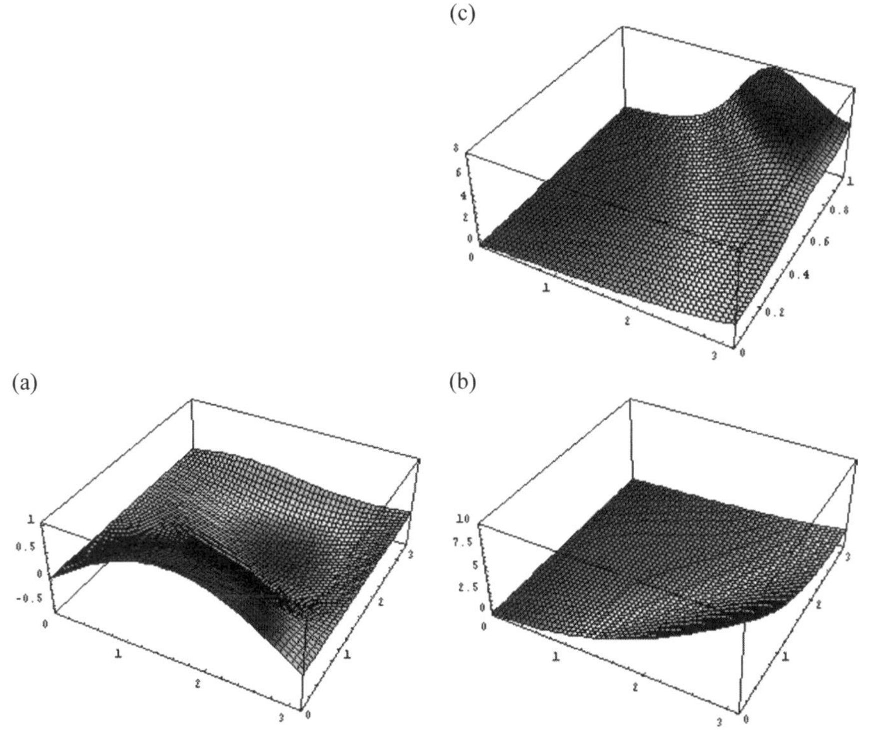

Fig. 9.8 (a) Inplane motion and 3D plot $X(x, t)$ (b) Antiplane motion and 3D plot $Z(t, z)$; (c) Inplane motion and 3D plot $X(x, t)$

The third case – 3D model

Finally under the load of shears $\{S_{32}, S_{31}\}$, we will consider a sequence of three consecutive solutions, as part of the model shown in Fig. 9.9. Under a given stress load, S_{ni}, a fracture process may develop in the 3D pattern: we may have a progress of the edge and screw dislocation motions along the planes perpendicular to directions {n, i}:

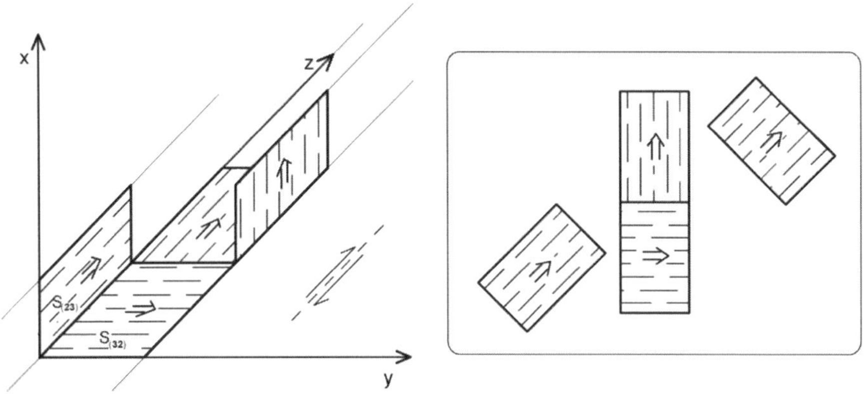

Fig. 9.9 The 3D model of fracture sequence and its projection used in presentations of simulations

for S_{32}, Δs_2 and b_3 we get

$$F_j = \varepsilon_{jk2} b_3 S_{32} \zeta_k \to \{F_3 = b_3 S_{32} \zeta_1,\ F_1 = -b_3 S_{32} \zeta_3\},$$

for S_{31}, Δs_1 and b_3 we get

$$F_j = \varepsilon_{jk2} b_3 S_{31} \zeta_k \to \{F_1 = -\, b_3 S_{31} \zeta_3,\ F_3 = b_3 S_{31} \zeta_1\}.$$

In simulations we will consider only the three consecutive solutions.

The first step

Equation (9.15) can be solved numerically with the assumed initial and boundary conditions:

$$X(x,0) = \sin x, \quad \frac{\partial}{\partial t} X(x,0) = 0, \quad X(0,t) = 0, \quad X(\pi,t) = 0.$$

An example of the obtained solution is presented in Fig. 9.10a in the rectangular domain $(x,t) \in [0,\pi] \times [0,\pi]$.

The second step

Equation (9.15), where unknown function is $Y(y,t)$, can be solved numerically with the assumed initial and boundary conditions:

$$Y(y,0)=X(\pi,y),\qquad \frac{\partial}{\partial t}Y(y,0)=\frac{\partial}{\partial x}X(x,y)\,|_{x=\pi}\,,$$
$$Y(0,t)=10(2\pi-t),\qquad Y(\pi,t)=2t.$$

An example of the obtained solution is presented in Fig. 9.10b in the rectangular domain $(y,t)\in[0,\pi]\times[0,\pi]$.

The third step

Equation (9.15), where unknown function is $Z(z,t)$, can be solved numerically with the assumed initial and boundary conditions:

$$Z(z,0)=Y(z,\pi),\quad \frac{\partial}{\partial t}Z(z,0)=\frac{\partial}{\partial y}Y(z,y)\,|_{y=\pi},\quad Z(0,t)=t.$$

An example of the obtained solution is presented in Fig. 9.10c in the domain $(z,t)\in[0,\pi]\times[0,1.29432\cdot10^{-7}]$.

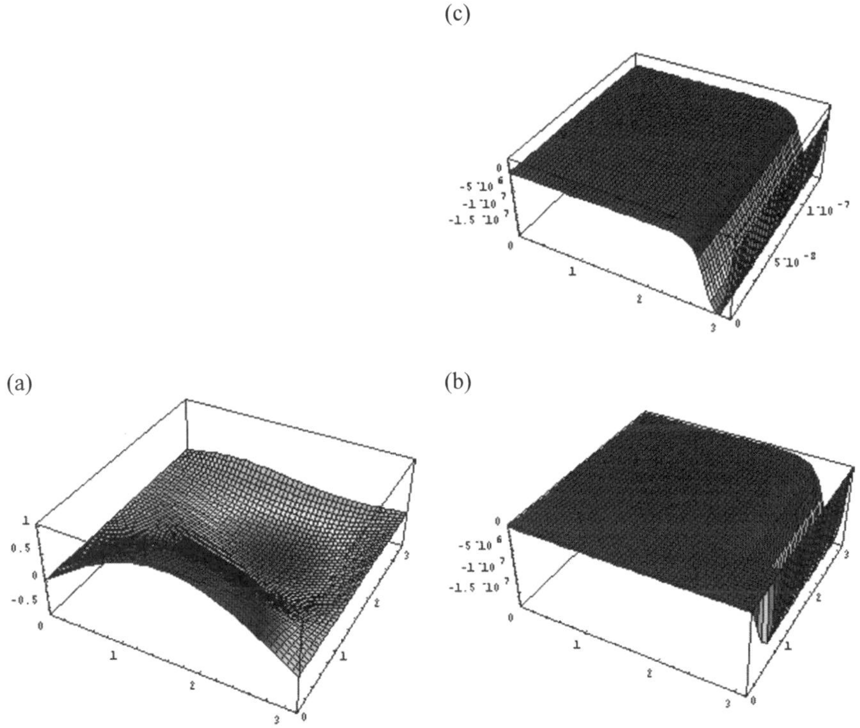

Fig. 9.10 (a) Inplane motion and 3D plot $X(x, t)$; (b) Inplane motion and 3D plot $Y(y, t)$; (c) Inplane motion and 3D plot $Z(z, t)$

9.5 Discussion

The presented approach permits to better understand a relation between the inplane and antiplane motions. A relation between these simultaneous motions on a common fault becomes evident. For a sequence of independent fault motions, extending in time to a neighboring faulting domain, we can take the final value of first solution as the initial condition for the equation describing the next fault motion. Such sequences shown in Figs. 9.3 and 9.6, can also describe an aftershock processes following the first solution – the main shock (shock and aftershock).

We have given some examples of the numerical simulation of the related edge and screw dislocation motions; the presented sequences related to a shear load give a fault propagation for both the 2D and 3D models. However, the governing equations used in numerical procedure lead very quickly to the instabilities; therefore, it will be important in future research to introduce the damping factors into the governing theory.

Finally, we shall note that the presented simulations are given only in a relative scale; moreover, there is a lack of proper proportions between the different constants in the governing equations. To improve this inadequacy, new experimental data are needed.

References

Boratyński W., Teisseyre, R., (2006), Continuum with Rotation Nuclei and Defects: Dislocation and Disclination Densities in "Earthquake Source Asymmetry, Structural Media and Rotation Effects" (eds. R. Teisseyre, M.Takeo, E. Majewski) Springer Berlin Heidelberg New York , 57-66.

Kossecka E., DeWitt R. (1977), Disclination kinematic, Arch. Mech. **29**, 633-651.

Mataga P.A., Freund L.B., Hutchison J.W. (1987), Crack Tipp plasticity in dynamic fracture, J.Phys. Chem. Solid **48**, 985-1005 .

Teisseyre R. (2001), Evolution, Propagation and Diffusion of Dislocation Fields, in "Earthquake Thermodynamics and Phase Transformations in the Earth's Interior" (eds. R. Teisseyre and E. Majewski), Academic Press (vol. **76** of International Geophysical Series), San Diego,San Francisco, NewYork, Boston, London, Sydney, Tokyo, 167-198.

Teisseyre R. and Yamashita T. (1999), Splitting stress motion equation into seismic wave and fault-related fields, Acta Geophys. Polon. **47,** 135-147.

10 Charged Dislocations and Various Sources of Electric Field Excitation

Krzysztof P. Teisseyre

Institute of Geophysics, Polish Academy of Sciences
ul. Księcia Janusza 64, 01-462 Warszawa, Poland
e-mail: kt@igf.edu.pl

10.1 Introduction

Before significant earthquakes, and during them, various anomalies in electric field and currents are often observed in the atmosphere and in the ground. The unusual phenomena include telluric current anomalies, EM radiation in radio frequencies, and light phenomena. An analysis of electric phenomena preceding and accompanying seismic events is out of the scope of this chapter. A considerable number of papers reviewing co- and preseismic electric anomalies, and possible mechanisms of their formation, has been published; we refer the reader to Gershenzon and Gokhberg (1992), Park et al. (1993), Freund (2002), Varotsos (2005), and to the book of Thanassoulas (2007). Varotsos with his co-workers have been developing a method of searching for, and recording the preseismic anomalies of potential differences; seismic electric signals (SES) were observed in certain sensitive areas, up to a distance of about 100 km from the epicentral area.

We present a review of possible mechanisms responsible for electric precursors in the lithosphere, concentrating on initial charge separation and interactions of dislocations and other crystal lattice defects in the presence of the stress field. We include certain complementary consideration to that presented by Teisseyre and Nagahama (2001), where some formulae have been derived for the energy transfer during the dislocation motions, charge separation and the electric/electromagnetic field emission. The interplay between the electric field, dislocations and asymmetric stresses is important for processes in the focal area.

Charge separation, large enough to give effect outside the focus, and some telluric and atmospheric currents evade easy explanation. There are many inspiring hypotheses, including the opinions that strong electric fields and currents play their own role in the fracture preparation and initiation. Hypothetical mechanisms governing tectono- and seismo-electric processes are quite numerous; some masking and, reversely, synergic amplification of effects may occur.

In the search of earthquake precursors, most promising is the phase of dilatation. Here open cracks nucleate at tips of a progressing dislocation array. This process, along with shear stresses, should promote rotations and deformations of grains, which change their polarization state. It is clear that the charge separation processes occur also at other times, and at the time of rupture they may be very intense in some parts of the focal area. The analysis of mylonite material along the faults points to the occurrence of pre- or coseismically generated high-intensity electric fields (see Ferré et al. 2005, and Freund et al. 2007). Pseudotachylite, i.e., the molten and subsequently cooled rock found in faults where a strong earthquake took place – is a glassy material that gives evidence for friction-driven melting. Partial melting causes lubrification; therefore, pseudotachylites are formed only in small quantities, usually as thin vein along the fault. Most of these materials are dark brown due to the magnetite content. Natural remnant magnetisation of the pseudotachylites reaches values 330 times higher than that of their host rock. This implies that currents which flowed in the faults when this molten material cooled below its Curie temperature (580°C) had to be on the order of 1 A/cm^2. Upon experimental basis, these authors state that such strong current densities may be produced in rocks subjected to stress, when the p-holes activation occurs.

We also touch unclear phenomena in the domain of triboelectricity, fractoemission and dielectrics conductivity or photoconductivity. Numerous articles have been devoted to these problems, mainly to the experiments. In geophysics, the basic physico-chemical processes are mostly hidden, and the problem of charge transfer through dielectric material is very complex. The small polaron model can be mentioned in this context, as well as concepts of electron tunnelling or charge transport via **hopping** vacancies in the O^{-2} sub-lattice which exists in many minerals. The latter approach, named p-holes conductance, has been developed in many papers by Freund and his co-workers; they present also a model of a large-scale electric circuits in and around the zone of earthquake preparation (Freund et al. 2007).

Principles of electrokinetic processes are discussed too. The discovery of ferroelectric structure of water is recalled here (Rusiniak 2000).

10.2 Effects of Varying and Transient Polarization Due to Mechanical Stimulation

10.2.1 Piezoelectric polarization

When classical piezoelectric material is subject to external force, the polarization occurs and the electric current forms; these entities are joined by the formulae:

$$\Pi = (\Sigma Q d)/V, \qquad \Pi = \mathbf{D} - \varepsilon_0 \mathbf{E}, \qquad \mathbf{j}_0^{\mathrm{pp}} = \frac{\partial \Pi}{\partial t}. \tag{10.1}$$

Polarization vector $\mathbf{\Pi}$ is the sum of the products of individual charges Q (of one sign) and the related dipole moments (separation vectors d) per volume unit, $\mathbf{D}$ is the electric induction vector, $\mathbf{E}$ is the vector of electric field, ε_0 is the electric permittivity of free space and $\mathbf{j}_0^{\mathrm{PP}}$ is the density of the current component at the source. Here V is the considered volume of rock.

In composite rocks which contain grains or movable dipoles of piezoelectric minerals, mechanical forces shift and rotate these elements; such a process may be mediated by dislocation system and the defects. The moving dislocation core may act both mechanically and electrostatically, as it carries an excess charge. Movable, polarized entities tend to orient parallel to the forming slip or microfracture plane; electrostatic forces and some transient excitations interfere with this process. In the same, deformation of polarizable grains causes their polarization or depolarization.

10.2.2 Transient stimulation

The transient stimulation is an electric polarization caused by change of external mechanical force. The polarization thus obtained decays quickly, which distinguishes it from real piezoelectric effect persisting under constant load. The electric current created in a transient process was also named the deformation-induced charge flow (Varotsos et al. 2001, Varotsos 2005). Experiments on very dry rock samples subjected to variable load revealed the ability of several rock types to transient electric stimulation; the observed potential curve was similar to the first time-derivative of load (Mavromatou and Hadjicontis 2001, Teisseyre K.P. et al. 2001). The results suggest that after each increment of load, the sample gets a new balance; this process may continue until the sample breaks. During each period of load stabilization, some complementary processes take place and manifest themselves as changes in relaxation.

Some experiments of Hadjicontis and Mavromatou have been later numerically simulated (Teisseyre K.P. 2002, Teisseyre K.P. et al. 2001). As the input data, the time-derivative of load was used; output had to simulate the voltage. From the shapes of the rising part of simulated voltage and the relaxation stage, we infer that there occurs a summation of some processes in the rock – each stimulus (when a certain threshold is exceeded) gives rise to a certain increase in potential difference. This effect drops immediately with exponential relaxation.

The concept of division of the observed potential differences into few parts of different relaxation times (Varotsos et al. 2001) enabled to achieve rough agreement of simulation curves with the experimental ones. A closer look at the experimental curves reveals that different rocks, for example granite and limestone, may also react to an abrupt break in the load increase episode, showing a small bay of opposite-oriented polarization. This effect is included in the simulation. Some other effects were introduced to the simulation algorithm, like the effect of second time-derivative of the load and the saturation factor. The latter, perhaps the most important one, corresponds to the observed load history: a strong stimulation exhausts the rock's charge carriers. Further repeating loads cause the polarization effects to become smaller, but after some time, this saturation effect disappears slowly.

A removal of load results in transient stimulation of reversed polarity, except for the LiF (a ionic crystal) which behaves differently – the removal of load gives no electric signal at all. It is to be noted that the case of granite is complex, as this rock contains grains of quartz which is a piezoelectric material. Polarization of these grains seems to be not important because of the chaotic dispersion of the axes ot these crystals.

Transient stimulation is generally attributed to the motion of dislocations caused by mechanical force. The dislocation core usually bears an excess electric charge and is surrounded by a cloud of dispersed charge of the opposite sign, borne by point defects. Mavromatou and Hadjicontis (2001), along with Varotsos (2005), attribute the observed effects to quicker response of the dislocation core system to the load stimuli. A complementary process – motion of opposite charges in point defects– causes the subsequent vanishing of the polarization, and also of the current. The key factor in the whole process is the appearance and variation of the **transient distance** *d* between the dislocation core and the center of the cloud related to it. When this charged cloud catches up the moving dislocation, the distance *d* disappears. Polarization caused by this ephemeral distance is the gradient polarization which we denote by $\nabla\Pi$ or $\tilde{\Pi}$. We assume that the generated component of electric current, $\mathbf{j}_0^{ts}$, is a time derivative of transient polarization and is proportional to density rate of dislocations through the volume of material (Teisseyre 1992, Teisseyre 2001, and Teisseyre and Nagahama 2001):

$$\tilde{\Pi} \equiv \nabla \Pi, \qquad \mathbf{j}_0^{\text{ts}} = \frac{\partial \tilde{\Pi}}{\partial t} \propto \frac{\partial \alpha}{\partial t}, \tag{10.2}$$

where α is the number of moving dislocations, e.g., counted on a certain cross-section perpendicular to the main direction of dislocation movement.

Let us consider an elementary episode of transient stimulation caused by the movement of a single dislocation. In a given moment, polarization of the rock volume is the sum of resultant charges of the dislocation core and the point-defects cloud: $\tilde{\Pi} = \Pi^d + \Pi^c$, the latter having an opposite sign. In the first stage of transient stimulation episode, the polarization rises. The compensation process is equivalent to the dislocation-caused polarization in the time-moment $t = b$ after which the polarization starts to relax. The following formula may be used to estimate the maximum polarization, $\tilde{\Pi}^x$:

$$\tilde{\Pi}^x = \Pi_b^d + \Pi_b^c = \int_{t=0}^{t=b} \frac{\partial (dq)}{\partial t} dt + \int_{t=0}^{t=b} \frac{\partial (dc)}{\partial t} dt,$$

from which we get

$$\tilde{\Pi}^x = \int_{t=0}^{t=b} \left(d\frac{\partial q}{\partial t} + q\frac{\partial d}{\partial t} \right) dt + \int_{t=0}^{t=b} \left(d\frac{\partial c}{\partial t} + c\frac{\partial d}{\partial t} \right) dt. \tag{10.3}$$

Here q is the charge at the dislocation core and c is the charge of the cloud of point defects (this has an opposite sign). The transient distance between the dislocation core centre and the centre of the point-defects cloud is symbolized by d. The time span between the start of the episode and the point of balance $t = b$ will be symbolized as τ_e. These entities are interrelated:

$$d = \tau_e \, (V_d - V_c), \tag{10.4}$$

where V_d is the velocity of dislocation core and V_c is the velocity of the cloud of charged point-defects associated to it (or the velocity of other compensating process, to be strict).

More precisely, this distance at the moment of balance (where polarization is at maximum) is

$$d = \int_{t=0}^{t=b} \frac{\partial C^d}{\partial t} dt - \int_{t=0}^{t=b} \frac{\partial C^c}{\partial t} dt, \tag{10.5}$$

where C^d is the position of the centre of dislocation core and C^c denotes the position of the centre of related cloud, both in relation to the direction of dislocations movement.

In the initial stage, when the stimulation rises, the velocity of the first process, V_d, should depend on external forces. Both this speed and the velocity of complementary process, V_c, should belong to a certain restricted range. Probably, the investigated stimulation occurs only for a certain range of load variation velocities.

After the start of each transient stimulation episode, the difference between both velocities varies and finally both processes stop, so that the core-cloud distance d diminishes and disappears. Meanwhile, other episodes may start in other points; for these episodes the separation of charges, expressed as d, may be different. The same concerns the time τ_e and the relaxation time. Summation of many simultaneous episodes of transient stimulation may result in an apparent split of the process into many portions, each with different time of relaxation, as mentioned in the description of experimental results. Further compensation processes may well have different physical nature; we cannot state that the only allowed mechanism is the pursuit of the dislocation by a cloud of point charges.

We may express the transient polarization in a certain volume of rock as a product of the moving dislocations density and the density, per dislocation length, of core-cloud dipole moments qd:

$$\tilde{\Pi} = \alpha \, qd, \tag{10.6}$$

where q is the mean charge of dislocation cores, d is the mean of distances defined above.

From this, we obtain the expression for density of the generated current component:

$$\mathrm{j}_0^{ts} = qd \frac{\partial \alpha}{\partial t} + \alpha q \frac{\partial d}{\partial t} + \alpha d \frac{\partial q}{\partial t}. \tag{10.7}$$

Assuming that time-variation of mean charge may be neglected, we equate the last term to zero:

$$\mathrm{j}_0^{ts} = \frac{\partial \tilde{\Pi}}{\partial t} = qd \frac{\partial \alpha}{\partial t} + \alpha q \frac{\partial d}{\partial t}. \tag{10.8}$$

For the distance *d* we put τV, where $V = (V_d - V_c)$ is the velocity difference (between dislocation motion and the compensation process, before the balance point) for a given group of transient stimulation episodes. Further, we get:

$$j_0^{ts} = q\tau V \frac{\partial \alpha}{\partial t} + \alpha q \frac{\partial (\tau V)}{\partial t} = q\tau V \frac{\partial \alpha}{\partial t} + q\tau\alpha \frac{\partial V}{\partial t} + q\alpha V \frac{\partial \tau}{\partial t},$$

or

$$j_0^{ts} = qV\left(\tau \frac{\partial \alpha}{\partial t} + \alpha \frac{\partial \tau}{\partial t}\right) + q\tau\alpha \frac{\partial V}{\partial t}. \tag{10.9}$$

As we see, the generated current depends both on the relative velocity V of the piezo-stimulating motion and on its acceleration; this may be related to the first and second time-derivatives of the load. The precise calculating of transient distance *d*, as in (10.5), would complicate the formula.

In this paper, we do not treat time τ as constant for a given material and conditions, because we are aware of concurrence of various relaxation processes which strongly implies that time τ shall vary also before the balance point when $V_c = V_d$. Moreover, the assumption that $\tau = const.$ will not significantly change the conclusion, and we get:

$$j_0^{ts} = q\tau V \frac{\partial \alpha}{\partial t} + q\tau\alpha \frac{\partial V}{\partial t} = q\tau\left(V \frac{\partial \alpha}{\partial t} + \alpha \frac{\partial V}{\partial t}\right).$$

So far, we assumed that the transient stimulation develops only in the described way, due to moving charged dislocation; $\mathbf{j}_0^{ts} = \mathbf{j}_0^{cd}$. We cannot exclude, however, the presence of other mechanisms. Mavromatou and Hadjicontis (2001) discuss the propagation of the electric excitation through a rod of rock, positioned very close to the sample subjected to above-mentioned experiments; this is a kind of ultra low frequency electromagnetic radiation. These authors present also some additional experiments; their findings give some light on transient stimulation mechanisms in LiF and granite. Irradiation of these minerals with γ rays before the experiments leads to different results: granite reacts to loads in the usual way, and LiF shows reduced transient excitation. For higher doses (above 10 Mrad), the transient stimulation and EM radiation completely disappear (still, in the case of LiF). Thermal annealing of irradiated samples removes the irradiation effect. These results point again to different mechanisms of transient stimulation in LiF in comparison to other investigated rocks. In another paper, Hadjicontis et al. (2004) indicate that the color centers (F-centers) in LiF crystal, produced by irradiation, act as traps for charges caused by moving microcracks. Whatever the mechanism underlying the transient stimulation might be, the speed of external load variation should bear a great importance.

10.2.3 Coalescence of opposite dislocations, microcrack coalescence

For a given material, parallel dislocations or microcracks carry an excess of charges of the same sign regardless of the dislocation sign in mechanical sense. Thus, when two parallel dislocations or microcracks approach each other, they are attracted through mechanical forces but in the same time repelled due to identical sign of electric charge. Coalescence leads to some mechanical energy release while excess charges would be expelled – these may be the exoelectrons, impurity ions or for example the O^- radicals (p-holes); these topics will be mentioned later. The generated current component is symbolized by $\mathbf{j}_0^{em}$.

10.2.4 The effects of large dielectric body under vibration or varying load

In the Earth's lithosphere, there occur many dielectrics and ferroelectrics – see Corry (1994). We assume that most of dipoles have the same orientation of crystallographic axes, and the material is not antiferroelectric. We may consider different situations:

A. Under mechanical forces, the electric field appears in the body; this may be either the piezoelectric effect or the gradient polarization. Therefore, the seismogenic oscillation gives rise to some pulses to the electric field, in the ULF range.
B. Electric polarization may also be modified due to rotation waves; for example, the polarized grains may be rotated. In a most common situation, the electric dipoles are oriented in such a way that they attenuate the external electric field; in this case, the rotational oscillations will produce an oscillating episode of enhanced propagation of this field. Strong electric fields, both of natural and industrial origin may be a valuable source of information.

10.3 Electrokinetics and the Properties of Water

The main interest of earthquake science in the electrokinetics lies in electric fields generated in rocks when aqueous solutions filtrate through them. An additional electric field on the ground or the change in ionic content of water are often referred to as preseismic electric anomalies. Problems of colloids and particulates seem to be less important, at least according to the present knowledge.

From laboratory research on liquid filtration through a porous (e.g. fibrous) material, the streaming potential is found and then its source, the so-called **zeta potential** ζ, is calculated.

When water contacts with a solid material and this material, bearing electric charge on its surface, comes into reaction with water (for example, the reaction of hydration), ions gather in the liquid on the solid-liquid interface and near it due to dissociation and electrostatic induction (see, e.g., Ishido and Mizutani 1981). At the contact zone, the **diffuse double layer** is formed. The content of this sheath varies from the solid to the liquid side as follows: to the solid surface, mainly ions of opposite sign are firmly attached, although water molecules are also present there; the outer part contains ions of again opposite sign (it means again the same sign as on the solid surface) and a higher content of water molecules adsorbed through hydration of ions. The border surface between liquid phase and this complex sheath of ions and molecules is called the **shear surface**, where slip motion can occur. Outside this slipping surface, there still are clouds of ions – this is the outer zone in diffuse double layer. Zeta potential ζ is the electric potential difference between the bulk of liquid and this slipping surface. The sign and value of zeta potential depend on physicochemical features of involved materials, and on specific conditions, among them the water pH (see, for example, Guichet et al. 2006).

The gravity and pressure cause the filtration processes. The observed potential difference – the **streaming potential** – is generally proportional to the applied pressure gradient. The streaming potential coefficient C is introduced here in the simplest formula:

$$C \equiv \frac{\Delta V}{\Delta p} = \frac{\varepsilon\zeta}{\eta\sigma}, \tag{10.10}$$

where ΔV is the streaming potential, p is the pressure, ε is the dielectric constant of liquid, ζ is its zeta potential, η the dynamic viscosity for shear motion and σ is the liquid's conductivity. The current density generated in a given direction is:

$$\mathrm{j}_0^{\mathrm{ce}} = \frac{\varepsilon\zeta}{\eta F^0}\nabla p - \frac{\sigma}{F}\nabla V, \tag{10.11}$$

(from Varotsos 2005) where F and F^0 are electric formation factors; the first term on the right-hand side is the convection current (in aqueous solutions treated in standard diffuse double layer theory, this is the current of ions from outside the slipping surface, swept along with water); the second term is the conduction current forced by the streaming potential ΔV (sum of the streaming potential gradient over an investigated way of fluid or over a rock sample).

The conduction current is the reaction to the charged fluid flow. To be strict, it also hampers (to some extent) this flow. In the steady fluid flow condition, absolute values of both terms on the right-hand side are equal; the first term, the convection current, is completely compensated by the conduction current. As a result, there is no electric current flowing, despite of the presence of streaming potential ΔV (it is known that the electrokinetic electric current appears due to a certain change on the way of fluid). For the case when $\mathbf{j}_0^{ce} = 0$, we come to the Helmholtz–Smoluchowski equation:

$$\frac{\Delta V}{\Delta p} = \frac{\varepsilon\zeta}{\eta\sigma}\frac{F}{F^0}, \tag{10.12}$$

from which we get the corrected expression for the streaming potential coefficient C:

$$C \equiv \frac{\Delta V}{\Delta p} = \frac{\varepsilon\zeta}{\eta\sigma}\frac{F}{F^0}. \tag{10.13}$$

Each of electric formation factors, F and F^0, includes permeability coefficient and the pores/cracks saturation factor. Factor F^0 does not depend on eventual conductance of the interface between solid and fluid, while factor F does depend:

$$F = \frac{F^0}{1 + \frac{2S_s}{r\sigma}}, \tag{10.14}$$

where S_s is the surface conductance (the conductance of those interfaces along the way of fluid in the rock), r is the pore radius (at cross section) and σ is again the fluid conductivity. If S_s is absent, we obtain $F = F^0$. This relation is strictly applicable for a very simple geometry of pores, e.g., for the set of identical parallel capillaries. In other cases, kind of integration will be needed, containing the spectrum of pore radii r together with certain geometrical coefficients; moreover, such an integration might involve the local S_s variation.

Electrokinetic relations proceed in three dimensions: the zeta potential forms perpendicular to solid surface, and the streaming potential, i.e., the experimentally obtained potential difference ΔV, emerges from integration of ζ and other relevant factors along the filtrating fluid way. We propose to express this dependence by the following linear integral form derived from (10.11):

$$j = \int_{x_0}^{x_N} \left(\frac{\partial \varepsilon}{\partial x} \zeta \frac{1}{\eta F^0} \nabla p + \frac{\partial \zeta}{\partial x} \varepsilon \frac{1}{\eta F^0} \nabla p + \frac{\partial \nabla p}{\partial x} \varepsilon \zeta \frac{1}{\eta F^0} - \frac{\partial F^0}{\partial x} \frac{1}{F^{0^2}} \varepsilon \zeta \frac{1}{\eta} \nabla p \right) \mathrm{d}x$$

$$- \int_{x_0}^{x_N} \left(\frac{\partial \sigma}{\partial x} \frac{1}{F} \nabla V + \frac{\partial \nabla V}{\partial x} \sigma \frac{1}{F} - \frac{\partial F}{\partial x} \frac{1}{F^2} \sigma \nabla V \right) \mathrm{d}x. \qquad (10.15)$$

Expression (10.15) includes changes of the fields and parameters along the path, due to variations, e.g., in temperature and porosity, when filtrating water enters the system of cracks/fractures. During dilatation phase, crack/fractures system increases its volume and, as a result, an increased ion content may appear in underground fluids or in the air above ground. Another precursor – change in ground self-potential – may appear even without any sight of electrokinetic currents.

We shall treat, however, such an analysis with care, because the electric behaviour of water is more complex than it was thought. As revealed by Rusiniak (e.g., Rusiniak 2000, 2004), electric permittivity of distilled water depends on many factors, and by no means it is constant even in stable temperature. Measured in higher frequency range, above 1 kHz, it has a value of about 80, provided that the sample is not too thin; in low frequency, e.g., 5 Hz, its electric permittivity ranges between 2×10^4 and 2×10^6. The electric permittivity appears to be depending on the sample thickness; it is higher in thin water samples. In the sample that is only 0.2 mm thick, the distilled water measured at 5 Hz had electric permittivity bigger than 3×10^5; for higher frequencies it dropped and fell below 100 at frequencies of $6\times10^4 - 10^5$ Hz.

Moreover, the electric permittivity of water grows with time after filling the measurement chamber; Rusiniak explains this effect by an increasing structurization – growth of the concentration of ordered domains in the water (we feel, however, that the time needed for such an arrangement of domains is too long to make this phenomenon important in the processes of filtration). Therefore, in very low frequencies, water (not only ice) exhibits ferroelectric properties.

In geophysical research, the relative dielectric permittivity of water is taken from classical, high frequency measurements (so, for room temperatures, the value of ca. 80 is adopted), while in the research on blood in medicine, the permittivity factors similar to those found by Rusiniak have already been applied for some time, but they were attributed to the water-bearing mixture (see Balan et al. 2004). In the light of these findings, we have to emphasize that the mechanism underlying streaming potential occurrence may depend also on the domain system in water. If there is a layer of dipoles, the domains firmly adjacent to the surface of solid, then the zeta-potential will retain its meaning. Possibly the natural waters – aqueous solutions – should be best treated as a fluid abundant with domains and ions, both having an influence on electrical properties. Anyhow, the realistic values of dielectric water permittivity should be taken into account.

Recent experiments with rocks samples dripped with distilled water and subjected to standardized strokes have shown that the pattern of post-stroke polarization (on which sides the signs "+" and "–" will appear) depends on the direction from which the stroke came (Rusiniak, 2007, personal communication). The experimenter explains his results as being due to progressing domain-type organization, that is polarization, along the impact wave.

During percolation and due to the progress of strains, the dislocation and crack net may vary. The magnetic and electric properties of rock skeleton may change too (Park et al. 1993).

Streaming potential, due to its dependence on the conditions, leads us towards spontaneous potentials sensitivity to local inhomogeneities (Gershenzon and Gokhberg 1992).

Widening the scope of considerations: fluids interact with electric and magnetic fields both in the preparation phase and during the rupture. Electrokinetics and magnetohydrodynamics, the domains important for Earth sciences, may be applied also to processes during rupture; not only to the water influence but also to the granulated rock material (see: Chap. 8), and to temporarily melted parts of fault walls (Freund et al. 2007).

Before the earthquake, circulation of fluids in the area of a future seismic event usually undergoes changes. The system of fractures and channels may be temporarily connected with deeper layers or magma chambers; this sometimes leads to spectacular pre- and co-seismic phenomena as geysers, water fountains, flames and so on (see Gold and Soter 1985). The above-mentioned hydraulic variations influence the electric/electromagnetic anomalies.

10.4 Less-Known Mechanisms of Charge Separation

10.4.1 The contact electrification and variation of rock properties

Electrification, in the micro-scale meaning, may rely on the already evoked dislocation reaction with stresses, and on dipoles arrangement, and the conductance typically relies on electron motions via a conduction band (as in metals) or a valence band (typical for semicondutors), or on the motion of ions. The last two topics are related to the reduction/oxidization problem. We came onto the process called contact electrification which Freund (2002) considers to be one of the fundamental processes occurring between electrically dissimilar materials contacting each other. We should recall here two basic terms which are in use in the contact electrification domain. When two materials are brought into contact (to make the things simpler, let both bodies be electrically neutral in macroscopic sense), transfer of certain amount of electrons from one to the other occurs. When these materials are later torn apart, they remain charged (see Horn and Smith 1992), although partial charge redistribution, via various processes, occurs during the time-interval when the distance between the two bodies increases (until it passes a critical distance). The force causing electrification (and in part also the redistribution of electrons) is related to the difference in **electrochemical potentials**. These potentials depend (see: Nernst equation) on temperature and other physico-chemical conditions; in the case of dissolved substances – on activities of the reduced and the oxydized agent. In contemporary research, another approach is in use – the attractiveness of various materials to the electrons is compared. The **work function** is the measure of attractiveness of the body to electrons.

Processes of electrification and the change of conductance may occur in a rock even without mediation of dislocations or the aqueous solutions. In the changing conditions, as pressure and temperature, the work functions of minerals, and portions of rock, change along. Change in distance or work functions is enough for producing variation in the intrinsic electric field. These variations affect the Fermi levels of electrons inside the body; therefore they lead to migration of electrons, seen as the charge motion. The related current density will be symbolized as $\mathbf{j}_0^{wf}$.

10.4.2 Electromagnetic processes caused by fracturing and rubbing

Plastic deformation and fracturing of rocks is sometimes accompanied by emission of electrons, molecules and ions and photons; the EM radiation also appears there. These phenomena may appear even under not very strong loads, if only the material of one kind is rubbed by another. The emission of electrons in these conditions is called triboelectricity, the emission of light is known as triboluminescence. In geophysics, the term triboelectricity has two meanings; it may mean charge separation caused by fracturing the crystal lattice or electrification by contact change or by rubbing that occurs between two materials when at least one of them is dielectric. The electric phenomena accompanying fracturing may be caused, at least in part, by polarization and/or transient polarization, discussed earlier.

When an advancing fracture separates charges, the freshly formed crack surfaces may contain many uncompensated charges; the electric dipoles are formed and this leads to electromagnetic radiation, especially if the newly formed fracture is forced to vibrate. This is one of the explanations to pre- and coseismic EM radiation. Charge separation related to formation of cracks and fracturing gives rise to still another current component $\mathbf{j}_0^{fr}$. It is to be noted that this kind of electrification process becomes especially difficult to understand when the distance between two bodies starts to increase; the processes of electron hopping (tunneling) between the defect surfaces may take place. The same concerns the reverse distribution of charges: besides the "back-tunneling", also corona discharges may occur. Electron hopping processes are responsible for charge movement in dielectrics and occur also in semiconductors. At the triboemission and fractoemission processes, electrons of low energy, up to 4 eV, are mainly released, but also ions and molecules. Such a low energy of the released electrons points to the excitation lower than the work function of the bulk material.

In addition to triboemission and fractoemission, we shall also mention pyroelectricity which means either thermal ionization or the emission of charged particles caused by heating of material (which often coexists with triboelectric process). Triboemission, fractoemission and pyroelectricity are sometimes treated jointly as exoemission. Dislocations-related exoemission, $\mathbf{j}_0^{ex}$, was discussed by Teisseyre and Nagahama (2001).

The traps from which exoemission can be released, for example the F-centers in alkali-halide crystal, called also the color centers, are a kind of defects in the transparent crystals. Such centers can absorb light of specific wavelengths. The crystal F-center is probably a vacancy in the negative-charged sub-lattice (vacancy in which an electron, or a few electrons, are trapped). The exoelectrons, exoions and exoatoms (the trapped electrons, ions or atoms) can be released at low energy impact, for example due to EM or thermal excitation. Poletaev and Shmurak (1984) explain dislocation-related exoelectron emission from colored alkali-halide crystals with F- centers as follows. Dislocations moved by plastic deformation capture electrons which were localized in an electronic trap (F-center). These electrons become quasi-free. Such an electron may move to the bottom of atoms' conduction band. This process causes movement of other electrons to higher energy level – further from the atomic nucleus; this is a kind of reverse Auger process. Later, due to thermal excitation, one of the electrons may be emitted. This would explain – in one of possible ways – the triboluminescence phenomena.

According to Prevenslik (2001, 2003), the repeating rubbing – which creates friction between metal and insulator – produces a small quantity of fine powder from the latter. This pulverized matter forms and gathers, under pressure, in a very narrow gap between metal and insulator. Such a small gap acts as QED – a trap for radiation energy. Regular low frequency EM radiation from molecules of powder (this is the radiation related to ambient temperature) is inhibited; therefore, the energy trapped in this way rises. An impulse of a higher frequency and thus a higher energy is emitted from the powder. This is an impulse of EM radiation in vacuum ultraviolet (VUV) range which in turn frees the electron from the metal surface. The metal acquires a locally positive charge and in the next episode of still closer contact with insulator, free electrons are transferred to the insulator. There remains a possibility that the VUV impulse, when absorbed by the insulator, raises its electrons from the valence to the conduction band, facilitating a subsequent charge transport. Prevenslik's articles throw also a new light on electric discharges.

Consideration of mechano-electric processes leads us to another phenomena: obscuring of polarization by adsorbed material, emission of this material and the pyroelectricity. All are found in the experiments such as reported by Fukao et al. (2003).

10.4.3 P-holes activation

Papers by Ferré at al. (2005) and Freund et al. (2007) bring experiments on various rocks, discussion on psedotachylite veins along faults, as well as new models of large natural electric circuits around an earthquake focal zone. An important mechanism is presented for electric precursory processes: excitation of the p-holes which abound in various rocks in the form of dormant p-hole pairs. Metal oxides, as MgO and silicates, contain vacancies and impurities in their crystal lattices, mainly the constituents of water which are built into a distorted lattice. When MgO crystallizes in the presence of water, some water molecules become only trapped and some incorporated into the crystal lattice:

$$(H_2O)_{\text{dissolved}} + (O_2)_{\text{structure}} \leftrightarrow (OH^-)_{\text{structure}} + (OH^-)_{\text{structure}}$$

These hydroxyl pairs can undergo redox conversion, splitting off molecular hydrogen and forming two O^- which in turn undergo self-trapping to a peroxy anion, O_2^{-2}, which is the positive hole pair (PHP):

$$(OH^-)_{\text{structure}} + (OH^-)_{\text{structure}} \rightarrow (H_2)_{\text{structure}} + (O_2^{-2})_{\text{structure}}$$

The presence of H_2 molecules and PHPs was confirmed by many methods. Such a PHP locates in the vacancy in Mg^{+2} sublattice; molecular hydrogen remains in the crystal lattice too. In silicates, even nominally anhydrous, a certain amount of water is present, which reacts with crystal lattice in a somewhat similar mode:

$$(H_2O)_{\text{dissolved}} + (O_3X/^{O}\backslash XO_3)_{\text{structure}} \leftrightarrow (O_3X/^{OH}{}_{HO}\backslash XO_3)_{\text{structure}}$$

where X is the ion of silicium, aluminium etc. Here the peroxy-pairs undergo redox transformation, by analogy to the situation in magnesium oxide:

$$(O_3X/^{OH}{}_{HO}\backslash XO_3)_{\text{structure}} \rightarrow (H_2)_{\text{structure}} + (O_3X/^{OO}\backslash XO_3)_{\text{structure}}$$

Peroxy-bonds in the rock may undergo dissociation in response, e.g., to mechanical force. If this occurs, the insulator turns into a semiconductor with p-type conductance, in the following way: one of O^- ions becomes a "mobile" carrier of positive charge – it can accept an electron to change into regular O^{-2}, later it can release an electron, and so on. In the sublattice of O^{-2} ions, these O^- radicals play role of holes $O^{\cdot}_{\cdot}$ The other $O^{\cdot}$ from the pair does not partake in the conductance. Compare the wider explanation in Malinowski (2001).

Positive charges, due to mutual repulsion, migrate to the surface of rock, until the charge stored there is compensated by negative charge which gathers inside the rock. These signs are relative, because O^- site is positive only in comparison to regular elements of O^{-2} sublattice. Anyhow, the accumulation of these oxygen radicals is ready to accept adequate number of electrons. The evidence that the defect electrons transfer may be the main mechanism in electric effects, comes from a series of experiments described by Freund (2002).

10.4.4 Exemplary experiments with friction and fracturing and their interpretation

The electric phenomena caused by friction at the stick-slip event are of considerable interest and were also subject of laboratory research. Takeuchi and Nagahama (2006) report results of tests performed upon quartz pegmatite cylinders pre-cut in such a way that in the middle of axis the cylinder was divided by oblique (45°) plane of cut. The sample was kept under confining pressure and loaded by piston from above through one of circular sample walls. Around the sample, the coil was placed, solenoid or toroid, in the plane of cut. During loading period, of about 100 s, each sample reacted with 2-4 stick-slip events. When a solenoid coil was used, one type of EM radiation was detected during the stick-slip events. The waveform of the induced voltage was composed of damped oscillations at frequencies $f \approx 300$, 700 and 9000 Hz that were almost in phase with the damped oscillations of the piston strain at $f \approx 300$, 700, 2300, 4400 and 9000 Hz. The toroidal coil also reacted when the sample underwent stick-slip event, but the shape of induced voltage was different – it was a sharp rise followed by an exponential decrease (decay time was about 2×10^{-4} s) upon which small-amplitude damped oscillations were superimposed (of frequencies $f \approx 300$, 2300, 4400 and 9000 Hz). These oscillations again were almost in phase with the damped oscillations of the piston strain (an external cause for these oscillations is rather improbable as the sample was kept in a metal vessel). Judging from the typical experimental results, the polarity of signal did not change despite the occurrence of these oscillations – the curve was above the zero level. Only the toroidal coil detected also a number of small EM impulse few seconds **before** the stick-slip. Such signals were accompanied by very small vibrations of the piston strain, but sometimes these vibrations were not detected, or were undistinguished from the noise. These small EM pulses had the same waveforms and polarity as the larger EM impulses during stick-slips. Occasionally, abnormally strong spike or burst of spikes was detected, irrespective of the type of coil used, their polarity varied even during the burst.

The coil produces voltage in response to magnetic field variations, and the type of coil is very important here. A solenoid coil is most sensitive to the magnetic impulses oriented perpendicularly to the plane of solenoid; it is less sensitive to oblique pulses and insensitive to variations proceeding in the plane of coil. A toroidal coil reacts to components in the plane of coil (i.e., in the plane of slip); its reaction on oblique magnetic pulses is smaller and perpendicular pulses cause no reaction. Therefore, the solenoid reacts mainly to electric field variations in its plane, while the toroidal coil is most sensitive to electric field oscillations in the perpendicular direction.

The results clearly show pronounced polarity of electric phenomena accompanying slip events in the sample and their preparatory stages. During slip, movement of electric charges had both perpendicular and in-plane component and repeatable directions, because all larger impulses were similar. Such a signal started with a sharp peak of voltage perpendicular to the slip plane. Sources of these larger signals may be various:

- rotation of polarized quartz grains,
- change of polarization state of quartz grains,
- transient stimulation in other minerals,
- triboelectric processes an/or p-holes currents.

Any of these mechanisms shall be directed by oriented system of moving dislocations or/and microcracks. Damped oscillations were most probably caused by some mechanical imperfection of the whole measuring arrangement (it appears to be prone to vibrations).

Smaller signals were probably the effect of charge transport through the pre-cut gap, due to triboelectric or p-hole conductance. These signals were sometimes accompanied by small mechanical vibrations; both electrical and mechanical disturbances indicate that stress-strain rearrangement goes on and stronger events, e.g., of slip, are expected. The origin of "atypical", very strong signals is in our opinion unclear. They may come either from some breaks or slips in other parts of the sample, not at the pre-cut, or they might be caused by rapid release of some excitation stored in the crystal lattices of minerals, e.g., the intense triboelectric process or/and release of many p-holes, probably at the pre-cut, caused by intense milling at some asperities, present despite previous polishing of the sliding surfaces.

In the recent paper, Rabinovitch et al. (2006) discuss emission of LF (20-25 MHz) electromagnetic waves from rocks during fractures/cracks formation and during provoked friction episodes. These authors wonder why the obtained signal is not uniform but distorted into wavelets, and they come onto conclusion that the story of crack formation is written in the signal shape. The crack is created as a junction of several microcraks, each of them starting with a short burst of EM radiation. As these microcracks merge, also the EM radiation pattern develops. These authors postulate the following mechanisms. Breaking the bonds by the moving fracture moves atoms (on both created sides) into non-equilibrium positions which starts their oscillations around steady state, if cracks are formed when the rock vibrates. In these oscillations, positive charges move together in opposite phase to the negative ones. Initial charge separation could be either transverse to the crack surfaces, or along them. Damping of these oscillations is attributed to interactions with bulk phonons. In this model, the crack surfaces and edges may be electrically neutral.

10.5 Pre-earthquake Stress Variations as the Source of Rotations and Electric Processes

We follow the paper by Teisseyre et al. (2006). The inner preparation processes depend on the external stresses configuration; in extreme cases, for external shear and compression, we are dealing with a difference in preparatory stage before fracture. Shear stresses in the focal zone produce a definite organization of the dislocation arrays and cracks formed; the twist motions, as oscillations of the off-diagonal shear axes, take place. Repeated spatially-organized variations in shear stress and the resultant motions may facilitate both the material crushing and dipoles rearrangement along the fracture planes. Open cracks nucleation facilitates also rotations of grains in a common sense along the main fracture being formed, but opposite along some perpendicular sub-fractures.

Under a compression load, the induced shears and rotations appear: the confining conditions force the shears of opposite orientations, while a common sense of rotations helps in rearrangement of dipole directions.

In both cases, the progressing dislocation arrays, cracks nucleation and rotational motions produce charge separation, rotation of dipoles and other motions, via the discussed processes. In the first case – that of external shears – the shear processes play main role in fragmentation and fracturing; therefore, various tribolectric phenomena should be expected; these are less probable in the second case. Release of significant portion of energy before the main rupture is also more probable for the shear case.

For the compression load, a small part of energy may be released before the rupture, so that the probability of any precursory phenomena is lower, and the shock will probably be strong. The rotational motions play an important role both in the preparatory phase and in the main episodes of fracturing. Due to common sense of rotation, we may expect an amplification rather than annihilation of the generated rotational (spin) waves.

In a more realistic approach, we treat the focus zone as a volume subjected both to shear and compression load and we suppose that short time-periods with rotational motions prevalence alternate with those with shears and twist prevalence. This applies both to the preparatory stage and the rupture (see Chap. 8). Hence, an alternation in the type of electric precursors should also be expected.

10.6 Charge Separation and the Rise of Current

Following the consideration presented above, we distinguish many components of current generated in the rupture preparation zone (cf. Nagahama and Teisseyre 1998 or Teisseyre and Nagahama 2001):

$\mathbf{j}_0^{pp}$ – current due to change of piezoelectric polarization in anisotropic crystals,

$\mathbf{j}_0^{ts}$ – current caused by transient stimulation; the charged dislocations: $\mathbf{j}_0^{ts} = \mathbf{j}_0^{cd}$,

$\mathbf{j}_0^{em}$ – charge emissions caused by coalescence of opposite dislocations,

$\mathbf{j}_0^{fr}$ – current caused by charge separation at fracturing episode,

$\mathbf{j}_0^{ex}$ – exoemission current after fracturing or rubbing,

$\mathbf{j}_0^{tr}$ – current resultant from surface triboelectrical reactions,

$\mathbf{j}_0^{hs}$ – release of p-holes: the O^- radicals.

The current caused by electrokinetic effect(s), $\mathbf{j}_0^{ce}$, has no direct relationship to dislocations. Also the currents caused by **spatial and simultaneously temporal variation** of materials' work function, $\mathbf{j}_0^{wf}$ do not fit to the class of dislocation-caused charge separations which may be symbolized by $\mathbf{j}_0^{d}$.

Summarizing, when mechanical stimulation induces charge separation, the generated current density at the source may be expressed as:

$$\mathbf{j}_0 = \mathbf{j}_0^{ce} + \mathbf{j}_0^{wf} + \mathbf{j}_0^{d}. \tag{10.16}$$

Of course, not all components listed occur in every case. It may happen, for a given direction and distance, that certain constituents will have different signs. The above-mentioned processes are generally sensitive to conditions, and the possibility must be taken into account that they react with electric fields and electromagnetic oscillations occurring in the focal volume. If the generated current has to flow further, it must meet a medium that can conduct and has connections with acceptors of charges. The current will depend on materials' impedances. Mechanisms of conduction may be various: it may be, for example, a wave of certain excitation.

10.7 Large-Scale Electric Circuits

An electric current caused by mechanical process is inevitably followed by local compensating processes, as, for example, back-discharges through the gap, electrons tunneling, polarization of neighbouring susceptible rock or conduction through it, not to mention ionic conduction. However, electric currents do flow in the lithosphere, which implies a large scale organization of electric currents and partial separation from compensating phenomena.

A probable solution to the large-scale currents mystery was proposed by Mavromatou and Hadjicontis (2001). According to their idea, in the rock subjected to asymmetric or changing stress, there occurs electric current, causing secondary compensating flow in surrounding rocks; probably this compensating current is the one which is registered on the surface.

We have to return to the article of Freund et al. (2007), as these authors present a different model of charges accumulation and electric connections over wide distances around the focal area. They propose that stresses of tectonic origin "push horizontally" the positive charges created in the process of p-holes activation. When the border surface of p-holes accumulation zone reaches a well-conducting vertical fault (containing, for example, partially molten rock), the circuit closes, and the deep, well conducting layers of the lithosphere are able to complete the lacking electrons. From this moment, the stressed rock starts to act like a battery during exploitation.

Here, we would like to sketch a more general scheme. The activated p-holes repel each other which causes positive polarization of sample surface in respect to the central part. In favourable conditions, this may cause current flow on large spatial scale, provided that massive acceptors/donors of charges come in contact with the stressed rock volume. Let us imagine that the rock in which the massive activation of p-holes takes place, contacts on one side with a certain conducting formation (rocks with conducting capillaries or tectonic fault, especially if it is deep-penetrating and contacts with relatively good conductors at depth) and on the other side it contacts with some other conducting formation, soil or the atmosphere. The direction in which the current starts to flow may be a matter of chance. But, if we take into account that the Earth surface has usually a negative charge in relation to the atmosphere, it may be that this charge is **caused** by the p-holes current! Movement of O^- sites, which are electron-deficient, is accompanied by opposite movement of electrons or transfer of some points of electron excess. For some unknown reason, the O^- sites might migrate downwards and laterally, while the excess electrons – toward the Earth surface. The above-mentioned contact with good conductors, as a mineral vein or fault, may locally alter this scheme and allow to canalize the current into a narrow path.

10.8 Final Remarks

Experiments on contact electrification, p-holes activation and electric processes accompanying fracturing and rubbing show that a very strong electric field may be generated on a rock surface, leading to electric discharge – underground or to the atmosphere. Whatever the mechanism of triboelectric process is, it is quite probable that it will take place in the earthquake preparatory zone, e.g., between grains of various contents rubbed in the stream of gas, or between sliding rock surfaces. Definitely, these phenomena are more probable during the earthquake itself. Triboelectric phenomena are often accompanied by EM emission, sometimes also by absorption. The problems of charged dislocations and charge processes are to be now considered on physicochemical and even quantum level, so the electric precursors and all earthquake-related phenomena appear more and more complicated.

The presence of water and its influence is always taken into account in the studies on physical/chemical processes in rocks. Therefore, the discovery of ferroelectric behaviour of water (Rusiniak 2000) is of great importance as it changes some views on the electricity in nature. Also the macroscopically observable pre- and coseismic hydraulic anomalies, and sometimes drastic changes in fluids flow regime, as for example unexpected outflows of liquids and gases or, reversely, drying of the area (see Gold and Soter 1985) – important per se, also influence the seismic, electric and electromagnetic precursory activities.

Among the records of potential differences on the ground, probably most puzzling are those where the variation of voltage on one electrode array mirrors vertically the results from the other array. Such opposite precursory signals were found by Thanassoulas (2007). The generation of such signals may be due to various reasons, as local differences in charge separation processes, or the resistivity changes due to variations in oriented crack systems – see Teisseyre, K.P. (1991).

Despite many observations and attempts to systematise the research results, and to devise a coherent theory, scientific world did not yet manage to resolve questions dealing with electric precursors to the earthquakes. Thus, it will be not surprising if quite other, exotic explanations emerge in the future, as for example some processes in the lower crust, or even below.

References

Balan C, Balut C, Gheorghe L, Gheorghe C, Gheorghiu E, Ursu G (2004) Electrical and rheological properties of blood in simple shear flow. Part I: Experimental determination of blood permittivity and conductivity. Clin Hemorheol Micro **30**: 359-364

Corry CE (1994) Investigation of ferroelectric effects in two sulfide deposits. J Appl Geophys **32**: 55-72

Ferré EC, Zechmeister MS, Geissman JW, MathanaSekaran N, Kocak K (2005) The origin of high magnetic remanence in fault pseudotachylites: Theoretical considerations and mplication for coseismic electrical currents. Tectonophysics **402**: 1/4, 125-139

Freund F (2002) Charge generation and propagation in igneous rocks. J Geodyn **33**: 4/5, 543-570

Freund FM, da Silva MAS, Lau BWS, Takeuchi A, Jones HH (2007) Electric currents along earthquake faults and the magnetization of pseudotachylite veins. Tectonophysics **431**: 1/4, 131-141

Fukao S, Ito Y, Yoshikado S (2003) Mechanism of electric charge emission from $LiNbO_3$ single crystal. Trans Tech Publications, Switzerland, www.ttp.net

Gershenzon NI, Gokhberg MB (1992) On the origin of electrotelluric disturbances prior to earthquake. In: Earthquake prediction. Il Cigno Galileo Galilei Edizioni di Arte e Scienza, Roma, pp 515-525

Gold T, Soter S (1985) Fluid ascent through the solid lithosphere and its relation to Earthquakes. Pure Appl Geophys **122**: 2/4, 492-530

Guichet XL, Jouniaux L, Catel N (2006) Modification of streaming potential by precipitation of calcite in a sand–water system: laboratory measurements in the pH range from 4 to 12. Geophys J Int **166**: 1, 445-460

Hadjicontis V, Mavromatou C, Ninos D (2004) Stress induced polarization currents and electromagnetic emission from rocks and ionic crystals, accompanying their deformation. Nat Hazard Earth Sys **4**: 5/6, 633-639

Horn RG, Smith DT (1992) Contact electrification and adhesion between dissimilar Materials. Science **256**: 5055, 362-364

Ishido T, Mizutani H (1981) Experimental and theoretical basis of electrokinetic phenomena in rock-water systems and its applications to geophysics. J Geophys Res **86**: B3, 1763-1775

Malinowski S (2001) Physical and chemical properties related to defect structure of oxides and silicates doped with water and carbon dioxide. In: Teisseyre R, Majewski E (eds) Earthquake thermodynamics and phase transformations in the Earth's interior. (International Geophysics Series vol 76), Academic Press, San Diego, pp 441-459

Mavromatou C, Hadjicontis V (2001) Laboratory investigation of the electric signals preceding the fracture of crystalline insulators. In: Teisseyre R, Majewski E (eds) Earthquake thermodynamics and phase transformations in the Earth's interior. (International Geophysics Series vol 76) Academic Press, San Diego, pp 501-517

Nagahama H, Teisseyre R (1998) Dislocation field evolution and dislocation source/sink Function. Acta Geophys Pol **46**: 1, 13-33

Park SK, Johnston MJS,. Madden TR,. Morgan FD, Morrison HF (1993) Electromagnetic precursors to earthquakes in the ULF band: A review of observations and mechanisms. Rev Geophys **31**: 2, 117-132

Poletaev AV, Shmurak SZ (1984) Dislocation exoelectron emission of colored alkali-halide crystals. Sov Phys JETP **60**: 376-379

Prevenslik TV (2001) Contact electrification by cavity QED. http:// www. angelfire.com /super/cavityqed/contact.htm

Prevenslik TV (2003) Flow electrification by cavity quantum electrodynamics. Conference on Electrical Insulation and Dielectric Phenomena, 19-22 Oct. 2003, Annual Report, pp 670-673

Rabinovitch A, Frid V, Bahat D (2007) Surface oscillations – A possible source of fracture induced electromagnetic radiation. Tectonophysics **431**: 1/4, 15-21

Rusiniak L (2000) Dielectric properties and structure of water at room temperature. New experimental data in 5 Hz – 13 MHz frequency range. Phys Chem Earth (A) **25**: 2, 201-207

Rusiniak L (2004) Electric properties of water. New experimental data in the 5 Hz – 13 MHz frequency range. Acta Geophys Pol **52**: 1, 63-76

Takeuchi A, Nagahama H (2006) Electric dipoles perpendicular to a stick-slip plane. Phys Earth Planet Int **155**: 3/4, 208-218

Teisseyre KP (1991) Simulation of anisotropic changes of resistivity in primarily anisotropic rocks under load. Acta Geophys Pol **39**: 3/4, 315-333

Teisseyre KP (2002) Anomalous piezoelectric effects fund in the laboratory and reconstructed by numerical simulation. Ann Geofis **45**: 2, 273-278

Teisseyre KP, Hadjicontis V, Mavromatou C (2001) Anomalous piezoelectric effect: analysis of experimental data and numerical simulation. Acta Geophys Pol **49**: 4, 449-462

Teisseyre R (1992) Earthquake premonitory processes: evolution of stresses and electric current generation. Terra Nova **4**: 4, 509-513

Teisseyre R (2001) Evolution, propagation, and diffusion of dislocation fields. In: Teisseyre R, Majewski E (eds) Earthquake thermodynamics and phase transformations in the Earth's interior. (International Geophysics Series vol 76), Academic Press, San Diego, pp 167-197

Teisseyre R, Górski M, Teisseyre KP (2006) Fracture-band geometry and rotation energy release. In: Teisseyre R, Takeo M, Majewski E (eds) Earthquake source asymmetry, structural media and rotation effects. Springer, Berlin, pp 169-183

Teisseyre R, Nagahama H (2001) Electric and electromagnetic fields related to earthquake formation. In: Teisseyre R, Majewski E (eds) Earthquake thermodynamics and phase transformations in the Earth's interior. (International Geophysics Series vol 76), Academic Press, San Diego, pp 535-552

Thanassoulas C (2007) Short-term earthquakc prcdiction. H Dounias & Co, Athens, 373 pp

Varotsos P (2005) The Physics of seismic electric signals. TERRAPUB, Tokyo, 338 pp

Varotsos PA, Hadjicontis V, Nowick AS (2001) The physical mechanism of seismic electric signals. Acta Geophys Pol **49**: 4, 415-424

11 Friction and Fracture Induced Anisotropy: Asymmetric Stresses

Roman Teisseyre

Institute of Geophysics, Polish Academy of Sciences
ul. Księcia Janusza 64, 01-452 Warszawa, Poland
e-mail: rt@igf.edu.pl

11.1 Introduction

An advanced approach to the theory of asymmetric continuum relates to our former studies (Teisseyre 2004, 2005) and to our former monograph (Teisseyre et al. 2006a); here we follow the standard asymmetric theory presented in Chapter 7.

This theory may be generalized for the case of medium anisotropy, but we shall be aware of the additional influence of fracturing and micro-processes. The asymmetry of strain and stresses follows from the considerations by Shimbo (1975, 1995) and Shimbo and Kawaguchi (1976) as related to the friction processes and rotations of grains. We shall take into account the new constant μ^* representing the rotation rigidity of bonds and related to the inner friction; further on we assume its equality to the rigidity constant μ. Further, we shall note that fracture processes develop usually along the main fault plane, giving rise to the initial asymmetry of the fracture pattern; such a process already causes a deviation from isotropy. The same concerns the micro-fracturing. As a result, also internal friction may become an anisotropic phenomenon. Then, due to friction, the rotation of grains adjacent to the micro-slip planes causes the appearance of antisymmetric part of stresses. Antisymmetric stresses relate to the internal rotation motion; these stresses become important in zones with higher dislocation densities, under high stresses, or in zones where micro-fracturings nucleate; in such zones we can expect the presence of rotation nuclei. Owing to the additional constitutive law between the antisymmetric parts of stresses and strains we may arrive at the asymmetric and anisotropic continuum.

We refer to the former chapters on Asymmetric Continuum Theory (Chaps. 7 and 8) and we will consider the deviations related to original material anisotropy, and those related to friction and fracture processes.

We follow the asymmetric standard theory entirely based on the displacement field (see: Chap. 7). Such a theory employs the asymmetric stresses and asymmetric deformations (symmetric strains and antisymmetric rotations) on the one hand, and the related constitutive laws and motion equations on the other. Stresses are asymmetric while strains remain symmetric and rotations antisymmetric (notation same as in Chap. 7)

$$S_{kl} = S_{(kl)} + S_{[kl]}, \qquad E_{kl} = E_{(kl)}, \qquad \omega_{kl} = \omega_{[kl]}. \tag{11.1}$$

Instead of the Kröner method with the self-stresses, strains and rotations (see: Teisseyre and Boratyński 2003), we introduce the new parameters (material structure indexes: e^0, χ^0) to join the deformations, in an independent way, with some reference displacement or rotation motions (u_k, Ω_s):

$$\begin{aligned} E_{kl} &= e^0 E^0_{kl} = e^0 \frac{1}{2}\left(\frac{\partial u_l}{\partial x_k} + \frac{\partial u_k}{\partial x_l}\right), \\ \omega_{kl} &= \chi^0 \omega^0_{kl} = \chi^0 \frac{1}{2}\left(\frac{\partial u_l}{\partial x_k} - \frac{\partial u_k}{\partial x_l}\right), \quad u_k = \in_{kns} \frac{\partial \Omega_{sl}}{\partial x_n} \end{aligned} \tag{11.2}$$

where further on we put for a solid elastic medium $\left|e^0\right| = 1$ and $\chi^0 = 1$, where e^0 means the phase shift between the motions E^0_{kl} and ω^0_{kl}.

Co-action of these independent fields (E_{kl}, ω_{kl}) may lead to some extreme phenomena and to structural defects; for fracture processes, the indexes e^0 and χ^0 may take various values.

For the deviatoric motions/deformations, as explained in the former chapters (7 and 8), we have the wave equations for the spin and twist fields:

$$\Delta\omega_{[s]} = \frac{1}{V^2}\frac{\partial^2}{\partial t^2}\omega_{[s]}, \qquad \Delta\omega_{(i)} = \frac{1}{V^2}\frac{\partial^2}{\partial t^2}\omega_{(i)}. \tag{11.3}$$

11.2 2D Uniform Anisotropy

Instead of the contemporary approach to anisotropy based on the eikonal equation, we may use a more direct geometrical approach; we refer the reader to the papers by Teisseyre (1955) related to the problem of ray theory (see: Yajima and Nagahama 2006, and Teisseyre 2006; we follow the latter paper with the important improvements and corrections).

For the anisotropy problems, and in particular for friction, microfracturing and fracture induced anisotropy, we will try to apply the geometry methods; we limit ourselves to the 2D problem. We will consider two cases: the 2D uniform skew system anisotropy and the 2D fracture skew system anisotropy; the first is useful when considering the problem of wave propagation, while the other applies mainly to the part of continuum in which the fracturing develops from the microscale to the macroprocesses.

It will be convenient to express the 2D anisotropy using the related skew coordinate system. The fracture processes trace the pre-slip structure of media revealed in formation of the mylonite layers; the mylonite zones form an anisotropic system in which the difference in the grain radii causes the inertia-related anisotropy determining the fracture ability and friction properties.

Considering the 2D fracture skew anisotropy, we observe the occurrence of a main fracture plane and a system of weaker skew-oriented auxiliary fracturings. The related anisotropy can be expressed by the 2D skew coordinate system X^K ($K, S, \ldots. = \{1, 2\}$) as related to the anisotropy system (with an angle φ between the axes X^1 and x_1 related to the rectangular system $\{x_1, x_2\}$, and angle ψ between the axes X^2 and x_2; see: Fig. 11.1).

The skew system of the privileged slip planes presents different abilities for slip motions; such an ability could depend on the different radii of grains. In the considered 2D skew coordinate system $\{X^S\} = \{X^1, X^2\}$, we recall the contravariant (parallel) and covariant (perpendicular) projections of any vector **L** and the base versors, $\mathbf{e}_1$ and $\mathbf{e}_2$, of the skew system.

According to the invariant rules, the wave equations for spin and twist (11.3) can be written, in the skew anisotropic system, as follows:

$$\begin{aligned}
&\frac{\partial^2}{\partial X^K \partial X_K}\omega_{[n]} - \frac{1}{V^2}\ddot{\omega}_{[n]} = 0\,,\\
&\frac{\partial^2}{\partial X^K \partial X_K}\omega_{[n]} = \frac{\partial^2}{g_{SK}\partial X^S \partial X^K}\omega_{[n]} - \frac{1}{V^2}\ddot{\omega}_{[n]} = 0,\\
&\frac{\partial^2}{\partial X^K \partial X_K}\omega_{(n)} - \frac{1}{V^2}\ddot{\omega}_{(n)} = 0\,,\\
&\frac{\partial^2}{\partial X^K \partial X_K}\omega_{(n)} = \frac{\partial^2}{g_{SK}\partial X^S \partial X^K}\omega_{(n)} - \frac{1}{V^2}\ddot{\omega}_{(n)} = 0
\end{aligned} \tag{11.4}$$

where the velocity for the anisotropic state shall be written as $V^2 = g_{ks} V^k V^s$.

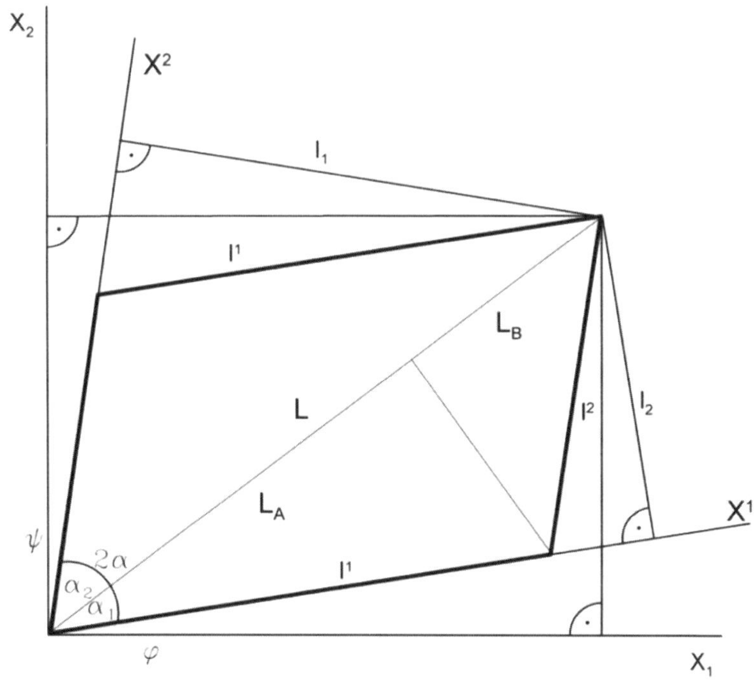

Fig. 11.1 The skew system of pre-slip or fracture planes

Referring to the 2D skew coordinate system $\{X^S\} = \{X^1, X^2\}$, (Fig. 11.1), we recall the contravariant (parallel projection) and covariant (perpendicular projection) components of any vector **L** and the base vectors, $\mathbf{e}_1$ and $\mathbf{e}_2$, of the skew system: a square of a finite length element L can be expressed with the help of the metric tensor g_{NS}

$$L^2 = g_{11} l^1 l^1 + g_{22} l^2 l^2 + 2 g_{12} l^1 l^2$$
$$= l^1 l^1 \cos^2 \alpha_1 + l^2 l^2 \cos^2 \alpha_2 + 2 l^1 l^2 \cos \alpha_1 \cos \alpha_2 \,.$$

where $g_{KS} = \mathbf{e}_K \mathbf{e}_S$ and $g_{11} = \cos^2 \alpha_1$, $g_{22} = \cos^2 \alpha_2$, $g_{12} = \cos \alpha_1 \cos \alpha_2$.

This relation can be applied to velocity relation in the anisotropic state

$$V^2 = g_{11} V^1 V^1 + g_{22} V^2 V^2 + 2 g_{12} V^1 V^2$$
$$= V^1 V^1 \cos^2 \alpha_1 + V^2 V^2 \cos^2 \alpha_2 + 2 V^1 V^2 \cos \alpha_1 \cos \alpha_2 \qquad (11.5)$$

leading at a constant density to

$$\mu = g_{11} \mu_1 + g_{22} \mu_2 + 2 g_{12} \sqrt{\mu_1 \mu_2}, \qquad g_{KS} = \mathrm{e}_K \mathrm{e}_S \qquad (11.6)$$

where we have assumed that the rigidity anisotropic moduli μ^{s}_{ik} are equal for both motions and they become reduced to two values, μ_1 and μ_2:

$$\mu^{3}_{[12]} = \mu^{3}_{(12)} = \mu_1 , \qquad \mu^{3}_{[21]} = \mu^{3}_{(21)} = \mu_2 . \tag{11.7}$$

We can write for the related solutions:

$$\omega_{[n]} = \omega^{0}_{[n]} \exp\left[i(X^K K_K - \varpi t) \right] = \omega^{0}_{[n]} \exp\left[i(g^{SK} X_K K_S - \varpi t) \right],$$
$$\omega_{(n)} = \omega^{0}_{(n)} \exp\left[i(X^K K_K - \varpi t) \right] = \omega^{0}_{(n)} \exp\left[i(g^{SK} X_K K_S - \varpi t) \right].$$

Our considerations are equivalent to the problem of stress moment anisotropy.

In addition to the basic isotropic elastic constitution relation, we will rely on the following anisotropic constitutive laws (cf. Eqs. 11.1 and 11.2):

$$\begin{aligned} S^{[12]} &= \mu_1 \omega_{[3]}, \quad S^{[21]} = \mu_2 \omega_{[3]} ; \\ S^{(12)} &= \mu_1 \omega_{(3)}, \quad S^{(21)} = \mu_2 \omega_{(3)} . \end{aligned} \tag{11.8}$$

We might note that for a sequence of such systems with variable angles φ and ψ between the consecutives skew systems and the initial orthogonal system, we may arrive at curvilinear 2D coordinates and related anisotropy.

11.3 2D Fracture/Friction Induced Anisotropy

In this system we keep the relations (11.7) only near the fracture planes. Thus, we can apply our consideration to the fracture related anisotropy compatible with the axes X^1 and X^2.

We assume the rigidity modulus anisotropy; in the near-fracture planes we put μ_1 and μ_2, and for farther regions we put μ_0

In this definition we included the isotropic constitutive law valid for regions far from the fracture planes and the definitions describing the low values of the rotation rigidity occurring closely to the fracture planes $X^1 = 0$ and $X^2 = 0$. These drops of rigidity may be of different rates and could be visualized by the different radii of grains in Fig. 11.2.

In this case we will arrive at different equations in the domains near to each of the fracture planes and also in the domain outside those zones:

$$\frac{\partial^2}{\partial X_k \, \partial X_k} \omega_{[n]} - \frac{1}{V^2_{ZONE}} \ddot{\omega}_{[n]} = 0 \quad \text{where} \quad V^2_{ZONE} = \{V_1^2, V_2^2, V_0^2\}. \tag{11.9}$$

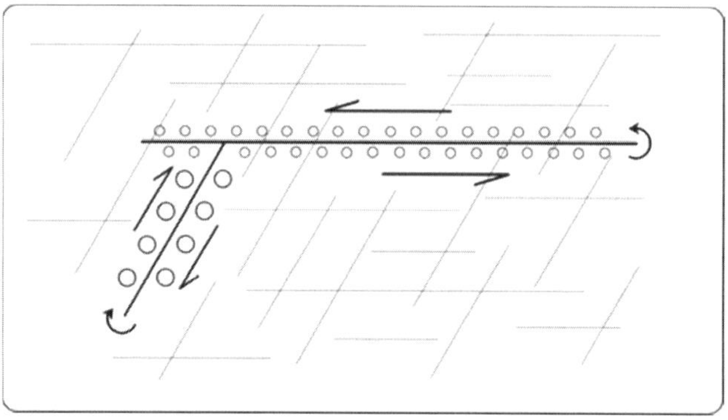

Fig. 11.2 2D fracturing skew system anisotropy

For both the spin and twist motions, we obtain anisotropy of the wave velocities along the axes X^1 and X^2 and an isotropic velocity for the domain outside the fracture zones:

$$V_1^2 = \frac{\mu_1}{\rho}, \qquad V_2^2 = \frac{\mu_2}{\rho}, \qquad V_0^2 = \frac{\mu_0}{\rho}. \tag{11.10}$$

The constitutive laws for a given zone can be presented as

$$S^{[KS]} = \mu_{ZONE}\omega_{[3]} \quad \text{and} \quad S^{(KS)} = \mu_{ZONE}\omega_{(3)}, \tag{11.11a}$$

$$S^{[K3]} = \mu_{ZONE} \in_{NK3} \omega_{[N]} \quad \text{and} \quad S^{(KS)} = \mu_{ZONE} \in_{NK3} \omega_{(N)}, \tag{11.11b}$$

where $E_{KS} = \frac{1}{2}\left(\frac{\partial u_S}{\partial x_K} + \frac{\partial u_K}{\partial x_S}\right)$, $\omega_{KS} = \frac{1}{2}\left(\frac{\partial u_S}{\partial x_K} - \frac{\partial u_K}{\partial x_{IS}}\right)$, and $E_{K3} = \frac{\partial u_3}{2\partial x_K}$, $\omega_{K3} = \frac{\partial u_3}{2\partial x_K}$.

For the dislocation field we obtain (cf. Chap. 7):

$$\alpha_{3L} = \varepsilon_{3MK}\frac{\partial^2 u_K}{\partial x_M \partial x_L} \quad \text{and} \quad \alpha_{33} = \varepsilon_{3MK}\frac{\partial^2 u_K}{\partial x_M \partial x_3}. \tag{11.12}$$

11.4 Conclusions

We have related our consideration on the antisymmetric stresses and rotations to the wave equations for 2D anisotropy: both the uniform 2D anisotropy and the skew system of fracture planes. We have applied to this problem the geometry methods related to covariant and contravariant components in the skew system of coordinates.

References

Shimbo M (1975) A geometrical formulation of asymmetric features in plasticity. Bull Fac Eng, Hokkaido Univ 77: 155-159

Shimbo M (1995) Non-Riemannian geometrical approach to deformation and friction. In: Teisseyre R (ed) Theory of earthquake premonitory and fracture processes. PWN, Warszawa, pp 520-528

Shimbo M, Kawaguchi M (1976) A note on the asymmetric fields. Bull Fac Eng Hokkaido Univ **80**: 75-79 (in Japanese)

Teisseyre R (1955) Optico-geometrical approximation for seismic waves in non-homogeneous media. Acta Geophys Pol **3**: 161-166

Teisseyre R (2004) Spin and twist motions in a homogeneous elastic continuum and cross-band geometry of fracturing. Acta Geophys Pol **52**: 2, 173-183

Teisseyre R (2005) Asymmetric continuum mechanics: deviations from elasticity and symmetry. Acta Geophys Pol **53**: 2, 115-126

Teisseyre R (2006) Asymetric continuum and anisotropy. Acta Geophys **54**: 3, 225-238

Teisseyre R, Boratyński W (2003) Continua with self-rotation nuclei: evolution of asymmetric fields. Mech Res Com **30**: 235-240

Teisseyre R, Takeo M, Majewski E (eds) (2006a) Earthquake source asymmetry, structural media and rotation effects. Springer, Berlin

Yajima T, Nagahama H (2006) Seismic ray theory for structural medium based on Kawaguchi and Finsler geometry. In: Teisseyre R, Takeo M, Majewski E (eds) Earthquake source asymmetry, structural media and rotation effects. Springer, Berlin, pp 329-336

12 Asymmetric Fluid Dynamics: Extreme Phenomena

Roman Teisseyre

Institute of Geophysics, Polish Academy of Sciences,
ul. Księcia Janusza 64, 01-452 Warszawa, Poland
e-mail: rt@igf.edu.pl

12.1 Introduction

The extreme wave phenomena in oceans, like soliton waves, are usually described by some differential non-linear equations, like soliton equations and other ones, not related directly to the fundamental Navier-Stokes equations.

However, a plausible physical basis for a more general approach to the mechanics of fluids can be based on the standard asymmetric continuum theory as developed recently (see: Chap. 7). In this theory, we assume that, besides the translational motions, there appear independent rotations of points (particles/grains) which supplement the former motions. Further on, we assume that a fluid may be described by the constitutive laws joining the time rates of the asymmetric stresses with time rates of the symmetric strains and antisymmetric rotations. Moreover, we shall include to the rotational motions not only the spin but also the twist rate motion related to oscillations of the strain shear rates.

Finally, we consider the extreme phenomena following directly from the interaction between spin and shear/twist rate motions.

12.2 Standard Asymmetric Fluid Theory

Asymmetric fluid theory has to account for deviations from the classical approach; it includes, among other things, the interaction between spin and shear rate motions. Some unusual wave phenomena, as e.g. extreme (soliton) ocean waves, lead us to formulate the theory in which the shear/twist rate motion related to viscosity is introduced and possible deviations from the Navier-Stokes equations may be also included. The asymmetric fluid theory could naturally explain the above-mentioned extreme wave phenomena and, using the system of the obtained relations, we can replace the nonlinear soliton equations as introduced without strict theoretical relation to the classical fluid dynamics.

Thus, our theory is based on two groups of relations, for the symmetric and antisymmetric stress and deformation field rates:

$$\tilde{S}_{kl} = \tilde{S}_{(kl)} + \tilde{S}_{[kl]}, \qquad \tilde{E}_{kl} = \tilde{E}_{(kl)}, \qquad \tilde{\omega}_{kl} = \tilde{\omega}_{[kl]} \tag{12.1}$$

and fluid properties are defined by the following constitutive relations:

$$\tilde{S}_{(ik)} = 2\eta \tilde{E}_{ik}, \qquad \tilde{S}_{[ik]} = 2\eta \tilde{\omega}_{ik} \tag{12.2}$$

where η is viscosity. Further on, we introduce the dynamical indexes (see: Chap. 7), $\left|\tilde{e}^0\right| = 1$ and $\tilde{\chi}^0 = 1$, related to the phase differences and joining these deformation fields, $\tilde{E}_{kl}$ and $\tilde{\omega}_{kl}$, with the reference displacement velocities:

$$\tilde{E}_{kl} = \tilde{e}^0 \tilde{E}^0_{kl} = \tilde{e}^0 \frac{1}{2}\left(\frac{\partial \upsilon_l}{\partial x_k} + \frac{\partial \upsilon_k}{\partial x_l}\right), \qquad \tilde{\omega}_{kl} = \tilde{\omega}^0_{kl} = \frac{1}{2}\left(\frac{\partial \upsilon_l}{\partial x_k} - \frac{\partial \upsilon_k}{\partial x_l}\right) \tag{12.3}$$

where υ_k is the displacement velocity.

For simplicity, we will consider a non-compressible fluid, $\tilde{E}_{ss} = 0$, and, following the Standard Asymmetric Continuum Theory (see: Chap. 7), we define the twist rate oscillations related to the deviatoric shear rate tensor, and the 4D invariant definition of twist as based on the off-diagonal $\tilde{E}^D_{ki}$ values).

The balance equations for field rates lead us to the wave equations for the related spin and twist rate fields, $\tilde{\omega}_{[s]}$ and $\tilde{\omega}_{(k)}$:

$$\varepsilon_{kps} \frac{\partial}{\partial x_p} \tilde{\omega}_{[s]} - \frac{1}{V}\frac{\partial}{\partial t} \tilde{\omega}_{(k)} = \frac{4\pi}{V} \tilde{J}_k, \qquad \varepsilon_{kps} \frac{\partial \omega}{\partial x_p} \tilde{\omega}_{(s)} + \frac{1}{V}\frac{\partial}{\partial t} \tilde{\omega}_{[k]} = 0. \tag{12.4}$$

The appearance of the coupled waves transversal to the transport motion brings physical background for diffraction in fluids as explained usually by the Huygens principle.

We may already note that a co-action of the rotation motions (spin rate: $\tilde{\omega}_{[ki]} = \varepsilon_{ski} \tilde{\omega}_{[s]}$, and twist rate: $\varepsilon_{ski} \tilde{\omega}_{(s)} = \tilde{E}^D_{ki}$) can lead to the extreme phenomena in fluids.

In 2D we can demonstrate this co-action in the following way; let us consider the following sum of these motions:

$$\tilde{E}_{12} + \tilde{\omega}_{12} = C\upsilon_{(2,1)} + \upsilon_{(2,1)} = C\frac{1}{2}\left(\frac{\partial \upsilon_k}{\partial x_i} + \frac{\partial \upsilon_i}{\partial x_k}\right) + \frac{1}{2}\left(\frac{\partial \upsilon_k}{\partial x_i} - \frac{\partial \upsilon_i}{\partial x_k}\right) \tag{12.5}$$

where for $C = 0$ we have only the spin motion, while a full coincidence will occur at $C = 1$.

The effect of such a superposition of the spin and twist motions is presented in Fig. 12.1.

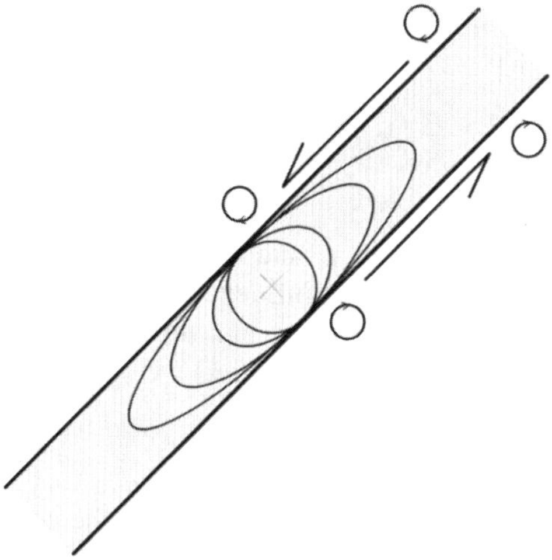

Fig. 12.1 Extreme motions: soliton wave represented by deformation of a circle for different parameters C [Amplitude plot (Mathematica 5.0)]

We shall define, as well, the possible dynamic discontinuities using the following integral along a closed circuit (see: Chap. 7):

$$\tilde{B}_l = \oint\left[\tilde{E}_{kl} + \tilde{\omega}_{kl}\right]\mathrm{d}l_k = \oint\left[\tilde{e}^0\tilde{E}^0_{kl} + \tilde{\omega}^0_{kl}\right]\mathrm{d}l_k. \tag{12.6}$$

In this way we arrive at some dynamic objects – v-dislocations:

$$\tilde{B}_l = \iint\left(\tilde{\alpha}_{pl} - \frac{1}{2}\tilde{\alpha}_{ss}\delta_{pl}\right)\mathrm{d}s_p \tag{12.7}$$

and with the constitutive relations (12.2) we arrive at the relation between the v-dislocation density and asymmetric stress rates:

$$\tilde{\alpha}_{pl} - \frac{1}{2}\tilde{\alpha}_{ss}\delta_{pl} = \frac{\varepsilon_{pmk}}{2\eta}\frac{\partial}{\partial x_m}\left(\tilde{S}_{(kl)} - \frac{\nu}{1+\nu}\delta_{kl}\tilde{S}_{ii} + \tilde{S}_{[kl]}\right). \tag{12.8}$$

In some extreme dynamic situations, there may start a bond breaking process between molecules/grains of fluid; just after it, we can expect the released rebound spin motion retarded in phase. Such a hypothesis is supported by the following solution of the homogeneous wave equations for the twist and spin rates (see: Chap. 7):

$$\tilde{\omega}_{(.)} = \mathrm{i}\tilde{\omega}_{[.]}, \quad \tilde{\omega}_{[.]} = \tilde{\omega}^0_{[.]} \exp\left[\mathrm{i}(kx - \varpi t)\right], \quad \tilde{\omega}_{(.)} = \tilde{\omega}^0_{(.)} \exp\left[\mathrm{i}(kx - \varpi t)\right] \tag{12.9}$$

where with the constants: $\tilde{\omega}^0_{[s]} = \mathrm{abs}(\tilde{\omega}^0_{[s]})\exp(\mathrm{i}\,\pi/2)$, $\mathrm{abs}(\tilde{\omega}^0_{(s)}) = \mathrm{abs}(\tilde{\omega}^0_{[s]})$ we can fulfill the field Eq. (12.4).

Here, an important role may be played by the turbulence related solution in the system $\{r, \phi, z\}$:

$$\tilde{\omega}_{(\varphi)}(r) = \mathrm{i}\tilde{\omega}_{[\varphi]}(\mathrm{r}), \qquad \Box\, \tilde{\omega}_{[\varphi]}(r) = 0. \tag{12.10}$$

Using the introduced waves, $\tilde{\omega}_{[s]} = \mathrm{i}\tilde{\omega}_{(s)}$, we arrive at a possibility to study the dynamic defect objects and to explain the synchronization of these dynamic processes due to an influence of the propagating waves. The conjugate solutions presented suggest that the spin rebound motion is delayed in phase by $\pi/2$; for the integral (12.6) we obtain

$$\tilde{B}_l = \oint\left(-\mathrm{i}\tilde{E}^0_{kl} + \tilde{\omega}^0_{kl}\right)\mathrm{d}l_k, \qquad \tilde{e}^0 = -\mathrm{i} \tag{12.11}$$

arriving at the conjugate dynamic disclosure and the conjugate dislocation objects (v-dislocations) connected to the stress and deformation in a complex manner.

The presented conjugate solutions (12.9 and 12.10) suggest that the spin rebound motion shall be delayed in phase by $\pi/2$; when the breaks of bonds between particles occur, the micro-spin motions become released; we may expect that a correlation between the recorded twist rate motions and spin rate motions shifted by $\pi/2$ in phase could exist in some wavelets.

Further, when we approach the extreme processes, we may observe the "accumulation" phase with the co-action of the twist and spin and formation of the above-mentioned dynamic objects: v-dislocations. For an incompressible fluid, the related dynamic disclosures, v-dislocations (12.7) may follow the conjugate solutions (12.9) and lead to the turbulence phenomena (12.10), or even to a foam formation.

12.3 Conclusions

We have presented new ideas how to construct the asymmetric fluid theory with the asymmetric stress rate field. We believe that this approach can help to explain some extreme fluid phenomena related to atmosphere and oceans including the soliton waves and turbulence motions.

13 Fracture Band Thermodynamics

Roman Teisseyre

Institute of Geophysics, Polish Academy of Sciences
ul. Księcia Janusza 64, 01-452 Warszawa, Poland
e-mail: rt@igf.edu.pl

13.1 Introduction

Basing on the fracture band model, we consider the thermodynamic conditions related to seismic energy release and the rotation counterpart in a fracture process.

We shall be aware of different nature and scale of the rotation processes that take part in such extremely complicated fracture phenomena in which the dynamic processes proceed together with the simultaneous changes of material properties (see: Teisseyre 1996). We shall recall a special role of rotations in the energy release effectiveness under different load conditions, and, further on, we shall include the rotation impact on the granulation processes accompanying the material crushing.

The constitutive laws undergo simultaneously considerable changes, from those describing the elastic, to plastic and, further, mylonite-type material. In the narrow zones adjacent to fracturing, the shear stresses break the molecular bonds and in the crashed rock material the stresses immediately drop to a much lower level, while together with the advancing material granulation, we shall include a rapid increase of the stress and strain rates. Finally, in that narrow zone adjacent to fracturing, the stresses and strains may be gradually neglected and progressively replaced by their time-rates. To describe these processes we shall simultaneously introduce the changes into the related constitutive relations. As a result, the rock properties in this zone may even approach those similar to fluid. Such conditions may permit to include in fracture description the transport Navier-Stokes relations. The fracturing transport, the bond breaking and granulation processes force us to include, into the fracturing description, the hypothesis that the twist-shear deformations leading to the bond breaking precede by $\pi/2$ in phase the rebound rotation motion; this means that the difference between the shear motion and spin motion shall reach minimum when the latter is shifted by $\pi/2$ in phase.

We shall underline that the considered conditions in the mylonite zone can serve as a basis to formulate the asymmetric fluid theory with the extreme motion phenomena and dynamic defect objects.

A counterpart of rotations and rotation energy release at fracture processes (e.g., in an earthquake source) explains fragmentation and spall processes and permits to estimate the efficiency of different fracturing modes. We shall again underline that in any theoretical approach, the elastic rotation energy can be considered only when assuming the constitutive law joining rotations with the antisymmetric stresses or stress moments.

Teisseyre et al. (2006) have reexamined Dieterich's compression experiments (Dietrich 1978) coming to the conclusion that, under the compression load, induced precursory shear stresses may arise at some centers in the source region; at the fracturing event, we arrive at the coseismic rebound compensation leading, also, to the release of induced stresses by rebound process. Similarly, the precursory rotations, associated with the newly formed dislocations or cracks, shall have an opposite orientation to that related to the coseismic process. At the precursory stage these repeated processes lead to micro-fractures, while during the seismic event there will occur, under compression load, the fracturing with the rock fragmentation and the rebound macro-rotations at the inner centers where the precursory induced stress had accumulated.

13.2 Earthquake Dislocation Theory

The Earthquake Dislocation Theory (Droste and Teisseyre 1959, Teisseyre 1961, 1964, 1970) has been based on the the elastic dislocation theory (e.g., Eshelby et al. 1951, Nabarro 1951, Kröner 1981). Dislocations, in a physical meaning, are related to disorder in a crystal lattice (e.g., slip along a certain glide plane): a lattice disorder appears only at the edges of dislocation plane; this notion differs from a geological dislocation – fault. We recall some basic ideas of this theory:

- Dislocated area with a constant slip value (the Burgers vector, $\Delta u = \lambda$, where λ is the lattice constant) is bounded at its edges by the dislocation line, called in physics just a dislocation.
- Elastic energy is concentrated around a dislocation line being a source of elastic deformation.
- Dislocations can have different signs (orientation of slip vector as related to the normal of the dislocated area): the dislocations of opposite signs attract each other to merge in the dislocated areas, while

dislocations of the same sign repel each other; a dislocated area is bounded at the dislocations of opposite sign.

- An external stress field acts on a dislocation line; when this load overpasses a stress resistance, the dislocations moves (the Koehler force).
- Under an external load the dislocations of the same sign form a dislocation array; this happens when the first leading dislocation is stopped by some barrier with a greater stress resistance.
- An array forms a concentration of stress field $S = n_A S_0$ (S_0 is the external stress load, n_A is the number of dislocations) at a first dislocation and when the concentration of stresses is very high, such an array can be approximated by a dislocation with the slip $n_A \lambda$.
- An array, with a sufficiently great number of dislocations, well approximates the field of a crack tip.
- A moving array, under shear load, can meet an opposite array; the external forces and additionally the attraction of the opposite arrays may break the barriers and the stresses attached to the two arrays will be mutually canceled with a release of their energies. Recall now the basic elements of the Earthquake Dislocation Theory (Droste and Teisseyre 1959, Teisseyre 1961, 1964, 1970):
- In an early premonitory time-domain, the dislocation arrays of different signs are formed; their interaction generates the micro-cracks (we come from quasi-static to dynamic processes and from stress resistance to friction).
- In an advanced premonitory time-domain, the interactions between the microcracks and also their mutual annihilations lead to the formation and expansion of bigger and bigger cracks: fragments of slip fracture develop with a rapid dynamic process, in which the fragmented parts merge together in a main fault.
- An earthquake process may develop along two perpendicular planes, but in reality, due to the fact that any geological space is extremely complex, one main fracture plane is usually formed (for such effect, see: Chap. 8).

In this Earthquake Dislocation Theory, an energy release is explained by the mutual annihilations of the strain energies of the opposite groups of defects: dislocation arrays, tips of micro-cracks and cracks and those of fragments.

13.3 Fracture Band Model

Next, we present some basic elements of fracture band model (Teisseyre 1996, 1997, Teisseyre et al. 2001):

- Under a situation with a high density of dislocations (earthquake preparation zones), besides the basic lattice we expect an existence of a super-lattice related to the interactions between the densely distributed dislocations (such a super-lattice may be associated with an internal structure and band of the old slip planes); the super-lattice constant Λ shall fulfill the condition: $\Lambda \gg \lambda$.
- In earthquake preparation zone the super-lattice may be very irregular; in order to make it more coherent, we define the vacant dislocations: the difference between the best fitted regular super-lattice and that of the really existing ones defines a number of the vacant dislocations.
- The thermodynamics of a super-lattice (Teisseyre 2001) is based on the Gibbs formation energy $\hat{g}^f$ for a vacant dislocation and relates to an equilibrium number of vacant dislocations:

$$\hat{n}_{eq} = \frac{1}{\Lambda^3} \exp\left(-\frac{\hat{g}^f}{kT} \right). \tag{13.1}$$

- For a fracture model we assume for simplicity a disc model of an earthquake volume $\Delta V = \pi R^2 D$.
- The slip bands are active at a premonitory time-domain and during a fracture process: dislocation and crack slips and fracture slip.
- A local shear stress before an earthquake can be estimated by the value of the related stress drop (see: Teisseyre 1996, Teisseyre and Wiejacz 1993, Teisseyre 2001):

$$S = A\Delta S, \quad b = \Delta u = N\lambda \quad \text{and} \quad N = B\Delta S \tag{13.2}$$

where the Burgers dislocation vector b (slip vector) can be given by a product of number of dislocations and lattice constant, while the number of dislocations entering an array can be estimated as proportional to the shear local load.

- We arrive at the expression for the slip vector:

$$b = \Delta u = B\lambda\,\Delta S = \frac{B}{A}\lambda S. \tag{13.3}$$

– For a band of shear planes, the slip vector shall be multiplied by the ratio of a band width D and a super-lattice constant Λ (Fig. 13.1):

$$b = \Delta u = B\lambda \Delta S \frac{D}{\Lambda} = \frac{B}{A}\lambda S \frac{D}{\Lambda}. \tag{13.4}$$

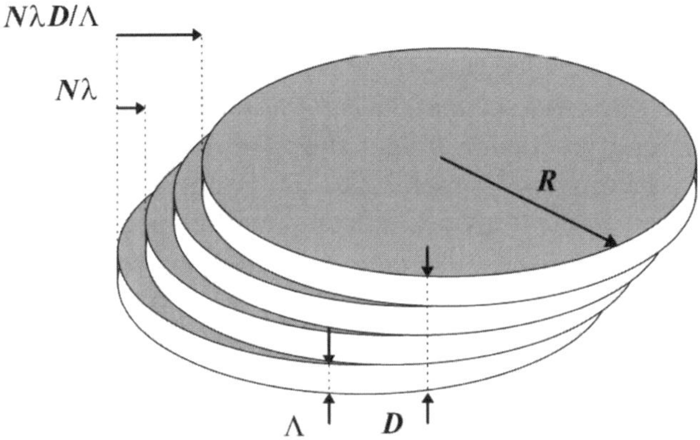

Fig. 13.1 Shear band model

13.4 Earthquake Thermodynamics

Basic thermodynamic relations for line defects (dislocations and vacant dislocations) are derived under the assumption of a dense network of defects forming a kind of superlattice (Teisseyre and Majewski 1990, 1995, 2001, and Majewski and Teisseyre 1997). The thermodynamic functions of line defects can be associated with defects in the superlattice. Let us confine our considerations to the irreversible (plastic) deformations of solids.

To distinguish the thermodynamic functions used here from those used under pressure conditions, we introduce the symbols with hat and we consider only a pure shear work under shear load $S_{(..)}$ and under induced friction stress moment $S_{[..]}$ with $\mathrm{d}E$ being an incremental strain and that of rotation $\mathrm{d}\omega$. We consider deformations under a constant volume; the work $\mathrm{d}\hat{W}$ done on a body (per unit of volume), the internal energy change $\mathrm{d}\hat{U}$, and the heat received in an exchange with the surrounding $\mathrm{d}Q$ become:

$$\mathrm{d}\hat{W} = S\mathrm{d}E = S_{(..)}\mathrm{d}E + S_{[..]}\mathrm{d}\omega \geq 0, \quad \mathrm{d}\hat{U} = \mathrm{d}Q + S\mathrm{d}E. \tag{13.5}$$

while for an internal energy stored in a medium we put (see: Chap. 7):

$$E = S_{(ks)} E_{ks} + S_{[ks]} \omega_{ks} .$$

For the Helmholtz free energy $\hat{F}$ and Gibbs free energy $\hat{G}$ we have:

$$\hat{F} = \hat{U} - T\tilde{S}, \quad \hat{G} = U - SE - T, \quad T\mathrm{d}\tilde{S} \geq \mathrm{d}Q, \quad \mathrm{d}\tilde{S} \geq 0 \tag{13.6}$$

where T is the absolute temperature and $\tilde{S}$ is the entropy; $\mathrm{d}\tilde{S}$ would be the entropy production due to the irreversible processes occurring inside the system.

The local formulation of the second law of thermodynamics requires that the entropy production be positive wherever an irreversible process occurs (Teisseyre and Majewski 2001). It is postulated that even outside the equilibrium, the entropy depends only on the same variables as at equilibrium. In order to derive the expression for the entropy production, Prigogine (1979) introduced some additional assumptions. Namely, it has been assumed that the entropy production can be determined for the conditions near the equilibrium.

Formation of a dislocation gives negative contribution to the Gibbs energy (Kocks et al. 1975) and therefore it is not possible to find a minimum of the Gibbs function in respect to the number of dislocations. Thus, the dislocation distribution cannot exist as the thermodynamically stable system, since the Gibbs free energy has no minimum of any equilibrium concentration of dislocations.

As we mentioned in the former section of this chapter, for a dense dislocation distribution there appear the repulsive interactions between dislocations, and a kind of dislocation supper-lattice could be considered (Teisseyre and Majewski 1990, 2001, Majewski and Teisseyre 1997).

These processes are accompanied by an internal friction related to displacement formed by dislocations and hence a spin motion appears as inherently present there (see: Chap. 3).

The Gibbs energy minimum can now exist as the equilibrium number of the vacant dislocations. We can consider the structure of a cross-zone consisting of bands of layerlets; such a structure favors the appearance of some macroscopic dislocations under conditions of shearing deformation. The particular values of the Burgers vector become related to particular layer thicknesses. In this sense we suppose that a fine boundary structure could play the role of a quantization factor; this problem is related to the earthquake shear band model.

Consider a continuum that contains a regular (cubic) supper-lattice of dislocation lines with a certain supper-lattice parameter Λ ($\Lambda \gg \lambda$ where λ

relates to a basic rock lattice). The notion of the supper-lattice is directly related to the shear band model of fracturing (Teisseyre and Majewski 2001), see Fig 13.1.

We associate the thermodynamic functions of line defects with the defects in a super-lattice; the Gibbs free energy may have a minimum corresponding to the equilibrium concentration of the vacant dislocations in the super-lattice. Many results can be now transferred from the thermodynamics of the point defects (Varotsos and Alexopoulos 1986). The regular super-lattice which includes the dislocations and vacant dislocations may be described in a very rough approximation by a characteristic distance Λ (super-lattice constant). For the ideal super-lattice (no vacant dislocations), the mean value of distances following from distribution of dislocations defines the reference dislocation density $\alpha^0 = \lambda/\Lambda^2$, while for a real body with n dislocations we may add to it other $\hat{n}$ vacant dislocations in such a way that the whole set $n + \hat{n} = N$ (dislocations and vacant dislocations) fits to a regular super-lattice with the smallest error. For the density of dislocations α, and vacant dislocations $\hat{\alpha}$, we can write (Teisseyre and Majewski 2001):

$$\alpha = \left(1 - \frac{\hat{n}}{N}\right)\frac{\lambda}{\Lambda^2}, \qquad \alpha = \frac{\hat{n}\lambda}{N\Lambda^2}\frac{\lambda}{\Lambda^2}\exp\left(-\frac{\hat{g}^f}{kT}\right) \tag{13.7}$$

where the number $\hat{n}$ can be identified with an equilibrium value in relation to the formation energy of vacant dislocation $\hat{g}^f$ per length of the crystal lattice λ.

The stress field and the resistance stress (e.g., the drag resistance in a dislocation motion and the friction stress in a crack motion) are defined as (Kocks et al. 1975):

$$S = \frac{\partial \hat{W}}{\partial E}, \qquad S_F \equiv \frac{\partial \hat{F}}{\partial E}$$

while the Gibbs function for a crystal containing the vacant dislocations can be written as

$$\hat{G} = \hat{G}^0 + \hat{n}\hat{g}^f - T\hat{S}_c$$

where $\tilde{S}_c$ is the configuration entropy.

Near the equilibrium state under a constant local shear S and temperature T, the Gibbs energy is close to its minimum and the equilibrium values can be found as follows:

$$\left.\frac{\partial \hat{G}}{\partial \hat{n}}\right|_{S,T} = 0, \quad \hat{n}^{\mathrm{eq}} = N \exp\left(-\frac{\hat{g}^{\mathrm{f}}}{kT}\right),$$
$$\hat{\alpha} = \frac{\lambda}{\Lambda^2} \exp\left(-\frac{\hat{g}^{f}}{kT}\right), \quad \tilde{S}_{\mathrm{c}} = \hat{n}\left(k + \frac{\hat{g}^{\mathrm{f}}}{T}\right) \tag{13.8}$$

while the Gibbs energy function becomes

$$\hat{G} = \hat{G}^0 + \hat{n}\hat{g}^f - T\tilde{S}_c \tag{13.9}$$

where $\hat{n}$, $\hat{g}^f$ and $\hat{S}_c$ mean the number of vacant dislocations, its formation energy and configurational entropy, respectively.

The equilibrium free energy is less than that for an ideal superlattice $\hat{G}^0$; the difference is kT per line vacancy, per length of crystal lattice.

For the point defect thermodynamics, Varotsos and Alexopoulos (1986) have introduced the so-called $CB\Omega$ theory approximating the contribution to the Gibbs energy from the formation of point defects.

For the line vacancies a change of the Gibbs energy depends on the stress level and resistance stress. Therefore, we postulate for the approximative value of such change per unit element (formation energy of vacant dislocation) the following expression defining the $C\mu b\lambda^2$ model:

$$\hat{g}^f = C\mu b\lambda^2, \quad \hat{n}^{eq} = N \exp\left(-\frac{C\mu b\lambda^2}{kT}\right) \tag{13.10}$$

where C is a constant; $\hat{g}^f$ becomes here independent of stress load and resistance, μ is the rigidity, b is the Burgers vector of dislocation.

Concluding, a body containing some number of dislocations cannot be in the state of equilibrium; there is no minimum of the Gibbs function, because, by reducing the number of dislocations, we always get a smaller value of the free energy. For a dense distribution of dislocations we can assume, due to their interaction, that there exists a certain supper-lattice composed of dislocations.

The equilibrium density of the vacant dislocations may be written now with the help of (13.8)

$$\hat{\alpha} = \frac{\lambda}{\Lambda^2} \exp\left(-\frac{C\mu b\lambda^2}{kT}\right) \tag{13.11}$$

and becomes useful, when looking for the most probable density value of defects after the energy release in a fracturing process. The density $\alpha^0=\lambda/\Lambda^2$ may be identified here with the reference density.

We can assume that before an earthquake a supper-lattice is almost completely filled in by dislocations ($n \approx N$ and $\hat{n} \approx 0$). The maximum number of dislocations in arrays could reach $(\Lambda/\lambda)^2$ per area Λ^2. The total moment for an area $\Delta s=N/\Lambda^2$ affected by the arrays of dislocation along the slip planes becomes:

$$\bar{M} = \mu\lambda\Delta s = \mu\lambda N\Lambda^2\left(\frac{\Lambda}{\lambda}\right) = \mu N\Lambda^3.$$

After an earthquake, the number of vacant dislocations $\hat{n}$ shall increase, probably to the equilibrium value (see: Eq. 13.11) and hence we can express the seismic moment by the number of coalescence processes related to surface element Λ^2 as equal to $\Delta\hat{n} = \hat{n}^{eq}(\Lambda/\lambda)$; the introduced factor expresses the maximum of concentration of dislocations in arrays. We obtain for the seismic moment

$$\bar{M}_0 = \bar{M}\Delta\hat{n} = \mu N\Lambda^3\left(\frac{\Lambda}{\lambda}\right)\Delta\hat{n} = \mu N\Lambda^3\left(\frac{\Lambda}{\lambda}\right)\exp\left(-\frac{C\mu\lambda\Lambda^2}{kT}\right) \tag{13.12}$$

where C is a constant for the given structure.

Using the expression for a change of the free energy values we may include the formation of dislocation arrays along the glide planes and we put

$$G = G^0 + \Delta\hat{n}\left(\frac{\Lambda}{\lambda}\right)kT.$$

According to these results, the total energy release ΔE and seismic moment become:

$$\Delta E = G - G^0 = \Delta\hat{n}\left(\frac{\Lambda}{\lambda}\right)kT = \left(\frac{\Lambda}{\lambda}\right)NkT\exp\left(-\frac{C\mu\lambda\Lambda^2}{kT}\right)$$
$$\text{and}\quad \bar{M}^0 = \mu\Lambda^3\frac{\Delta E}{kT}. \tag{13.13}$$

This formula is an important relation between the energy release density and seismic moment density; for instance, for a given ΔE the elementary seismic moment $\bar{M}^0$ decreases with temperature. A free energy related to defect formation, $\hat{g}^f$, is proportional to $\mu\lambda\Lambda^2$ being constant for a given structure; for a greater value of Λ the seismic moment becomes greater too.

Neglecting the term related to the formation entropy, we can write for the entropy density change:

$$\Delta\tilde{S} = kN\left(\frac{\Lambda}{\lambda}\right)\left(1+\frac{C\mu b\lambda^2}{kT}\right)\exp\left(-\frac{C\mu\lambda\Lambda^2}{kT}\right).$$

All these relations concern the quantities referred to the multiple of the cubic volume $N\Lambda^3$ as multiple of parallelepipeds; thus, we can correct these quantities to that related to a given source volume by the factor $\pi R^2 D / N\Lambda^3$:

$$M_0 = \mu\pi R^2 D\left(\frac{\Lambda}{\lambda}\right)\exp\left(-\frac{C\mu\lambda\Lambda^2}{kT}\right),$$

$$\Delta E = \pi R^2 D\frac{kT}{\Lambda^2\lambda}\exp\left(-\frac{C\mu\lambda\Lambda^2}{kT}\right) \quad \text{and} \quad \frac{M_0}{E^{\text{rad}}} = \frac{\mu\Lambda^3}{\eta kT}, \tag{13.14}$$

$$\Delta\tilde{S} = \pi R^2 D\frac{k}{\Lambda^2\lambda}\left(1+\frac{C\mu\lambda\Lambda^2}{kT}\right)\exp\left(-\frac{C\mu\lambda\Lambda^2}{kT}\right)$$

where $\eta\Delta E = E^{\text{rad}}$, η is the seismic efficiency; E^{rad} is the radiated energy.

13.5 Conclusions

In the above consideration we took into account the processes related to slip and rotation; a total released energy includes that related to wave radiation and that to heat caused by friction.

The basic elements of the earthquake thermodynamics are based on the former results related to dislocation interactions and fracture processes.

A role of micro-fracturing in the bond breaking process is similar under both the confining pressure and external shears; however, we observe the essential differences for rotations in the larger scales. A confining condition leads to formation of induced opposite arrays of dislocations, resulting in fragmentation processes and chaotically oriented macro-rotations, leading therefore to a rotation release process, while a shear condition leads to more concentrated fracturing along some planes, high shear strain release and the correlated rotations.

Considerations on the fracture band model and earthquake thermodynamics show how we can better understand physics and geometry of fracture and energy release processes.

References

Dieterich JH (1978) Preseismic fault slip and earthquake prediction. J Geophys Res **83**: B8, 3940-3954

Droste Z, Teisseyre R (1959) The mechanism of earthquakes according to dislocation theory. Sci Rep Tohoku Univ Ser 5, Geophys **11**: 1,55-71

Eshelby JD, Frank FC, Nabarro FRN (1951) The equilibrium of linear arrays of dislocations. Philos Mag **42**, 351-364

Kocks UF, Argon AS, Ashby MF (1975) Thermodynamics and kinetics of slip. Pergamon Press, Oxford, New York, 288 pp

Krőner E (1981) Continuum theory of defects. In: Balian R, Kléman M, Poirer JP (eds) Physics of defects (Les Houches, Session XXXV, 1980). North Holland, Amsterdam, pp 217-315

Majewski E, Teisseyre R (1997) Earthquake thermodynamics. Tectonophysics **277**: 1-3, 219-233

Nabarro FRN (1951) The synthesis of elastic dislocation fields. Philos Mag **42**: 1224-1231

Prigogine I (1979) Irreversibility and randomness. Astrophys Space Sci, **65**: 2, 371-381

Teisseyre R (1961) Dynamic and time relations of the dislocation theory of earthquakes. Acta Geophys Pol **9**: 1-2, 3-58

Teisseyre R (1964) Some remarks to the dislocational model of energy release in the earthquakes. Acta Geophys Pol **12**: 2, 89-98

Teisseyre R (1970) Crack formation and energy release caused by the concentration of dislocation along fault planes. Tectonophysics **9**: 6, 547-557

Teisseyre R (1996) Shear band thermodynamical earthquake model. Acta Geophys Pol **44**: 3, 219-236

Teisseyre R (1997) Shear band thermodynamical model of fracturing with a compressional component. In: Gibowicz S, Lasocki S (eds) Rockburst and seismicity in mines. A.A.Balkema, Rotterdam, Brookfield, pp 17-21

Teisseyre R (2001) Shear band thermodynamic model of fracturing. In: Teisseyre R, Majewski E (eds) Earthquake thermodynamics and phase transformations in the Earth's interior. (International Geophysics Series vol 76). Academic Press, San Diego, pp 279-292

Teisseyre R, Górski M, Teisseyre KP (2006) Fracture-band geometry and rotation energy release. In: Teisseyre R, Takeo M, Majewski E (eds) Earthquake source asymmetry, structural media and rotation effects. Springer, Berlin, pp 169-184

Teisseyre R, Majewski E (1990) Thermodynamics of line defects and earthquake processes. Acta Geophys Pol **38**: 4, 355-373

Teisseyre R, Majewski E (1995) Earthquake thermodynamics. In: Teisseyre R (ed) Theory of earthquake premonitory and fracture processes. PWN, Warszawa, pp 586-590

Teisseyre R, Majewski E (2001) Thermodynamics of line defects and earthquake thermodynamics. In: Teisseyre R, Majewski E (eds) Earthquake thermodynamics and phase transformations in the Earth's interior. (International Geophysics Series vol 76), Academic Press, San Diego, pp 261-278

Teisseyre R, Wiejacz P (1993) Earthquake sequences: stress diagrams. Acta Geophys Pol **41**: 2, 85-100

Varotsos PA, Alexopoulos KD (1986) Thermodynamics of point defects and their relation with bulk properties. North-Holland, Amsterdam, New York

14 Interaction Asymmetric Continuum Theory

Roman Teisseyre

Institute of Geophysics, Polish Academy of Sciences
ul. Księcia Janusza 64, 01-452 Warszawa, Poland
e-mail: rt@igf.edu.pl

14.1 Introduction

We present a logical development of the Standard Asymmetric Continuum Theory in which the interaction terms are introduced to the deformations: strains and rotations. In a concise theory, we shall not only include the relations between stresses and dislocation fields and some aspects of fracture processes, but also take into account various interactions related to thermal, electric, magnetic and other source fields.

There are different ways how to include such interactions. One way relies on the special choice of the constitutive relation in order to join the stress field with strain and other fields; equivalently we can use the Kröner method with the total, elastic and self fields; however, our aim is to include the interactive fields into the Standard Asymmetric Theory. To do so, we may present some equivalences between the Kröner fields and those used in the asymmetric theory.

An appropriate choice of the constitutive laws is a typical approach to influence of the thermal, electric and other fields on the stress changes; for the asymmetric stresses we may write:

$$\begin{aligned} S_{(kl)} &= 2\mu\left(E_{kl} + e^1\delta_{kl}F + e^2F_{(kl)}\right), \\ S_{[kl]} &= 2\mu\left[\omega_{kl} + \chi^1\varepsilon_{kls}G_s + \chi^3G_{[kl]}\right], \end{aligned} \tag{14.1}$$

where F, $F_{(kl)}$, G_s, $G_{[kl]}$ are the non-mechanical fields influencing stresses; and e^1, e^2, χ^1, χ^3 are the constants and according to the Standard Asymmetric Theory, we may write for strain and rotation (see: Chap. 7):

$$E_{kl} = e^0E^0_{kl}, \qquad \omega_{kl} = \chi^0\omega^0_{kl}.$$

In the Kröner method the physically significant elastic fields, S_{ks}, E_{ks}, ω_{ks}, are given by the differences between the total fields, S^0_{ks}, E^0_{ks}, ω^0_{ks}

(related directly to the displacement differentials), and the self fields, S^S_{ks}, E^S_{ks}, ω^S_{ks} (related to internal interaction nuclei):

$$S_{ks} = S^0_{ks} - S^S_{ks}, \qquad E_{ks} = E^0_{ks} - E^S_{ks}, \qquad \omega_{ks} = \omega^0_{ks} - \omega^S_{ks}. \tag{14.2}$$

Only the total field preserves the usual symmetry properties, the elastic and self fields may be asymmetric.

In the Standard Asymmetric Theory we have the following relations (Chap. 7)

$$S_{ks} = S^0_{ks} + S_{[ks]}, \qquad E_{ks} = e^0 E^0_{ks}, \qquad \omega_{ks} = \chi^0 \omega^0_{ks}.$$

Here, stresses are asymmetric, while strains remain symmetric, and rotations are antisymmetric. Comparison between these fields and those in the Kröner theory may be valid only for the restrained values of the self fields:

$$S_{[ks]} \Leftrightarrow -S^S_{ks}, \qquad (e^0 - 1)E^0_{ks} \Leftrightarrow -E^S_{ks}, \qquad \omega^0_{ks}(\chi^0 - 1) \Leftrightarrow -\omega^S_{ks}. \tag{14.3}$$

A comparison with the micromorphic theories brings the following relations joining the strains, $e_{(ks)}$, and the micro-strains (micro-rotations), φ_{ks} (ϕ_{lk}), with the rotation field, ω_{ks}:

$$e_{(ks)} \Leftrightarrow E^0_{ks}, \qquad \varphi_{ks} = \phi_{lk} \Leftrightarrow \omega_{ks} = \chi^0 \omega^0_{ks}. \tag{14.4}$$

Here the self fields, S^S_{ks}, E^S_{ks}, ω^S_{ks}, are restricted to antisymmetric self-stresses, symmetric self-strain and antisymmetric self-rotations.

However, in the asymmetric theory the stress and the deformation tensors would remain asymmetric:

$$S_{ks} = S_{(ks)} + S_{[ks]}, \qquad D_{ks} = E_{ks} + \omega_{ks} = e^0 E^0_{ks} + \chi^0 \omega^0_{ks} \tag{14.5}$$

and then we can write the equivalence relations between the asymmetric and self fields of the Kröner theory:

$$D_{ks} = E_{ks} + \omega_{ks} = e^0 E^0_{ks} + \chi^0 \omega^0_{ks} \Leftrightarrow E^S_{ks}\left(\frac{1}{e^0} - 1\right)^{-1} + \omega^S_{ks}\left(\frac{1}{\chi^0} - 1\right)^{-1}. \tag{14.6}$$

To formulate the interactive linear asymmetric theory we could replace the definition (14.5) by the following ones

$$\begin{gathered} D_{ks} = E_{ks} + \omega_{ks}, \qquad E_{kl} = e^0 E^0_{kl} + e^1 \delta_{kl} + e^2 F_{(kl)}, \\ \omega_{kl} = \chi^0 \omega^0_{kl} + \chi^1 \varepsilon_{kls} G_s + e^3 G_{[kl]}. \end{gathered} \tag{14.7}$$

For the elastic solids: $e^0 = 1,\ \left|\chi^0\right| = 1$, and we take the simplified versions of the constitutive relations (14.1) for asymmetric stresses:

$$S_{(kl)} = 2\mu\left(E^0_{kl} + \delta_{kl}F + F_{(kl)}\right), \quad S_{[kl]} = 2\mu\chi^0\left[\omega^0_{kl} + \varepsilon_{kls}G_s + G_{[kl]}\right]. \tag{14.8}$$

14.2 Thermal Interaction Field

We introduce a thermal field, $F = -\alpha^{\text{ther}}(T - T_0)$, to the corresponding elastic fields in the asymmetric theory:

$$E_{kl} = E^0_{kl} - \delta_{ik}\alpha^{\text{ther}}(T - T_0), \qquad \omega_{kl} = \chi^0\omega^0_{kl}. \tag{14.9}$$

The related compatibility condition becomes

$$\varepsilon_{ikm}\varepsilon_{jtn}\frac{\partial^2}{\partial x_k \partial x_t}\left(E_{mn} - \alpha^{\text{ther}}(T - T_0)\right) =$$

$$= \varepsilon_{ikm}\varepsilon_{jtn}\frac{\partial^2 E_{mn}}{\partial x_k \partial x_t} - \left(\delta_{ij}\frac{\partial^2}{\partial x_k \partial x_k} - \frac{\partial^2}{\partial x_i \partial x_j}\right)\alpha^{\text{ther}}T = 0.$$

For the thermal nucleus, we can take the temperature distribution

$$T = \frac{\vartheta}{r},$$

where ϑ is a source intensity; this expression follows from the stationary case, $\nabla^2 T = 0$. However, for a wide temperature range the thermal heat transfer is not constant and the stationary case can be described as

$$\frac{\partial}{\partial x_i}q_i = -\frac{\partial}{\partial x_i}\left(k_{is}\frac{\partial}{\partial x_s}T\right) = 0, \tag{14.10}$$

where **k** is the tensor of heat transfer (heat conduction) which may be anisotropic.

14.3 Dislocation Related Polarization: Polarization Gradient Theory

The classical piezoelectric effect appears in the anisotropic crystals, piezoelectric dielectrics (see: Toupin 1956, Mindlin 1972); the piezo-electric constants are discussed for the different crystalographic classes by Nowacki (1983).

However, some experiments indicate that in different isotropic bodies the anomalous piezoelectric effects are observed - laboratory tests proved that an electric response is then proportional to time change of the applied load (see: e.g., Hadijcondis and Mavromatou 1994, 1995). Mindlin (1972) generalized the Toupin (1956) theory assuming that internal energy depends also on polarization gradient. The respective mechanism relates to displacement of a dislocation core (electrically charged) under applied load in respect to its surrounding cloud of defects (having the opposite, compensating charges) and, thus, to the formation time dependent dipoles (see: Chap. 10). Dislocation core-cloud shift (due to motion of the dislocation system) follows the Mindlin theory (1972).

For isotropic bodies we assume that the internal energy is a function of strains, electric polarization and gradient of polarization.

We define the polarization and polarization gradient fields :

$$\Pi_i = D_i - \varepsilon E_i,$$
$$\Pi_{ij} = \Pi_{(ij)} + \Pi_{[ij]} = \frac{\partial \Pi_i}{\partial x_j} = \frac{1}{2}\left(\frac{\partial \Pi_j}{\partial x_i} + \frac{\partial \Pi_i}{\partial x_j}\right) + \frac{1}{2}\left[\frac{\partial \Pi_j}{\partial x_i} - \frac{\partial \Pi_i}{\partial x_j}\right], \tag{14.11}$$

where ε is the permittivity of vacuum.

The local internal energy $U(E_{ij}, \Pi_i, \Pi_{ij})$ is now depending on elastic deformation, strains E^0_{ks} and rotations (14.6):

$$D_{ks} = E_{ks} + \omega_{ks} = E^0_{ks} + \chi^0 \omega^0_{ks}, \qquad e^0 = 1 \tag{14.12}$$

and on polarization gradient field.

For isotropic bodies we may put for the internal energy:

$$U = \frac{1}{2} S_{ij} D_{ij} + k \Pi_{ss} E^0_{kk} + n \Pi_{(ij)} E^0_{ij} + m \chi^0 \Pi_{[ij]} \omega_{[ij]} + U^{\Pi}, \tag{14.13}$$

where $U^{\Pi} = a \Pi_{ss} \Pi_{kk} + b \Pi_{(ij)} \Pi_{(ij)} + c \Pi_{[ij]} \Pi_{[ij]}$, or with $S_{ns} = \frac{1}{3} S_{ii} \delta_{ns} + S^D_{ik}$ and $S^D_{ns} = S_{ns} - \frac{1}{3} S_{ii} \delta_{ns}$.

$$U = \frac{1}{3} S_{ii} E^0_{kk} + \frac{1}{2} S^D_{ik} E^0_{ik} + \frac{1}{2} \chi^0 S^D_{ik} \omega^0_{ik} + k \Pi_{ss} E^0_{kk} \\ + n \Pi_{(ij)} E^0_{ij} + m \chi^0 \Pi_{[ij]} \omega_{[ij]} + U^{\Pi} \tag{14.14}$$

and hence for stresses we obtain:

$$S_{ij} = \frac{\partial U}{\partial D_{ij}} = \lambda \delta_{ij} E_{ss}^0 + 2\mu E_{ij}^0 + 2\mu \chi^0 \omega_{ij}^0 + k\Pi_{ss} + 2n\Pi_{(ij)} + 2m\Pi_{[ij]}. \tag{14.15}$$

There appears a 2D current related to gradient polarization:

$$J_{ij} = \rho \Pi_{ij}. \tag{14.16}$$

The current amplitude in a given direction can be found from the distance from center to the appropriate point on the surface:

$$J = J_{ij} x_i x_j, \quad \text{where} \quad x_s x_s = 1.$$

Depending on measurement system, we can find the different components of this current.

14.4 Conclusion

In the frame of asymmetric continuum theory, we have presented a new approach to the interaction problem. We considered the isotropic cases related to thermal excitation and gradient polarization. The new result we obtained is the relation between the asymmetric stresses and strains, rotations and gradient polarization.

References

Hadjicontis V, Mavromatou C (1994) Transient electric signals prior to rock failure under uniaxial compression. Geophys Res Lett **16**: 1687-1690

Hadjicontis V, Mavromatou C (1995) Electric signals recorded during uniaxial compression of rock samples: their possible correlation with preseismic electric signals. Acta Geophys Pol **43**: 1, 49-61

Mindlin RD (1972) Elasticity, piezoelectricity and crystal lattice dynamics. J Elasticity **2**: 4, 217-282

Nowacki W (1983) Efekty elektromagnetyczne w stałych ciałach odkształcalnych. PWN, Warszawa

Toupin RA (1956) The elastic dielectrics. J Rat Mech Anal **5**, 849-915

15 Fracture Physics Based on a Soliton Approach

Eugeniusz Majewski

Institute of Geophysics, Polish Academy of Sciences
ul. Księcia Janusza 64, 01-452 Warsaw, Poland
e-mail: emaj@igf.edu.pl

15.1 Introduction

From soliton theory there is a natural route to fracture physics, by way of energy carriers and interactions. Solitons are quanta of energy (Majewski 2006a, b, c, d, e). Energy is the most decisive factor in fracture processes (Teisseyre and Majewski 2001, 2002). Solitons are rooted in the theory of nonlinear differential equations and their integrability conditions. Many of extreme and fracture processes are described in the language of nonlinear wave equations. Solitons are exact solutions of the nonlinear wave equations. Recently, it has become more evident that solitary equations such as, for instance, the Klein-Gordon equation or nonlinear Schrödinger equation, are applicable in many physical problems where interactions play an important role. Since interactions constitute an essential feature of many physical phenomena, it seems reasonable to attempt to construct some mathematical models that will enable us to illuminate characteristic features of nonlinear wave processes in interacting media. A possible way to describe such nonlinear models is to apply solitons.

15.2 The Dilaton Mechanism

According to Polchinski (2000), *"dilaton is the massless scalar with gravitational-strength couplings, found in all perturbative string theories. An exactly massless dilaton would violate limits on nongravitational interactions, but a mass for the dilaton is not forbidden by any symmetry and so dynamical effects will generate one in vacua with broken symmetry (the same holds for other moduli). The string coupling constant is determined by the value of the dilaton field.*" Thus, the dilaton is a particle in a scalar field that is associated with gravity. Dilaton is to a scalar field as photon is to the electromagnetic field. Polchinski (2000) also mentioned that in the strong interactions, large fluctuating gauge fields in the vacuum are responsible for quark confinement.

The idea of fluctuating fields was adopted in fracture physics as well. For example, Zhurkov (1983) investigated a dilaton mechanism of the strength of solids. He assumed that negative density fluctuations affect the strength of solids. He referred to these fluctuations as dilatons. He viewed them as short-living objects. The good feature of dilatons is that they can absorb energy from their neighborhood in the surrounding medium. They can accumulate the energy like capacitors in electricity only up to a certain threshold value. Next, they release their energy, and the dilaton breaks, nucleating a crack. Kozák and Šilený (1985) considered a similar idea, when they described seismic energy release in earthquake processes. Fracture in solids usually occurs along some inhomogeneities as asperities, cracks, inclusions, and so on. Kozák and Šilený (1985) investigated a 2D homogeneous medium with a linear inhomogeneity compressed in one direction. They were modeling an earthquake process along a tectonic fault. In such a case the directions of stresses were induced by the geometry of the fault. The fault area is determined by the value of stresses and is increasing up to the fracture level. When the fracture takes place, then the seismic energy is released. Gusev (1988) considered two similar dilatancy-based models to explain coda-wave precursors and P/S spectral ratio. He assumed that the mechanism of seismic wave propagation is related to fast variations of the dilatancy. Kozák and Šilený (1985) and Gusev (1988) were carrying their considerations in terms of linear waves. Nikolaev (1989) applied nonlinear approach to this problem and took into account dissipation as well. Engelbrecht and Khamidullin (1988) and Engelbrecht (1997) used the concept of dilaton to modeling the Earth's crust. Namely, they idealized the crust as a hierarchy of elastic blocks with a thin interface layers between them. These layers were viewed as dilatons, i.e., the inhomogeneities that absorbed, stored and released energy. The authors derived an equation that took into account the dilaton mechanism by including the related body force.

15.3 The Nonlinear Klein-Gordon Equation

In order to describe the dilaton mechanism, we start with the nonlinear Klein-Gordon equation (KG) in the following form

$$\frac{\partial^2 \omega}{\partial t^2} - \frac{\partial^2 \omega}{\partial \phi^2} + \frac{\partial}{\partial \omega}\Psi(\omega) = 0\,, \tag{15.1}$$

where Ψ is a general nonlinear function of the spin ω that may be expressed in different ways as a function of the rotational angle ϕ. We may introduce here a lot of forms of nonlinearity. For example, in case of an exponential nonlinearity

$$\frac{\partial \Psi(\omega)}{\partial \omega} = \exp(\omega), \tag{15.2}$$

Eq. (15.1) can be used as a modeling equation for various physical processes. In case of a cubic nonlinearity

$$\frac{\partial \Psi(\omega)}{\partial \omega} = \omega^3 - \omega, \tag{15.3}$$

Eq. (15.1) also can model many physical phenomena (cf., Rajaraman 1982). For our purpose, we shall consider the nonlinear Klein-Gordon equation in the following form

$$\frac{\partial^2 \omega}{\partial t^2} - \frac{\partial^2 \omega}{\partial \phi^2} = f(\omega). \tag{15.4}$$

Here, the function $f(\omega)$ is the body force connected to the so-called dilatancy mechanism and it takes the form

$$f(\omega) = C\omega - D\omega^3, \tag{15.5}$$

where C and D are positive constants.

The soliton solution of this equation is as follows

$$\omega(\phi) = \pm\left(\frac{C^{1/2}}{D}\right)\tanh\left[\frac{C^{1/2}(\phi - \phi_1)}{\sqrt{2}}\right]. \tag{15.6}$$

Figure 15.1 displays a soliton solution $\omega(\phi)$ of the Klein-Gordon equation (15.4) and the energy density $E(\phi)$ of the soliton. It is noteworthy that the energy curve has a pulse-like profile. The solution above the ϕ-axis plotted in Fig. 15.1 is called the "kink" and the one beneath the ϕ-axis the "antikink." The invariance of the solution with respect to translation is clearly visible, since a change in ϕ_1 merely moves the solution along the ϕ-axis. Solutions to Klein-Gordon and the sine-Gordon equations have "kink" profiles. It should be mentioned that if we wish to obtain soliton graphs in a form of pulse profiles, we should plot derivatives of the solutions.

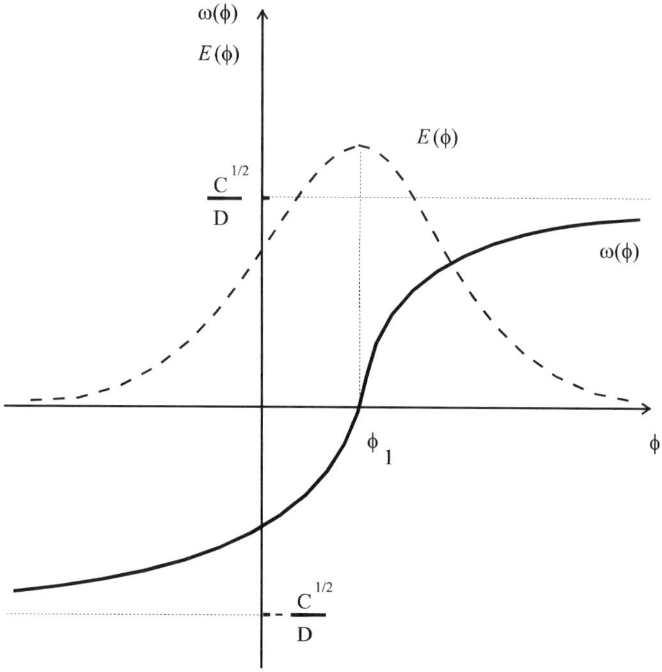

Fig. 15.1 The function $\omega(\phi)$ represents the static kink solution of the Klein-Gordon equation (15.2). The function $E(\phi)$ describes the energy density of the kink (modified from Rajaraman 1982)

15.4 Coupled Klein-Gordon Equations Applied to Modeling a Two-Layer Model

We discuss here a nonlinear model consisting of two layers of grains (or blocks) with linear spin (twist) interactions between the grains and nonlinear spin (twist) interactions between the layers. We assume the so-called degenerated medium introduced by Teisseyre (2004) where only rotational motions exist and translational motions vanish. So, we assume that grains may rotate and their angular position is denoted by ϕ_i. The moment of inertia of any grain of the upper layer is I_u and that of the lower layer is I_d. A distance between centers of adjacent grains in each layer is denoted by a. Spin (twist) interactions between the adjacent grains in the layer are represented by the spin (twist) interaction coefficients $\Re_u$ and $\Re_d$ for the upper and lower layer, respectively. The function representing interactions between the layers is expressed as function of spins (twists) of corresponding grains of the upper and lower layers. The form of this function depends on the interface structure.

We introduce the following notation: ω_i and ϕ_i are the spins (or twists) of the i-th grains of the upper and lower layer, respectively, and $\Im(\omega_i, \phi_i)$ is the spin (twist) energy of interactions between these grains. Processes of spin (twist) interactions in the two layers of grains are characterized by the following system of equations

$$I_u \frac{\partial^2 \omega}{\partial t^2} - \Re_u \left(\omega_{i+1} - 2\omega_i + \omega_{i-1} \right) = -\Im_{\omega_i} \left(\omega_i, \varphi_i \right), \tag{15.7}$$

$$I_d \frac{\partial^2 \varphi}{\partial t^2} - \Re_d \left(\varphi_{i+1} - 2\varphi_i + \varphi_{i-1} \right) = -\Im_{\varphi_i} \left(\omega_i, \varphi_i \right). \tag{15.8}$$

In order to simplify the above set of equations, we adopt the following dimensionless variables

$$\bar{t} = \frac{V_u t}{a}, \quad \bar{\phi} = \frac{\phi}{a}, \quad \bar{\omega} = \frac{\omega}{a}, \quad \bar{\varphi} = \frac{\varphi}{a} \sqrt{\frac{I_d}{I_u}}, \quad \bar{\Re} = \frac{\Re}{I_u V_u^2}, \tag{15.9}$$

$$V = \frac{V_d}{V_u} = \sqrt{\frac{\Re_d I_u}{\Re_u I_d}}, \qquad V_u^2 = \frac{\Re_u a^2}{I_u}, \qquad V_d^2 = \frac{\Re_d a^2}{I_d}. \tag{15.10}$$

Following a procedure presented by Khusnutdinova and Silberschmidt (2003) in the context of translational motions, we apply the force function in the form $f\left(\bar{\omega}, \bar{\varphi}\right) = -\bar{\Re}\left(\bar{\omega}, \bar{\varphi}\right)$. After employing the long-wave approximation, we obtain from Eqs. (15.7) and (15.8) the system of coupled Klein-Gordon equations for spins (twists)

$$\frac{\partial^2 \omega}{\partial t^2} - \frac{\partial^2 \omega}{\partial \phi^2} = f_\omega \left(\omega, \varphi \right), \tag{15.11}$$

$$\frac{\partial^2 \varphi}{\partial t^2} - V^2 \frac{\partial^2 \varphi}{\partial \phi^2} = f_\varphi \left(\omega, \varphi \right), \tag{15.12}$$

where the overbar is omitted and the force function $f(\omega, \phi)$ represents the spin (twist) interactions between the layers of grains. A similar set of coupled Klein-Gordon equations for translational motions was derived by Khusnutdinova and Silberschmidt (2003) in the framework of lattice modeling of nonlinear waves in a bi-layer with delamination.

15.5 The Generalized Korteweg-de Vries (KdV) Equation

Porubov et al. (2003) investigated selection of localized nonlinear seismic waves. They considered a longitudinal wave of displacements. On the contrary, we consider a spin wave. They took into account the phenomena associated with the dilaton mechanism. These phenomena are characterized by fluctuations of energy field. We should consider the role of the microstructure of the material. Let us consider the Korteweg-de Vries (KdV) (cf., Korteweg and de Vries 1895) equation for a spin solitary wave in the following modified form

$$\frac{\partial \omega}{\partial t} + \sigma \frac{\partial \omega}{\partial \phi} + \omega \frac{\partial \omega}{\partial \phi} + D \frac{\partial^3 \omega}{\partial \phi^3} = \varepsilon f(\omega), \tag{15.13}$$

where $\sigma = 0$ or $\sigma = 1$, and $\omega = \omega(\phi, t)$ is the spin at the angular position ϕ and time t, and ε is a small parameter; here the quantities of spin and time were scaled, and the function $f(\omega)$ is the body force related to the dilatancy mechanism and has the form

$$f(\omega) = -A\omega + B\omega^2 - C\omega^3, \tag{15.14}$$

where A, B and C are constants.

If the RHS of Eq. (15.13) equals zero, then this equation transforms into the KdV equation. The nonlinearity is modeled in this equation by the third term that is proportional to the spin while the dispersion is described by the fourth term that is proportional to the third derivative of the spin. The interplay between these two terms is crucial for the formation of solitons. However, a soliton is characterized by its stable shape and constant propagation velocity. According to the dilaton mechanism, the internal energy is stored in a solid body. When the spin soliton propagates through the body, then the locked-in internal energy can be released. Moreover, an additional energy influx can amplify the spin wave. Following Porubov et al. (2003), and assuming $\sigma = 0$ and $f(\omega) = 0$, we can write the solution of Eq. (15.13) as follows

$$\omega(\phi,t) = 12D\lambda^2 \cosh^{-2}\left[\lambda\left(\phi - 4D\lambda^2 t\right)\right], \tag{15.15}$$

where λ is a free parameter.

In addition, Porubov et al. (2003) attempted to find an asymptotic solution of the Eq. (15.13) for the case when $f(\omega)$ is expressed by Eq. (15.14).

15.6 The Spin and Twist Strain Solitons

Samsonov (1988) investigated some exact solutions of nonlinear longitudinal wave equations with dispersion and dissipation and referred to them as to the DDE equations. In the following years, Samsonov (2001) considered theoretical and experimental aspects of strain solitons in solids and how to construct them. He considered longitudinal waves, but we apply his equations to describe rotational waves. We write his nonlinear equation in terms of spin and rotational angle in order to describe the spin and twist strain solitons (Majewski 2006b)

$$\frac{\partial^2 \omega}{\partial t^2} - \frac{\partial^2 \omega}{\partial \phi^2} = \varepsilon \frac{\partial^2}{\partial \phi^2}\left(6\omega^2 + a\frac{\partial^2 \omega}{\partial t^2} + b\frac{\partial^2 \omega}{\partial \phi^2} + g\frac{\partial \omega}{\partial t}\right), \tag{15.16}$$

where ω is the rotational field, ϕ is the rotational angle, ε is a small parameter, g is the dimensionless 'viscosity', and a and b are constant coefficients. This equation is called the double dispersive equation (DDE), describing dispersion and dissipation and having exact solitary solutions. This equation can describe the spin and twist strain solitons. Let us consider this equation in the framework of the theory of a degenerated asymmetric continuum, in which the displacement motions vanish, and only rotational motions are retained. Both the spin motions characterized by antisymmetric fields and the bend-twist motions characterized by symmetric fields have been considered (cf. Teisseyre 2004). Now we write equations for spin and twist strain waves in the form of the DDE equations:

$$\frac{\partial^2 \omega_{[s]}}{\partial t^2} - V^2 \frac{\partial^2 \omega_{[s]}}{\partial \phi^2} = \varepsilon^2\left(C_1 \frac{\partial^4 \omega_{[s]}}{\partial \phi^2 \partial t^2} + C_2 \frac{\partial^4 \omega_{[s]}}{\partial \phi^4} + C_3 \frac{\partial^2}{\partial \phi^2}\left(\omega_{[s]}^2\right)\right),$$

(15.17)

$$\frac{\partial^2 \omega_{(t)}}{\partial t^2} - V^2 \frac{\partial^2 \omega_{(t)}}{\partial \phi^2} = \varepsilon^2\left(D_1 \frac{\partial^4 \omega_{(t)}}{\partial \phi^2 \partial t^2} + D_2 \frac{\partial^4 \omega_{(t)}}{\partial \phi^4} + D_3 \frac{\partial^2}{\partial \phi^2}\left(\omega_{(t)}^2\right)\right),$$

where $\omega_{[s]}$ and $\omega_{(t)}$ denote the spin and twist components of the rotational strain field, respectively; the velocities for spin and twist strain waves are equal, $V^2 = 2\mu^*/\rho$, where μ^* is the rotation rigidity, ρ is the material density, ε is a small parameter, and C_i and D_i ($i = 1, 2, 3$) are constant coefficients. In our case, the spin strain solitons take the following form (see Samsonov 2001, Majewski 2006b)

$$\omega_{[s]} = \frac{\Psi}{2}\cosh^{-2}\sqrt{\frac{\Psi}{\Psi(1+C_1)+C_2}}\left(\phi \pm t\sqrt{1+\Psi}\right), \tag{15.18}$$

where $\Psi = 6(V^2 C_1 + C_2)/C_3$.

Correspondingly, the twist strain solitons can be expressed in the form

$$\omega_{(t)} = \frac{\hat{\Psi}}{2}\cosh^{-2}\sqrt{\frac{\hat{\Psi}}{\hat{\Psi}(1+D_1)+D_2}}\left(\phi \pm t\sqrt{1+\hat{\Psi}}\right), \tag{15.19}$$

where $\hat{\Psi} = 6(V^2 D_1 + D_2)/D_3$.

15.7 Splitting the Spin Strain Solitons Propagating along the Fracture Surface into the Fracture-Zone Related Part and the Elastic Part

We follow here a splitting procedure proposed by Teisseyre and Yamashita (1999) who split the stress motion equations into seismic wave and fault-related fields (see Chapter 8 in Teisseyre et al. 2006). We apply this method to Eq. $(15.17)_1$ and split the equation for elastic spin strain solitons into elastic soliton and fracture-zone related soliton equations. A total spin strain ω can now be presented as the sum $\tilde{\omega}+\bar{\omega}$ of the elastic field $\tilde{\omega}$ and the self-spin (or plastic spin) part, which is assumed to rapidly decrease away from the fracture surface (Teisseyre 2001, Boratyński and Teisseyre 2006). We can identify these parts with an elastic field $\tilde{\omega}$ and a fracture-zone related field $\bar{\omega}$. After splitting, the first equation for the elastic strain soliton takes the form

$$\frac{\partial^2 \tilde{\omega}_{[s]}}{\partial t^2} - V^2 \frac{\partial^2 \tilde{\omega}_{[s]}}{\partial \phi^2} = \varepsilon^2\left(C_1 \frac{\partial^4 \tilde{\omega}_{[s]}}{\partial \phi^2 \partial t^2} + C_2 \frac{\partial^4 \tilde{\omega}_{[s]}}{\partial \phi^4} + C_3 \frac{\partial^2}{\partial \phi^2}\left(\tilde{\omega}^2_{[s]}\right)\right) \tag{15.20}$$

The fracture-zone related equation for the self-spin strain soliton is as follows

$$\frac{\partial^2 \bar{\omega}_{[s]}}{\partial t^2} - V^2 \frac{\partial^2 \bar{\omega}_{[s]}}{\partial \phi^2} = \varepsilon^2\left(C_1 \frac{\partial^4 \bar{\omega}_{[s]}}{\partial \phi^2 \partial t^2} + C_2 \frac{\partial^4 \bar{\omega}_{[s]}}{\partial \phi^4} + C_3 \frac{\partial^2}{\partial \phi^2}\left(\bar{\omega}^2_{[s]}\right)\right) \tag{15.21}$$

The above two equations have soliton solutions in the form of the DDE type solitons determined by Eq. (15.18). These solitons are spin strain waves. The latter equation describes the fracture-zone related soliton that

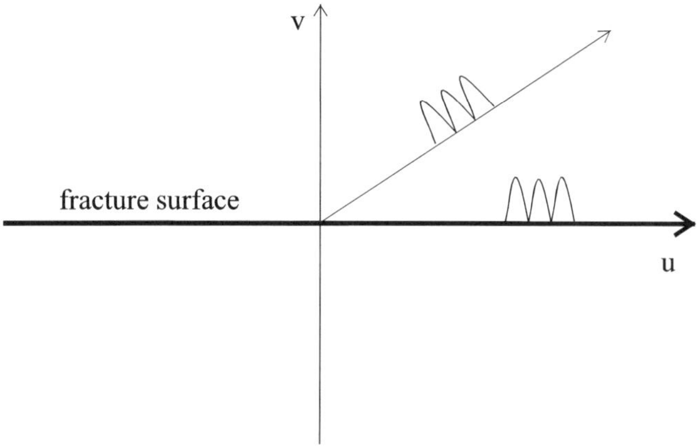

Fig. 15.2 Illustration of the splitting the spin strain solitons propagating along the fracture surface into the fracture-zone related part and the elastic part

can be generated by past fracture events and may propagate along the fracture surface to trigger new fracture events. Figure 15.2 depicts the splitting the total spin strain soliton equation into an elastic spin strain soliton and fracture-zone related soliton equations.

15.8 The Sine-Gordon Model of Moving Dislocations

The sine-Gordon equation (SG) is a special case of the Klein-Gordon equation (KG). It was employed in the modeling of many physical processes such as: motion of crystal dislocations in the periodic Peierls potential, motion of charged particles, biological processes like DNA dynamics, magnetic flux in the theory of Josephson junction ladders, Bloch wall motion in magnetic crystals, and two-dimensional models of elementary particles in the unified field theory (Rajaraman 1982). More information concerning soliton applications can be found in Barone et al. (1971), Scott et al. (1973), and Infeld and Rowlands (2000). The sine-Gordon equation can be expressed in the following form

$$\frac{\partial^2 \omega}{\partial x^2} - \frac{\partial^2 \omega}{\partial t^2} = \sin \omega(x,t). \tag{15.22}$$

The motion of dislocations in the periodic Peierls potential was analyzed by Hashizume (1993) starting from a kink-antikink solution of the sine-Gordon equation. Hc assumcd that the movement of the dislocation climbing the potential peak occurs by bending of a part of the dislocation line toward the potential ridge, formation of double kinks and their propagation in the opposite directions along the dislocation line. Small effects of external stress, friction and random force for the moving dislocation line were calculated by Hashizume (1993) by using soliton perturbation theories directly based on the inverse scattering transform.

15.9 Soliton Ratchets

Salerno and Quintero (2001) investigated the so-called soliton ratchets. They used a generalized double sine-Gordon equation in order to describe a ratchet system, i.e., a periodically forced Brownian particle moving in an asymmetric potential in presence of damping and periodic driving. The soliton ratchet is a unidirectional motion of the soliton in an asymmetric potential with damping and periodic forcing. Salerno and Quintero (2001) showed that the asymmetry of the potential yields a spatially asymmetric internal mode on the soliton profile that can be generated using the periodic force. Due to the damping mechanism, this mode is able to exchange energy with the translational mode. In a result, the soliton has a resultant unidirectional motion in spite of the action of an ac force. Here the damping can be viewed as a coupling mechanism between the internal and translational modes of the soliton. The job of a periodic force is to disturb the system and drive it out of equilibrium. An associated thermal energy can induce the conversion of the ac force into effective work that results in a unidirectional motion of the soliton. Salerno and Quintero (2001) found that a kind of resonance occurs when the internal mode and the external driving force are phase locked. In such a case the solitons moves at the maximum velocity. A similar resonance occurs when the frequency of the external ac force is varying and suddenly perfectly matching the damping value that is fixed together with other parameters of the system.

Salerno and Quintero (2001) considered the generalized double sine-Gordon equation

$$\frac{\partial^2 \omega}{\partial t^2} - \frac{\partial^2 \omega}{\partial x^2} = -\sin(\omega) - \lambda \sin(2\omega + \theta) = -\frac{d\mathfrak{A}(\omega)}{d\omega}, \tag{15.23}$$

where the asymmetric potential is expressed as follows

$$\mathfrak{A}(\omega) = C - \cos(\omega) - \frac{\lambda}{2}\cos(2\omega + \theta). \tag{15.24}$$

Here λ denotes the asymmetry parameter, θ denotes a phase constant, and C denotes a constant which determines the zero of the potential.

The soliton ratchet system can be applied to describe motions of dislocations during deformational and fracture processes.

15.10 The Generalized Sine-Gordon Model of Rock Fracture

Bykov (1999) presented a mathematical model for the unstable deformation of rock along an existing fracture zone. He showed that this deformation process is really nonlinear and the sine-Gordon equation can be applied as the best modeling tool. He was carrying an analysis on the influence of geometric irregularities and friction of the contact surfaces on the evolution of the velocity of solitary slip waves that are generated by local deformation effects and propagate along the fracture zone.

Moreover, Bykov (2006) considered solitary waves in crustal faults and their application to earthquakes. His starting point was a generalized sine-Gordon equation in the form

$$\frac{\partial^2 U}{\partial \xi^2} - \frac{\partial^2 U}{\partial \tau^2} = \sin U + \alpha \frac{\partial U}{\partial \tau} + \gamma(\xi)\delta(\xi - L)\sin U + \sigma(\tau), \tag{15.25}$$

where $U = 2\pi u/a$, $\xi = \pi x/ap$, $\tau = \pi\omega_0 t/p$, $p^2 = a^2 D/4mgh$, $\omega_0^2 = D/m$, $\alpha \approx a\mu/d\Delta\rho(gh)^{1/2}$, $\gamma = H/L$, and U is the displacement of blocks situated periodically along the fracture zone, a is the distance between the block centers, d is the diameter of the circular contact of the blocks, D is the tangential contact stiffness, g is the gravity acceleration, h is the distance between the block centers of the adjacent block layers, m is the mass of the block, μ is the viscosity of the layer between the blocks, Δ is the layer thickness, ρ is the density of the block material, α and γ are the parameters of friction and inhomogeneity, respectively, H and L are the height of the asperities and the distance between them normalized to ap/π, $\delta(\xi)$ is the Dirac delta-function, and $\sigma(\tau)$ is the function which reflects the external load at the contact of the fault surfaces.

Bykov (2006) gave a physical interpretation to all the terms on the RHS of the above equation. Namely, he interpreted the first term as the "restoring" force, the second term is the friction force, the third term is responsible for corrections for inhomogeneities, the last term describes the initiation external load on the fault. After numerical treatment of this equation, he found the displacement function in the form of a kink soliton and the slip velocity function in the form of a pulse soliton.

15.11 Links Between Solitons and Moving Cracks

Sharon et al. (2001) considered rapidly moving tensile cracks in a brittle material. In 3D, they form a singular front. The front has to interact with asperities or material inhomogeneities on the fracture surface. The cracks propagate as single coherent entities, despite random interactions with asperities. Sharon et al. (2001) revealed that perturbations to a crack front in a brittle material yield lasting and localized solitary waves. They referred to these waves as to 'front waves.' The front waves propagate along the crack front at the speed of sound.

15.12 Fracture Solitons in Polymer Chains

Manevitch et al. (2003) noted that a large part of polymer chain energy is stored in localized excitations of soliton-like packets. These solitons take active part in the process of mechanical degradation and fracturing of polymer chain. In a result of their experimental and molecular-dynamic modeling of the process of mechanical degradation of polymer chain, the discovered high-energy molecular products of mechanical degradation generate a chain branching process yielding an explosive nucleation of submicrocracks.

15.13 Chaos of Soliton Systems

Lou et al. (2001) described chaos in two different soliton systems and discussed special Lax pairs for chaos systems. They considered the (2+1) Davey-Stewartson (DS) model and the asymmetric Nizhnik-Novikov-Veselov (ANNV) model. They derived the Lorenz system and its general form from the DS equation and the ANNV equation. They considered integrability of these two models using the Lax pairs or IST (Inverse Scattering Transformation). They found that the both systems have some special chaotic solutions. Lou et al. (2001) used the fact that these two models are related to the Kadomtsev-Petviashvili (KP) equation whereas the self-dual Yang-Mills (SDYM) equation is the generalization of the DS and KP equations.

15.14 The Soliton Complexes

Bogdan et al. (1999) proposed the idea of soliton complexes in a nonlinear dispersive medium. They suggested that strongly interacting solitons constitute bound soliton complexes that propagate without any loss of energy.

Usually the complexes consist of three or more identical solitons. The soliton complexes can attain their excited states. They represent solutions of nonlinear dispersive equations with the fourth and higher spatial derivatives. Bogdan et al. (1999) investigated the dispersive sine-Gordon (dSG), double sine-Gordon (dDSG) and triple sine-Gordon (dTSG) solitons. They described the classification of the excited states of soliton complexes. They used also the two-soliton ansatz approximation to find the existence condition for the soliton complex in the dSG and dDSG equations. They described these soliton complexes using appropriate Lagrangians and Hamiltonians. The soliton complexes can be used to form the nonlinear dynamics of dislocations.

Fig. 15.3 Illustration of spin soliton complexes. The triple spin soliton complexes move downward

15.15 The Soliton Arrays

Solitons can form discrete arrays that move together, for instance, as an array of dislocations or disclinations (see Fig. 15.3).

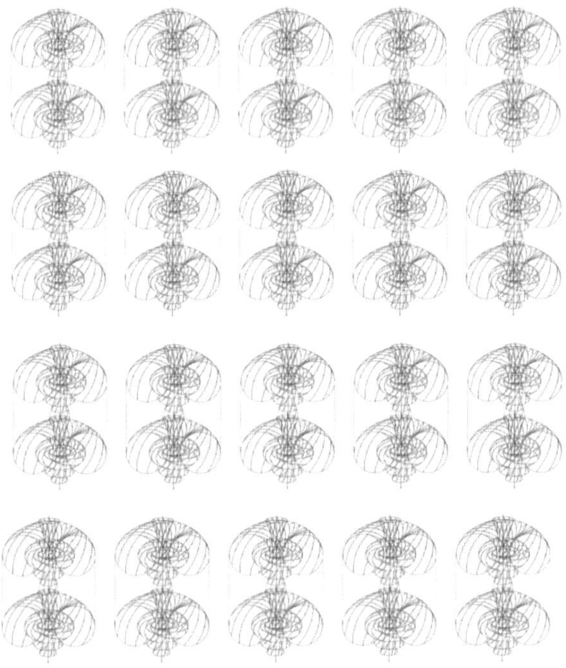

Fig. 15.4 Illustration of the discrete array of spin solitons moving downward

15.16 Conclusions

This chapter briefly presented a review of the most characteristic examples of soliton applications to problems of fracture physics. We started with the dilaton mechanism and its applications. The two-layer model of spin (twist) interactions was proposed. We applied the DDE equations to describe the spin and twist strain solitons. In addition, we reviewed some results concerning rock fracture processes and relations of solitons with moving cracks and dislocations. We mentioned about chaotic soliton systems and about soliton complexes and arrays.

References

Barone A, Esposito F, Magee CJ, Scott AC (1971) Riv Nuovo Cim **1**: 227

Bogdan MM, Kosevich AM, Maugin GA (1999) Soliton-complex dynamics in strongly dispersive medium. ArXiv:patt-sol/9902009 v1:1-32

Boratyński W, Teisseyre R (2006) Fault dynamics and related radiation. In: Teisseyre R, Takeo M, Majewski E (eds) Earthquake source asymmetry, structural media and rotation effects, Springer-Verlag, Berlin, Heidelberg, pp 77-89

Bykov VG (1999) Mathematical modeling of slip in a rock along the uneven fracture. J Mining Science **36** (3): 1062-7391

Bykov VG (2006) Solitary waves in crustal faults and their application to earthquakes. In: Teisseyre R, Takeo M, Majewski E (eds) Earthquake source asymmetry, structural media and rotation effects, Springer-Verlag, Berlin, Heidelberg, pp 241-253

Engelbrecht J (1997) Nonlinear wave dynamics: complexity and simplicity. Kluwer, Dordrecht

Engelbrecht J, Khamidullin Y (1988) On the possible amplification of nonlinear seismic waves. Phys Earth Planet Inter **50**: 39-45

Gusev AA (1988) Two dilatancy-based models to explain coda-wave precursors and P/S spectral ratio. Tectonophysics **152**: 227-237

Hashizume Y (1993) Sine-Gordon model of moving dislocation. J Phys Soc Japan **62**, 7: 2241-2247

Infeld E, Rowlands G (2000) Nonlinear waves, solitons and chaos. Cambridge University Press, Cambridge

Khusnutdinova KR and Silberschmidt VV (2003) Lattice modeling of nonlinear waves in a bi-layer with delamination. Proc Estonian Acad Sci Phys Math **52**, 1: 63-75

Korteweg DJ, de Vries G (1895) On the change of form of long waves advancing in a rectangular canal and on a new type of long stationary waves. Philos Mag (5) **39**: 422-443

Kozák J, Šilený J (1985) Seismic events with non-shear component. I. Shallow earthquakes with a possible tensile source component. Pure Appl Geophys **123**: 1-16

Lou S, Tang X, Zhang Y (2001) Chaos of soliton systems and special Lax pairs for chaos systems. ArXiv:nlin/0107029v2

Majewski E (2006a) Soliton physics. In: Teisseyre R, Takeo M, Majewski E (eds) Earthquake source asymmetry, structural media and rotation effects, Springer-Verlag, Berlin, Heidelberg, pp 113-128

Majewski E (2006b) Seismic rotation waves: spin and twist solitons. In: Teisseyre R, Takeo M, Majewski E (eds) Earthquake source asymmetry, structural media and rotation effects, Springer-Verlag, Berlin Heidelberg, pp 255-272

Majewski E (2006c) Seismic rotation waves in the continuum with nonlinear microstructure. In: Teisseyre R, Takeo M, Majewski E (eds) Earthquake source asymmetry, structural media and rotation effects, Springer-Verlag, Berlin Heidelberg, pp 293-300

Majewski E (2006d) Tectonic solitons propagating along the fault. In: Teisseyre R, Takeo M, Majewski E (eds) Earthquake source asymmetry, structural media and rotation effects, Springer-Verlag, Berlin Heidelberg, pp 301-309

Majewski E (2006e) Complexity of rotation soliton propagation. In: Teisseyre R, Takeo M, Majewski E (eds) Earthquake source asymmetry, structural media and rotation effects, Springer-Verlag, Berlin Heidelberg, pp 255-272

Manevitch LI, Zarkhin LS, Enikolopian NS (2003) Nonlinear dynamics and the problem of polymer fracture. J Appl Polymer Science **39** (11-12): 2245-2258

Nikolaev AV (1989) Scattering and dissipation of seismic waves in presence of nonlinearity. Pure Appl Geophys **131**: 687-702

Polchinski J (2000) String theory, vol I and II, Cambridge University Press, Cambridge

Porubov AV, Gursky VV, Maugin GA (2003) Selection of localized nonlinear seismic waves. Proc Estonian Acad Sci Phys Math **52**, 1: 85-93

Rajaraman R (1982) Solitons and instantons: an introduction to solitons and instantons in quantum field theory. North-Holland, Amsterdam

Salerno M, Quintero NR (2001) Soliton ratchets, arXiv:nlin/0107011v1 [nlin.PS] 6 Jul, pp 1-4

Samsonov AM (1988) On some exact solutions of nonlinear longitudinal wave equations with dispersion and dissipation. In: Dispersive Waves in Dissipative Fluids. Proc EUROMECH Coll 240, Bologna, DG Crighton and F Mainardi (eds) Tecnoprint, Bologna, pp 56-57

Samsonov AM (2001) Strain solitons in solids and how to construct them. Chapman & Hall/CRC, London

Scott AC, Chiu FYF, Mclaughlin DW (1973) Proc IEEE **61**: 1443

Sharon E, Cohen G, Fineberg J (2001) Propagating solitary waves along a rapidly moving crack front. Nature **410**: 68-71

Teisseyre R (2001) Deformation dynamics: continuum with self-deformation nuclei. In: Teisseyre R, Majewski E (eds), Earthquake thermodynamics and phase transformations in the earth's interior. Academic Press, San Diego

Teisseyre R (2004) Spin and twist motions in a homogeneous elastic continuum and cross-band geometry of fracturing. Acta Geophys Pol **52**: 173-183

Teisseyre R, Majewski E (eds) (2001) Earthquake thermodynamics and phase transformations in the earth's interior. Academic Press, San Diego

Teisseyre R, Majewski E (2002) Physics of earthquakes. In: Lee WHK, Kanamori H, Jennings PC, Kisslinger C (eds) International handbook of earthquake & engineering seismology, Part A. Academic Press, San Diego, pp 229-235

Teisseyre R, Takeo M, Majewski E (eds) (2006) Earthquake source asymmetry, structural media and rotation effects, Springer-Verlag, Berlin Heidelberg

Teisseyre R, Yamashita T (1999) Splitting stress motion equation into seismic wave and fault-related fields. Acta Geophys Pol **47**: 135-147

Zhurkov SN (1983) Dilaton mechanism of the strength of solids. Sov Phys Solid State **25**: 1797-1800

16 Canonical Approach to Asymmetric Continua

Eugeniusz Majewski

Institute of Geophysics, Polish Academy of Sciences
ul. Księcia Janusza 64, 01-452 Warsaw, Poland
e-mail: emaj@igf.edu.pl

16.1 Introduction

This chapter presents a canonical approach to the theory of asymmetric continua in terms of Lagrangians and Hamiltonians. The Lagrangian is expressed in terms of the spin structure and spin rates. Consequently, the Hamiltonian is described as a function of the spin structure and angular momenta (Majewski 2006a). Landau and Lifshitz (1958, 1960) gave a clear exposition of the Lagrangian and Hamiltonian formulations of classical mechanics. In their formulation for a system of material points, the Lagrangian is a function of generalized positions and generalized velocities of the material points. They viewed Hamiltonian as a function of the generalized positions and generalized momenta. Maugin (2003) formulated the canonical mechanics of a nonlinear elastic material. Kleinert (1988, 1989, 2008) elucidated an action approach to gravitational and electromagnetic fields. A Lagrangian formulation of an asymmetric elastic continuum was applied by Majewski (2006b) in the context of rotational seismic waves and accompanying them spin and twist solitons. Moreover, Majewski (2006c) used a Lagrangian approach in the framework of the gauge field theory of an elasto-plastic continuum with dislocations. In the gauge field theory the fundamental equations are obtained by variations of the gauge invariant Lagrangian.

The goal of the canonical approach in physics is to start from a single Hamilton's Principle of Least Action and to find an extremum of an integral describing energy in order to derive differential equations of motion, called in mechanics Lagrange's equations. In the variational calculus they are called Euler's equations for the general mathematical problem of determining the extrema of an integral. The Lagrangian of a mechanical system represents the difference between its kinetic and potential energies and allows us to derive the equations of motion of the system. They relate coordinates, velocities and accelerations.

Mathematically speaking, the number of equations of motion is equal to the number of generalized coordinates, which are treated as unknown functions. Let us say that we have n generalized coordinates. In such a case, the set of equations of motion comprises a set of n second-order differential equations for n unknown functions—generalized coordinates. The general solution has $2n$ arbitrary constants. In order to find these constants and to describe completely the motion of the system in question, we have to determine the initial conditions. By the initial condition we mean the initial numerical values of the coordinates and velocities. They define the state of the system at the initial time.

We should be aware that the construction of the Lagrangian is connected with a choice of variables. Consequently, the same variables will reappear in the derived Lagrange's equations of motion. Classically, the generalized coordinates and velocities are used to construct the Lagrangian that describes the difference between the kinetic and potential energies of the mechanical system. We should emphasize that this is not the only possible choice of variables. Sometimes, particularly, when we need to know the total energy of the system, it is more convenient to describe the system in terms of the generalized coordinates and momenta. The best way to transfer from one set of variables to another set is to use the Legendre's transform. In order to find the energy of a Lagrangian system, it is convenient to construct the so-called Hamiltonian. Using the Hamiltonian, we can obtain a new set of first-order differential equations of motion. These equations of motion are called Hamilton's equations or canonical equations. In these equations, the unknown functions are the generalized coordinates and the generalized momenta of the system. They can be treated as evolution equations for the generalized coordinates and generalized momenta.

As far as we know, the canonical approach was applied to translational motions. Our goal, here, is to show how the canonical approach can be applied to rotational motions. Rotational motions can be realized in asymmetric media. Thus, our contribution consists in the application of a different set of variables: spin structure and spin rates to create the Lagrangian, and the spin structure and angular momenta to form the Hamiltonian.

16.2 Hamilton's Principle

The fundamental variational principle of mechanics is Hamilton's Principle that can be viewed as a Principle of Least Action on intuitive grounds. Hamilton's Principle states that from all possible paths of motion of a mechanical system of material points, in reality, the path of motion is realized for which the following integral

$$\int_{t_a}^{t_b} \mathrm{d}t(K-U) \tag{16.1}$$

attains an extremum, where K and U are the kinetic and potential energies of the system of material points, respectively.

Hamilton's Principle, formulated for a mechanical system of material points, can be easily generalized to apply to a material continuum. In such a case, we have to consider an elementary reference volume of the continuum, and instead of the kinetic and potential energies of material points, we should use the kinetic and potential energy densities per elementary reference volume of the continuum (cf., Maugin 2003). In order to generalize Hamilton's Principle to apply to electromagnetic or gravitational fields, one should consider an extremum of field action (cf., Kleinert 1988, 1989, 2008).

16.3 Action of Spin and Twist Fields

Here, for the purpose of clarity, we confine ourselves to spin and twist motions. We neglect translational motions at this stage of our considerations. Teisseyre (2004) considered the theory of a degenerated continuum, in which there exist only the spin and twist axial motions but displacement motions vanish (see also Teisseyre et al 2006, Teisseyre and Boratyński 2006). Thus, the main variable in this asymmetric theory of spin and twist motions is the rotation angle $\phi^{\kappa}(x^{\kappa})$ about the x^{κ} axis $\kappa = 1, 2, 3$. The direction of the pseudo-vector ϕ^{κ} coincides with the rotation axis x^{κ} of the continuum elementary volume and its length is equal to the angle of rotation. The angle of rotation is dual to the spin structure field $\omega_{\mu\nu}(x^{\kappa})$, thus, $\omega_{\mu\nu} = \varepsilon_{\mu\nu\kappa}\phi^{\kappa}$, where μ, ν, κ=1, 2, 3, and $\varepsilon_{\mu\nu\kappa} = \varepsilon^{\mu\nu\kappa} = 0, +1, -1$ is the totally antisymmetric 3D permutation symbol called Eddington's epsilon, $\varepsilon_{\mu\nu\kappa} = +1, -1$ if $\mu\ \nu\ \kappa$ is an even and an odd permutation of (1, 2 ,3), respectively, otherwise $\varepsilon_{\mu\nu\kappa} = 0$.

Landau and Lifshitz (1960) gave an excellent exposition of the Lagrangian and Hamiltonian formulations of classical mechanics. From now on, we follow the general line of reasoning from their exposition. However, there are some differences. Their exposition is for a system of material points and their Lagrangian is formulated in terms of the generalized coordinates and generalized velocities, and their Hamiltonian is expressed in terms of the generalized coordinates and generalized momenta. On the contrary, our presentation deals with an elementary reference volume of the continuum and we use a different set of variables. For the Lagrangian, we use the spin structure and spin rates, but for the Hamiltonian, we use the spin structure and angular momenta.

The most elegant approach to the formulation of the law governing the spin and twist fields is the principle of an extremum of field action. This principle requires that the spin and twist field should be determined by a Lagrange function $L(\omega,\dot{\omega},t)$, where ω is the spin (twist) field, $\dot{\omega}$ is the spin (twist) rate field, and t is time. The action of the spin and twist fields satisfy a certain condition.

Here nine quantities $\omega_{\mu\nu}(\mu,\nu = 1, 2, 3)$ completely defining the spin structure of a system are called spin coordinates of the system, and the spin rates $\dot{\omega}_{\mu\nu}$ are called its angular velocities. These quantities define the position of the system in the so-called state space Σ. Let the positions of the system in the state space Σ at times t_a and t_b be determined by the spin coordinates $\omega^{(a)}$ and $\omega^{(a)}$, respectively. Thus, we assume that the curve representing a path in the state space Σ has both ends a and b fixed. Then, in order to move the system between these positions in the state space Σ, the action of the spin (twist) field must satisfy the condition that the following integral

$$A = \int_{t_a}^{t_b} \mathrm{d}t\, L(\omega(t),\dot{\omega}(t),t) \tag{16.2}$$

attains an extremum. The integrand L is called the Lagrangian of the system concerned, and A is called the action of the spin (twist) field. Note that the mechanical state of the system is completely defined when the spins and spin rates are given.

16.4 The Euler-Lagrange Equations

Let us compare the action for the path $\omega = \omega(t)$ with that of another path $\omega' = \omega(t) + \delta\omega(t)$, where the small function $\delta\omega(t)$ is called a variation of function $\omega(t)$ or a variation of the path. The variation of the action A when ω is replaced by $\omega + \delta\omega$ is

$$\delta A = \int_{t_a}^{t_b} \mathrm{d}t\, L(\omega+\delta\omega,\dot{\omega}+\delta\dot{\omega},t) - \int_{t_a}^{t_b} \mathrm{d}t\, L(\omega,\dot{\omega},t). \tag{16.3}$$

The end points are fixed, thus, the variations of the path vanish at the end points

$$\delta\omega(t_a) = \delta\omega(t_b) = 0. \tag{16.4}$$

Thus the condition for attaining an extremum is as follows

$$\delta A = \delta \int_{t_a}^{t_b} \mathrm{d}t\, L(\omega(t), \dot{\omega}(t), t) = 0. \tag{16.5}$$

The extremum condition is that the time derivative equals zero

$$\frac{\partial A}{\partial t} = \frac{\partial}{\partial t} \int_{t_a}^{t_b} \mathrm{d}t\, L(\omega, \dot{\omega}, t) = 0. \tag{16.6}$$

Now, we are in a position to take the variation in Eq. (16.5), and in a result, we obtain

$$\delta A = \int_{t_a}^{t_b} \mathrm{d}t \left(\frac{\partial L}{\partial \omega} \delta\omega + \frac{\partial L}{\partial \dot{\omega}} \delta\dot{\omega} \right) = 0. \tag{16.7}$$

Note that the variation of the time derivative is equivalent to the time derivative of the variation, i.e., $\delta\dot{\omega} = \mathrm{d}\delta\omega / \mathrm{d}t$. In other words, the variation of the path commutes with the time derivative. Employing this fact and integrating the second term by parts, we get

$$\delta A = \left[\frac{\partial L}{\partial \dot{\omega}} \delta\omega \right]_{t_a}^{t_b} + \int_{t_a}^{t_b} \mathrm{d}t \left(\frac{\partial L}{\partial \omega} - \frac{\mathrm{d}}{\mathrm{d}t} \frac{\partial L}{\partial \dot{\omega}} \right) \delta\omega = 0. \tag{16.8}$$

From the fixed ends described by Eq. (16.3), it follows that the boundary term in Eq. (16.8) must vanish. The remaining integral also must vanish for all variations $\delta\omega$, thus the spin structure $\omega(t)$ must fulfill the following Euler-Lagrange equation of motion

$$\frac{\mathrm{d}}{\mathrm{d}t} \left(\frac{\partial L}{\partial \dot{\omega}} \right) - \frac{\partial L}{\partial \omega} = 0. \tag{16.9}$$

Due to the fact that the spin structure $\omega(t)$ has nine tensor components, nine different functions $\omega_{\mu\nu}(t)$ must be varied independently according to Hamilton's principle of least action. Actually, we get a set of nine equations of motion of the system

$$\frac{\mathrm{d}}{\mathrm{d}t}\left(\frac{\partial L}{\partial \dot{\omega}_{\mu\nu}}\right)-\frac{\partial L}{\partial \omega_{\mu\nu}}=0, \qquad (\mu,\nu=1,2,3). \tag{16.10}$$

These are second-order differential equations for the spin structure $\omega(t)$. In order to solve this set of equations, we have to supplement it with the initial conditions, which determine the state of the system at an initial time, i.e., the initial values of all the components of the spin structure and spin rates.

16.5 Additive Decomposition of the Lagrangian

If a mechanical system is a combination of three independent fields X, Y, and Z, e.g., elastic field, defect field, and electromagnetic field, and each of these fields can be described by its own Lagrangian L_X, L_Y and L_Z, respectively, then, the Lagrangian of the whole system can be decomposed as the sum of all partial Lagrangians

$$L = L_X + L_Y + L_Z\,. \tag{16.11}$$

The additive decomposition of the Lagrangian yields consequences for the form of the equations of motion of the above-mentioned fields. Due to the independence of these fields from each other, i.e., due to the lack of any interactions between these fields, the motion equations should be free from quantities pertaining to the other fields. For example, Majewski (2006c) applied a Lagrangian formulation using the gauge field theory of an elasto-plastic continuum with dislocations, and decomposed the total Lagrangian into two parts: (i) the part describing the deformational energy of the material, (ii) the part describing the energy of moving dislocations.

16.6 The Canonical Equations (Hamilton's Equations)

The Lagrangian in the previous section was expressed in terms of the spin structure and spin rates. However, sometimes, it is useful to describe the system by the spin structure and the angular momenta. In order to find the energy of a Lagrangian system, it is convenient to construct the so-called Hamiltonian H. It can be obtained by Legendre's transform

$$H = M_{\mu\nu}\dot{\omega}^{\mu\nu} - L, \tag{16.12}$$

where

$$M_{\mu\nu} \equiv \frac{\partial L}{\partial \dot{\omega}_{\mu\nu}} \tag{16.13}$$

is called the canonical angular momentum.

Note that the Hamiltonian expresses the energy of the system, whereas the Lagrangian represents the difference between the kinetic and potential energies. In order to elucidate Legendre's transform here, we have to express the total differential of the Lagrangian in terms of the spin structure and spin rate, i.e.

$$\mathrm{d}L = \frac{\partial L}{\partial \omega_{\mu\nu}} \mathrm{d}\omega^{\mu\nu} + \frac{\partial L}{\partial \dot{\omega}_{\mu\nu}} \mathrm{d}\dot{\omega}^{\mu\nu} . \tag{16.14}$$

By virtue of Eq. (16.13), the above relation can take the form

$$\mathrm{d}L = \dot{M}_{\mu\nu}\, \mathrm{d}\omega^{\mu\nu} + M_{\mu\nu}\, \mathrm{d}\dot{\omega}^{\mu\nu} , \tag{16.15}$$

because it follows from Lagrange's equations that $\partial L / \partial \omega_{\mu\nu} = \dot{M}_{\mu\nu}$. If we reshape the above relation as follows

$$\mathrm{d}L = \dot{M}_{\mu\nu}\, \mathrm{d}\omega^{\mu\nu} + \mathrm{d}(M_{\mu\nu}\dot{\omega}^{\mu\nu}) - \dot{\omega}^{\mu\nu}\, \mathrm{d}M_{\mu\nu} , \tag{16.16}$$

then, we move the expression $\mathrm{d}(M_{\mu\nu}\dot{\omega}^{\mu\nu})$ with the opposite sign to the LHS of Eq. (16.16), we obtain

$$\mathrm{d}(M_{\mu\nu}\dot{\omega}^{\mu\nu} - L) = -\dot{M}_{\mu\nu}\mathrm{d}\omega^{\mu\nu} + \dot{\omega}^{\mu\nu}\, \mathrm{d}M_{\mu\nu} . \tag{16.17}$$

Let us now take a closer look at the argument of the differential. It has a physical interpretation as the energy of the Lagrangian system. It is called the Hamilton's function or Hamiltonian of the system

$$H(M, \omega, t) = M_{\mu\nu}\dot{\omega}^{\mu\nu} - L. \tag{16.18}$$

Note that the Hamiltonian is expressed in terms of the spin structure and angular momentum. The energy determined by Eq. (16.18) is the foundation of the Hamiltonian theory. Using the Hamiltonian, from Eq. (16.17), we get

$$\mathrm{d}H = -\dot{M}_{\mu\nu}\, \mathrm{d}\omega^{\mu\nu} + \dot{\omega}^{\mu\nu}\, \mathrm{d}M_{\mu\nu} . \tag{16.19}$$

It should be emphasized that the above differential consists of two independent variables, i.e., the spin structure and angular momentum. Now, from the structure of the differential, we infer that

$$\dot{\omega}_{\mu\nu} = \frac{\partial H}{\partial M_{\mu\nu}}, \qquad \dot{M}_{\mu\nu} = -\frac{\partial H}{\partial \omega_{\mu\nu}}. \tag{16.20}$$

These equations of motion are called Hamilton's equations or canonical equations. It is a set of first-order differential equations. The unknown functions are here the components of the angular momentum $M_{\mu\nu}(t)$ and the components of the spin structure $\omega_{\mu\nu}(t)$. They can be treated as evolution equations for the spin structure and angular momentum.

In a particular case, when the Hamiltonian is not an explicit function of time, we get $\mathrm{d}H/\mathrm{d}t = 0$. This relationship expresses the law of conservation of energy.

So far, we were describing the mechanical system using the dynamical variables $\omega, \dot{\omega}$ or ω, M. In order to enrich our description, we can introduce to the Lagrangian and to the Hamiltonian some other parameters or variables which will give a better description of the mechanical system in question. Thus, in order to enrich our Lagrangian, we introduce the variable Ψ. In such a case, Eq. (16.15) takes the form

$$\mathrm{d}L = \dot{M}_{\mu\nu}\,\mathrm{d}\omega^{\mu\nu} + M_{\mu\nu}\,\mathrm{d}\dot{\omega}^{\mu\nu} + \frac{\partial L}{\partial \Psi}\mathrm{d}\Psi. \tag{16.21}$$

Consequently Eq. (16.19) yields

$$\mathrm{d}H = -\dot{M}_{\mu\nu}\,\mathrm{d}\omega^{\mu\nu} + \dot{\omega}^{\mu\nu}\,\mathrm{d}M_{\mu\nu} - \frac{\partial L}{\partial \Psi}\mathrm{d}\Psi. \tag{16.22}$$

Applying here partial differentiations, we keep other quantities constant. Following Landau and Lifshitz (1958, 1960), we can obtain

$$\left(\frac{\partial H}{\partial \Psi}\right)_{M,\omega} = -\left(\frac{\partial L}{\partial \Psi}\right)_{\dot{\omega},\omega}, \tag{16.23}$$

where the suffixes to the derivatives indicate the quantities which are to be kept constant during the differentiation.

We can illustrate this relationship in a very simple way. If we disturb the Lagrangian by adding a small value ΔL, then, the new Lagrangian will be of the form $L = L_1 + \Delta L$. Next, we disturb the Hamiltonian by corresponding addition of ΔH, and the new Hamiltonian will be $H = H_1 + \Delta H$, then, we obtain

$$\left(\Delta H\right)_{M,\omega} = -\left(\Delta L\right)_{\dot{\omega},\omega}. \tag{16.24}$$

Thus, we can compare this situation to the relationship between the potential and kinetic energies in the pendulum motion. When the potential energy of a swinging bob increases by a particular amount, then the kinetic energy decreases by the same amount. Thus, the total mechanical energy can be viewed as continuously shifting between the kinetic and potential forms.

It should be emphasized that time t can be treated as one of the parameters. So, by analogy to Eq. (16.23), the partial time derivatives of L and H are related by (Landau and Lifshitz 1958, 1960)

$$\left(\frac{\partial H}{\partial t}\right)_{M,\omega} = -\left(\frac{\partial L}{\partial t}\right)_{\dot{\omega},\omega}. \tag{16.25}$$

The Hamiltonian for a spin (twist) motion for a unit volume of mass of density ρ in the orthogonal Cartesian coordinates x^1, x^2, x^3 takes the form

$$H = \frac{1}{2\rho}\left(M_{x^1}^2 + M_{x^2}^2 + M_{x^3}^2\right) + U\left(x^1, x^2, x^3\right), \tag{16.26}$$

16.7 Conclusions

This chapter briefly presented the main results concerning the canonical approach to asymmetric media. Hamilton's Principle was the starting point of our considerations. Based on this principle, the equations of motion were formulated in terms of the spin structure and spin rates. On the basis of different variables taken into considerations, we pointed out that the Lagrangian and Hamiltonian approaches deal with the energy of the mechanical system and adding a new variable is equivalent to shifting energy between the Lagrangian and the Hamiltonian.

References

Kleinert H (1988) Lattice defect model with two successive melting transitions. Phys Lett **A130**: 443

Kleinert H (1989) Gauge Fields in Condensed Matter. World Scientific, Singapore

Kleinert H (2008) Gravity with torsion (in preparation)

Landau LD, Lifschitz EM (1958) The Classical Theory of Fields, Addison-Wesley, Reading, MA

Landau LD, Lifschitz EM (1960) Mechanics, Addison-Wesley, Reading, MA

Majewski E (2006a) Rotational energy and angular momentum of earthquakes. In: Teisseyre R, Takeo M, Majewski E (eds) Earthquake source asymmetry, structural media and rotation effects, Springer-Verlag, Berlin Heidelberg, pp 217-225

Majewski E (2006b) Seismic rotation waves: spin and twist solitons. In: Teisseyre R, Takeo M, Majewski E (eds) Earthquake source asymmetry, structural media and rotation effects, Springer-Verlag, Berlin Heidelberg, pp 255-272

Majewski E (2006c) Tectonic solitons propagating along the fault. In: Teisseyre R, Takeo M, Majewski E (eds) Earthquake source asymmetry, structural media and rotation effects, Springer-Verlag, Berlin Heidelberg, pp 301-309

Maugin GA (2003) Nonlinear wave mechanics of complex material systems. Proc Estonian Acad Sci Phys Math **52**: 5-11

Teisseyre R (2004) Spin and twist motions in a homogeneous elastic continuum and cross-band geometry of fracturing. Acta Geophys Pol **52**: 173-183

Teisseyre R, Boratyński W (2006) Deviations from symmetry and elasticity: asymmetric continuum mechanics. In: R Teisseyre, M Takeo, E Majewski (eds.), Earthquake source asymmetry, structural media and rotation effects, Springer-Verlag, Berlin Heidelberg, pp 31-41

Teisseyre R, Białecki M, Górski M (2006) Degenerated asymmetric continuum Theory. In: R Teisseyre, M Takeo, E Majewski (eds.), Earthquake source asymmetry, structural media and rotation effects, Springer-Verlag, Berlin Heidelberg, pp 43-55

PART III

DEFORMATIONS IN RIEMANNIAN GEOMETRY

17 Continuum Theory of Defects: Advanced Approaches

Hiroyuki Nagahama[1] and Roman Teisseyre[2]

[1] Department of Geoenvironmental Sciences, Graduate School of Science Tohoku University, Aoba-ku, Sendai 980-8578, JAPAN
e-mail: nagahama@dges.tohoku.ac.jp

[2] Institute of Geophysics, Polish Academy of Sciences
ul. Księcia Janusza 64, 01-452 Warszawa, Poland
e-mail: rt@igf.edu.pl

Theoretical descriptions of defect field based on differential geometry (Kröner 1981) or gauge theory (Edelen and Lagoudas 1988, Yamasaki and Nagahama 2002) are referred to as the continuum theory of defects. In the last few years, continuum theory of defects has created considerable interest in application to space and planetary sciences such as cosmic strings (e.g., Katanaev and Volovich 1999), Einstein-Cartan gravity (e.g., Hehl and Kröner 1965, Puntigam and Soleng 1997), seismicity (e.g., Teisseyre 1995c, Takeo and Ito 1997) and geodesy (Yamasaki and Nagahama 1999). In this chapter, we introduce the continuum theory of defects.

17.1 Geometry of Deformation

In the Riemannian space, a holonomic system of coordinates ξ^k (the Greek letter is for a variable, the Latin one for index) can be introduced following Teisseyre (1995a). In practical application, we will assume that the sequence of tangent spaces forms an Euclidean space, and that further transformation is possible from the coordinates ξ^k to coordinates describing the state of undeformed medium x^k (its natural state). Then we can see that the generalization of the problem of medium deformation leads to interesting geometric and topologic analogies.

Of utmost imporfance for us is to introduce the deformation described by the field of dislocational displacements. This means that when we choose a certain arbitrary, closed path in the real medium and proceed along it through the particular elements brought to the natural state, then we obtain in general an open contour.

Now, let us return to the question which interests us here, that is, to the real state of the medium. In the geometric sense, to the real state of the medium there corresponds a certain manifold. To this manifold, we will be able to introduce an affine connection field as well as a metric tensor field. In every point of our manifold, an Euclidean space tangent to it can be introduced. We demand that the sequence of those spaces form a Riemannian space and thus we define the medium in its intermediate state. In this intermediate state, it becomes possible to introduce a holonomic coordinate system ξ^k.

Thus, for real state elements, we can introduce in every point a local tangent Euclidean space, with a basis vector i_μ, where the Greek index letters refer to the real state of the medium's element. Obviously, the integrability of the basis vector field cannot be assumed in advance. The arc length for an element of a medium in the real state can be expressed by

$$dS = i_\mu d\xi^\mu, \tag{17.1}$$

where the local increments of the coordinates $d\xi^\mu$ and the corresponding increments in the tangent space are connected by the relationship

$$d\xi^\mu = c_k^\mu d\xi^k, \qquad d\xi^k = \overline{c}_\lambda^k d\xi^\lambda, \tag{17.2}$$

where

$$c_k^\mu \overline{c}_\lambda^k = \delta_\lambda^\mu, \qquad c_k^\mu \overline{c}_\mu^i = \delta_k^i, \tag{17.3}$$

Owing to the medium's inhomogeneity, the deformation (Eq. 17.2) is, in general, nonholonomic, but we postulate that the increments ξ^k form a holonomic system. The changes in basis vectors at passing to the adjacent point can be expressed by the relation

$$di_\mu = \Gamma_{\mu\lambda}^{\cdot\cdot\nu} d\xi^\lambda i_\nu = \Gamma_{\mu\lambda}^{\cdot\cdot\nu} i_\nu c_i^\lambda d\xi^i. \tag{17.4}$$

This relation defines the coefficients of affine connection $\Gamma_{\mu\lambda}^{\cdot\cdot\nu}$. Here the Greek indices refer to the curvilinear coordinates and the Latin ones to the Cartesian coordinates; Einstein's summation convention is used in respect to the indices appearing twice in one expression. In general, relations (17.2) and (17.4) are not integrable along the contour passing through the sequence of elements in the real state of the medium. The transformation c_i^ν is in general nonholonomic, and $(\partial / \partial x^{[k})c_{i]}^\nu \neq 0$. Therefore, the torsion tensor defined $S_{\delta\rho}^{\cdot\cdot\nu}$ by

$$S_{\delta\rho}^{\cdot\cdot\nu} = \Gamma_{[\delta\rho]}^{\cdot\cdot\nu} \tag{17.5}$$

differs from zero (Kondo 1955). Using the condition of holonomity for the ξ^k system, we get from Eq. (17.1)

$$\frac{\partial^2 S}{\partial \xi^k \partial \xi^i} - \frac{\partial^2 S}{\partial \xi^i \partial \xi^k} = 0, \tag{17.6}$$

and hence

$$\frac{\partial}{\partial \xi^k}\left(i_\mu c_i^\mu\right) - \frac{\partial}{\partial \xi^i}\left(i_\mu c_k^\mu\right) = 0. \tag{17.7}$$

Now, from Eqs. (17.4) and (17.5), we can get

$$S_{\delta\rho}^{\cdot\cdot\nu} = \Gamma_{[\delta\rho]}^{\cdot\cdot\nu} = -\frac{\partial}{\partial x^{[k}} c_{i]}^{\nu}\, \overline{c}_\rho^{k}\, \overline{c}_\delta^{i}. \tag{17.8}$$

The linear connection is as follows (Schouten 1954):

$$\Gamma_{\nu\lambda}^{\cdot\cdot\alpha} = \left\{ {\alpha \atop \nu\lambda} \right\} + S_{\nu\lambda}^{\cdot\cdot\alpha} - S_{\nu\cdot\lambda}^{\cdot\alpha} - S_{\cdot\lambda\nu\cdot}^{\alpha},$$
$$\left\{ {\alpha \atop \nu\lambda} \right\} \equiv \frac{1}{2} g^{\alpha\mu}\left(\frac{\partial g_{\mu\lambda}}{\partial \xi^\nu} + \frac{\partial g_{\mu\nu}}{\partial \xi^\lambda} - \frac{\partial g_{\nu\lambda}}{\partial \xi^\mu} \right). \tag{17.9}$$

$\left\{ {\alpha \atop \nu\lambda} \right\}$ is the Christoffel symbol and $g^{\alpha\mu}$ is the contravariant metric tensor in of a Riemannian (or non-Riemannian) space; it is usual to recall the Riemann-Cartan space time of the Eistein-Cartan theory (e.g., Hehl and Kröner 1965, Sabbata and Gasperini 1984, Sabbata and Sivaram 1994, Puntigam and Soleng 1997, Kawai 2000) or moving dislocation theory (Yamasaki and Nagahama 1999, 2002) when an affine asymmetric connection is introduced.

For a holonomic system, we have $S_{\delta\rho}^{\cdot\cdot\nu} = 0$, and the symmetry of the coefficients of connection $\Gamma_{[\delta\rho]}^{\cdot\cdot\nu} = 0$. If the torsion tensor vanishes, the connection coefficients are therefore equal to the Christoffel symbols (Riemannian connection).

$$R_{\nu\mu\lambda}^{\cdot\cdot\cdot\kappa} = 2\left(\partial_{[\nu}\Gamma_{\mu]\lambda}^{\kappa} + \Gamma_{[\nu|\rho|}^{\kappa}\Gamma_{\mu]\lambda}^{\rho}\right) + 2\Omega_{\nu\mu}^{\rho}\Gamma_{\rho\lambda}^{\kappa}, \tag{17.10}$$

where $\partial_\nu = \partial/\partial x^\nu$, $R_{\nu\mu\lambda}^{\cdot\cdot\cdot\kappa}$ is the Riemann-Christoffel curvature tensor, $\Omega_{\nu\mu}^{\rho}$ is the anholonomic object and $[\nu|\rho|\mu]$ means alternation in regard to μ and ν alone.

The deformation state can be defined by fifteen parameters. In the theory of elastic deformation, the full characteristic of the state is given by six parameters – components of the symmetric tensor $g_{\mu\rho}$. In the discussed general case of a continuous field of inhomogeneities (dislocations) we have, in addition to the six components of tensor $g_{\mu\nu}$, also nine components of the torsion tensor $S_{\delta\rho}^{\cdot\cdot\nu}$ which is antisysmetric in the indices ρ and δ (or – which is equivalent – nine coefficients of a nonholonomic deformation c_i^μ). Let us now consider a small closed circuit in a Riemannian space (no torsion) described by the differentials $d\xi^k$ (intermediate state). Applying now the nonholonomic transformation (Eq. 17.2), we obtained an open circuit in system $d\xi^\mu$ (real state).

Now we will express the quantities related to those transformations, that is:

- the resultant change $\Delta\xi^\mu$ which we obtain in proceeding along a closed contour in the real state of the medium;
- the resultant change $\Delta\,\mathbf{i}_\mu$ in basis vectors.

Those quantities express the previously mentioned inhomogeneity of the field of internal stresses related to the dislocation field. Finally, integrating relation (17.2) over a closed circuit in the x coordinate system we get the vector $\Delta\boldsymbol{\xi}$ for the full displacement in the form, according to Kondo (1955).

$$\Delta\boldsymbol{\xi} = \left(S_{\beta\alpha}^{\cdot\cdot\mu} + R_{\cdot\beta\nu\alpha}^{\mu}\xi^\nu\right) d\xi^\alpha d\xi^\beta \mathbf{i}_\mu . \tag{17.11}$$

The authors believe that here the correspondence between the nonholonomic transformation introduced by plastic distortions and the Riemannian-Christoffel curvature **R** and torsion **S** leads to the physically adequate definitions for dislocations and rotational dislocations (Kondo 1955, Teisseyre 1995a); after deformation for the disclosure of any closed (before deformation) circuit **L**, we get the increment of the displacement vector u^μ along a closed contour in generalized space describing the state of stress in the form:

$$[u^{\mu}]_L = \frac{1}{2} \in^{\nu\alpha\beta} \left(S_{\alpha\beta}^{\cdot\cdot\mu} + R_{\alpha\beta\lambda}^{\cdot\cdot\cdot\mu} u^{\lambda} \right) ds_{\nu}, \tag{17.12}$$

where $S_{\alpha\beta}^{\cdot\cdot\mu}$ and $R_{\alpha\beta\lambda}^{\cdot\cdot\cdot\mu}$ are the tensors of torsion and curvature of a specified non-Riemann space, and $d\mathbf{s}$ is the surface element related to the circuit.

Here we introduced a product of two antisymmetric tensors, so the division by 2 is required. Dividing the dislocation field into fields of Burgers type and rotational dislocations and introducing the density of dislocations α and density of rotational dislocation η, we obtain the density of Burgers dislocations (Kondo 1955, Teisseyre 1995a, Minagawa 1971, 1979):

$$[u^{\mu}]_L = \in^{\nu\alpha\beta} S_{\alpha\beta}^{\cdot\cdot\mu} ds_{\nu}, \qquad \alpha^{\mu\nu} = \in^{\nu\alpha\beta} S_{\alpha\beta}^{\cdot\cdot\mu}, \tag{17.13}$$

as well as the density of rotational dislocations (Teisseyre 1969)

$$[u^{\mu}]_L = \in_{\sigma\lambda}^{\mu} \eta^{\sigma\nu} u^{\lambda} ds_{\nu}, \qquad \eta^{\tau\nu} = \frac{1}{4} \in^{\tau\lambda\mu} \in^{\nu\alpha\beta} R_{\alpha\beta\lambda\nu}. \tag{17.14}$$

Here we shall note that in our definition the first index in dislocation density relates to its Burgers vector and the second one to the direction of its line (contrary to some other authors). The tensor $\eta^{\tau\nu}$ corresponds in the theory of Kröner (1958) to the incompatibility tensor I_{mn}.

The difference $\Delta\xi$ is therefore a measure of displacement of points of the medium in proceeding along a closed line, and constitutes the total value of the dislocation displacements. The density of those displacements as related to a surface unit is, of course, given by the expression in parentheses in Eq. (17.11). Density $S_{\alpha\beta}^{\cdot\cdot\mu}$ expresses the density of Burgers type dislocations, and $R_{\alpha\beta\lambda}^{\cdot\cdot\cdot\mu} \xi^{\nu}$ that of rotation type dislocations (disclinations). In general, those expressions correspond to the respective expressions for the Volterra dislocations:

$$u^k = b^k + \in^{kim} \omega_i x_m. \tag{17.15}$$

17.2 Deformation Measures and Incompatibility

In our approach, plastic distortions are the only sources of the incompatibilities. However, the plastic part of any field may be chosen arbitrarily; the definitions of defects, like dislocations and disclinations, shall be physically plausible and reasonable. Making the linear approximation after Kröner (1958), we get (Teisseyre 1995a):

$$\eta_{ik} = \in_{isr} \in_{kab} \frac{\partial^2}{\partial x_b \partial x_r} \left(\beta_{sa}^{PL} + \beta_{as}^{PL} \right),$$
$$\eta_{ik} = \in_{iab} \frac{\partial \alpha_{ak}}{\partial x_b} + \in_{kab} \frac{\partial \alpha_{ai}}{\partial x_b}, \tag{17.16}$$

$$\frac{\partial \alpha_{ik}}{\partial x_k} = 0, \qquad \eta_{ik} = \eta_{ki}, \qquad \frac{\partial \eta_{ik}}{\partial x_k} = 0. \tag{17.17}$$

Our definition of rotational dislocation indicates also that its line corresponds, for the particular case of distortion field $\beta_{ik}{}^{PL} = \beta\delta_{ik}$, to the dilatancy/compression line as formed by the surrounding field of the edge dislocations (Teisseyre 1969). In our approach the plastic distortions are the only sources of incompatibilities and the above definition of disclination density leads to its direct relation with the incompatibility tensor (Teisseyre 1995b):

$$\eta_{ij} = -2I_{ij}. \tag{17.18}$$

Some authors take a more general assumption that the plastic strain and the plastic (incompatible) rotation (to be more exact: the plastic equivalent of gradient of rotation) are the independent sources of incompatibility (DeWit 1971, Kossecka and DeWit 1977a, b); in our approach (Teisseyre 1995b), plastic strain and plastic rotation are related to plastic distortions

$$\varepsilon_{ik}^{PL} = \frac{1}{2} \left(\beta_{ik}^{PL} + \beta_{ki}^{PL} \right),$$
$$\omega_{ik}^{PL} = \frac{1}{2} \left(\beta_{ik}^{PL} - \beta_{ki}^{PL} \right). \tag{17.19}$$

In the definition by Kossecka and DeWit (1977a) the dislocations and disclinations are independent sources of incompatibilities. Their results are more general and apparently too complicated, but explain the relation between dislocation density α_{ik} and disclination density θ_{pq}, the latter being a source of dislocation lines:

$$\frac{\partial \alpha_{ik}}{\partial x_k} + \in_{ijk} \theta_{jk} = 0, \qquad \frac{\partial \theta_{ik}}{\partial x_k} = 0. \tag{17.20}$$

Note that in our convention the first tensor indexes both in dislocation density and disclinations refer to the Burgers vector and the second ones to vectors of density lines.

Earlier, Anthony et al. (1968) and Minagawa (1971, 1979) obtained the same relation by introducing the incompatible rotations for the Cosserat continuum (deformation in this continuum is characterized by displacement vector and micro-rotation tensor), where an independent role of plastic rotation increment, in this generalized continuum, justifies such an approach completely. As shown by Kossecka and DeWit (1977a), the general linear relation between the dislocations, disclinations and incompatibility is the following:

$$I_{ij} = \frac{1}{2}\left(- \in_{iab} \frac{\partial \alpha_{aj}}{\partial x_b} - \in_{jab} \frac{\partial \alpha_{ai}}{\partial x_k} + \theta_{ij} + \theta_{ji} \right). \tag{17.21}$$

Preserving the definition of rotation dislocations (17.15), we get

$$I_{ij} = \frac{1}{2}\eta_{ij} + \frac{1}{2}(\theta_{ij} + \theta_{ji}). \tag{17.22}$$

Independently, we shall note that the consideration by Kossecka and DeWit (1977a, b) on the loop dislocation density relate to its wrong definition. The definition of the disclination loop density is usually taken after Nabarro (1967), who probably introduced by a simple mistake the following definition:

$$\iint \alpha_{ik} dS_k = \oint \Delta_{ik} dx_k, \qquad \alpha_{is} = \in_{smk} \frac{\partial \Delta_{ik}}{\partial x_m}. \tag{17.23}$$

It has been shown that this definition is wrong (Teisseyre 1995c); the loop density Δ shall be derived from the integral relation

$$\iint \Delta_{ik} dS_k = \oint \alpha_{ik} dx_k \tag{17.24}$$

where on the left-hand side the loop dislocations are summed over a surface element (this corresponds to the well known synthesis of the dislocation fields presented earlier by Nabarro 1951) and the right-hand side represents the resulting dislocation line density encircling that surface. Further we get (Teissyre 1995c):

$$\Delta_{is} = \in_{smk} \frac{\partial \alpha_{ik}}{\partial x_m}. \tag{17.25}$$

Thus, it is not possible to identify the plastic distortion $\boldsymbol{\beta}^{PL}$ with the loop dislocation tensor $\boldsymbol{\Delta}$ as proposed by Kossecka and DeWit (1977a).

In our approach (plastic distortions $\boldsymbol{\beta}^{PL}$ are the only sources of incompatibilities and $\boldsymbol{\theta} = 0$) we can derive the field of rotational dislocations from the dislocation density by relation (17.16) and similarly to Eq. (17.25) we might define the loop density of disclinations from density of disclination lines.

Another question which we would like to rise is the use of scalar density function for dislocation fields as proposed by Aifantis (1982, 1984, 1985, 1986, 1987). Scalar density can be used either for the specified unique component of dislocations or for isotropic distribution of dislocations. Dislocations move and evolve under the influence of an external stress field and under mutual interactions; only for the case of pressure load we can believe that the distribution of all dislocation density tensor components is isotropic.

17.3 Evolution Equations for Stresses and Dislocations

The system of evolution equations (Teisseyre 1990, Teisseyre and Czechowski 1993, Czechowski et al. 1994, 1995) consists of the continuity condition for line dislocations:

$$\frac{\partial \alpha_{ik}}{\partial t} + \frac{\partial}{\partial x_j}(\alpha_{ik} V_j) = \Pi_{ik}, \tag{17.26}$$

where $\boldsymbol{\Pi}$ denotes the dislocation source/sink function; $\mathbf{V} = \mathbf{v}/c$ is the relative velocity; $\mathbf{v}$ is the dislocation velocity and c is the shear wave velocity.

This relation shall be supplemented by the expression for relative velocity of motion of dislocation density field (Magata et al. 1987), written by Teisseyre (1996) in the tensorial form (S_{in} and R_{in} are the stresses and stress resistances, respectively):

$$V_j = \in_{jnk} \frac{\alpha_{ik}}{|\alpha_{ik}|} + \frac{S_{in} - R_{in}}{\sqrt{\left|R_{in}{}^2 + \left(S_{in} - R_{in}\right)^2\right|}} \tag{17.27}$$

and by the differential relation between stresses and dislocation density tensor:

$$\alpha_{mj} - \frac{1}{2}\alpha_{ss}\delta_{mj} = -\frac{1}{2\mu} \in_{jtn} \frac{\partial}{\partial x_t}\left(S_{mn} - \frac{\nu}{1+\nu} S_{ss}\delta_{mn}\right), \tag{17.28}$$

where μ is the rigidity modulus and ν is the Poisson ratio.

However, we shall note that a more general relation for the stresses and dislocations shall include the asymmetric stresses (equivalent to stress moments (see: Chap. 7).

The form of Eqs. (17.26) and (17.27) permits, with the help of (17.28), to integrate them (eliminating dislocation density) in order to get the differential evolution equation for stresses (Czechowski et al. 1994); in another approach we can eliminate stresses (Teisseyre 1997) and get a system describing the evolution of dislocation density.

17.4 Source/Sink Functions of Dislocation Density

On the other hand, in his approaches to the choice of a source/sink function by using the dislocation density tensor, Mura (1963) proposed the following local balance law

$$\frac{\partial \alpha_{ik}}{\partial t} + \frac{\partial}{\partial x_l}(V_{ikl} - V_{ilk}) = 0 \tag{17.29}$$

using the notation $V_{ikl} = c\alpha_{ik}V_l$.

The same form is discussed by Teisseyre (1990, 1995c) with the additional source/sink term

$$\frac{\partial \alpha_{ik}}{\partial t} + c\frac{\partial}{\partial x_l}\left(\alpha_{ik}V_l - \alpha_{il}V_k\right) = \Pi_{ik}. \tag{17.30}$$

Upon assuming $\partial V_k / \partial x_l = 0$ (here index k relates to line of dislocation) and noting that $\partial \alpha_{il} / \partial x_l = 0,$ this equation leads to (17.26).

For the elasto-plastic continuum body, Teodosiu (1970) derived the equivalent continuity condition of the defect density which is expressed by the "dislocation flux", $J_{it} = c \in_{qnt} \alpha_{iq} V_n$, as follows:

$$\frac{\partial \alpha_{ik}}{\partial t} + \in_{kjt} \frac{\partial}{\partial x_j} J_{it} = 0. \tag{17.31}$$

His definition of **J** does not account for plastic velocity of dislocations; Kossecka and DeWitt (1977a, b) proposed the dislocation current tensor defined as difference between time and space derivatives of plastic distortion and plastic velocity as follows:

$$J_{it} = \frac{\partial \beta_{it}^{PL}}{\partial t} - \frac{\partial}{\partial x_t} v_i^{PL}. \tag{17.32}$$

However this equation has not accounted for local creation or annihilation of dislocations.

Lardner (1969a, b) has considered effects of local creation or annihilation of dislocations as in Eq. (17.16) with the source/sink function of a form which could be motivated by the work of Webster (1966a, b). However, they have not proposed any concrete tensor form of this function; the same remark concerns the works by Werne and Kelly (1978) and Caglioti and Bottani (1988).

Teisseyre (1990, 1995c) proposed proportionality P of the source/sink function to the dislocation current:

$$\Pi_{ik} = P\frac{\partial}{\partial x_t}\left(\alpha_{ik}V_l - \alpha_{il}V_k\right), \tag{17.33}$$

while Teisseyre and Nagahama (1997) proposed another formula for Π:

$$\Pi_{ik} = P_0\delta_{ik} + P_l\alpha_{ik} + P_2 \in_{kjs} \alpha_{is}V_j\,. \tag{17.34}$$

This form is supplemented with the production and coalescence rates of dislocations with the respective probability coefficients P_0, P_1 and P_2, which in general may depend on the space coordinates (due to material properties) and on time (due to creep damage processes). The last form of source/sink function is similar to that of the scalar rate equation of dislocation densities from the work of Webster (1966a, b). These tensor rate equations of dislocation densities, (17.33) and (17.34), have been proposed under the assumption of the distant parallelism criterion which means the absence of disclination.

A new form of the source/sink function is discussed by Teisseyre (1997) and is linked to the rebound mechanism of earthquakes. Moreover, the numerical simulations of an interacting model with the source/sink of dislocations reproduce the self-organized intermittency of dislocation movements (Miguel et al. 2001).

Anthony et al. (1968) obtained the relation between the dislocations and disclinations as presented in Eq. (17.20); considering the Cosserat continuum he considered the dislocation density tensor and disclination density tensor. When dealing with dislocations alone, Eq. (17.20) should reduce to Eq. (17.17) and therefore the disclination term is compared with the distribution of sources or sinks of dislocation lines (more on the sources or sinks of dislocation lines: Minagawa 1971). Similarly, DeWitt (1971) and Harris and Scriven (1971) have also considered the relation between disclinations and sources of dislocation line from the crystallographic point of view.

Consideration of the Riemann-Christoffel curvature tensor of a Riemannian (or non-Riemannian) space, under the assumption of non-distant parallelism criterion, may lead us to the relation between the incompatible tensor and the Riemann-Christoffel curvature tensor in the form suitable for exploring the defect properties in the material:

$$\eta^{\lambda\chi} + \theta^{\lambda\chi} = \frac{1}{4} \in^{\lambda\nu\mu} \in^{\pi\chi\rho} R_{\nu\mu\tau\rho} . \tag{17.35}$$

Under this assumption, Minagawa (1971, 1979) expressed the density of the sources of dislocations given by

$$R^{\chi} = \frac{1}{2} \in^{\lambda\nu\mu} R_{\lambda\nu\mu}^{\cdot\cdot\cdot\chi} . \tag{17.36}$$

By substituting Eq. (17.36) into Eq. (17.35) we obtain (cf. Minagawa 1971, 1979):

$$R_{\chi} = \in_{\lambda\chi\mu} (\theta^{\lambda\mu} + \eta^{\lambda\mu}). \tag{17.37}$$

This equation means that the sources of dislocations are converted into a distribution of disclination and line centers (rotational density). This role of disclinations has also been discussed by DeWitt (1971) from the crystallographic point of view.

17.5 Virtual Tearing (Kondo 1964)

At any rate, any real plastic manifold with imperfections can be torn into a teleparallelism (a distant parallelism) by virtual tearing (Kondo 1964). Kondo's (1964) concept of a teleparallelism by virtual tearing is introduced below.

An important case of untearability or imperfect tearing is observed in the Riemannian approach to of all plastic disturbances including dislocations or disclinations into the so-called incompatibility tensor. It should first be referred to *a-posteriori* continuum without tearing with a structure which is entirely Riemannian. Its curvature is, therefore,

$$K_{\nu\mu\lambda}^{\cdot\cdot\cdot\kappa} = 2\left(\frac{\partial}{\partial x^{[\nu}} \left\{ {\kappa \atop \mu]\lambda} \right\} + \left\{ {\kappa \atop [\nu|\rho|} \right\} \left\{ {\rho \atop \mu]\lambda} \right\} \right). \tag{17.38}$$

By a virtual tearing, the manifold can be brought to a teleparallelism so that the non-Riemannian curvature tensor, formally defined by $R_{\nu\mu\lambda}^{\cdot\cdot\cdot\kappa}$ as above, needs to vanish

$$R_{\nu\mu\lambda}^{\cdot\cdot\cdot\kappa} \equiv K_{\nu\mu\lambda}^{\cdot\cdot\cdot\kappa} + N_{\nu\mu\lambda}^{\cdot\cdot\cdot\kappa} = 0, \tag{17.39}$$

where $N_{\nu\mu\lambda}^{\cdot\cdot\cdot\kappa}$ is originated by the tearing. This equation represents an equation of compatibility in dislocation theory.

In three dimensions, nothing is lost from the non-Riemannian or incompatible metric characteristic represented completely by replacing it by a physically more significant

$$G_{\mu\lambda} = K_{\mu\lambda} - \frac{1}{2} K g_{\mu\lambda}, \tag{17.40}$$

where $K_{\mu\lambda}$ is the contracted curvature tensor, or Ricci tensor, and K the scalar curvature defined respectively by

$$K_{\mu\lambda} = K_{\kappa\mu\lambda}^{\cdot\cdot\cdot\kappa} \quad \text{and} \quad K = g^{\lambda\mu} K_{\mu\lambda} \tag{17.41}$$

(see McConnel 1931). Equation (17.40) has been derived by Turski (1966) from a variational principle or gauge theory (Utiyama 1956).

Hence, the contracted equation

$$G_{\mu\lambda} = -M_{\mu\lambda} \tag{17.42}$$

can be substituted for Eq. (17.39), where

$$\begin{aligned} M_{\mu\lambda} &= N_{\mu\lambda} - \frac{1}{2} g_{\mu\lambda} N, \\ N_{\mu\lambda} &= N_{\kappa\mu\lambda}^{\cdot\cdot\cdot\kappa}, \quad N = g^{\lambda\mu} N_{\mu\lambda}. \end{aligned} \tag{17.43}$$

The tensor $M_{\mu\lambda}$, as well as $N_{\kappa\mu\lambda}^{\cdot\cdot\cdot\kappa}$, depends on the tearing which needs to be perfect. If it is assumed as above, Eq. (17.42) is used for determining the internal strain expressed by the tensor $\varepsilon_{(\mu\lambda)} \equiv \frac{1}{2}(g_{\mu\lambda} - \delta_{\mu\lambda})$ included in $G_{\mu\lambda}$. Hence, the latter is properly called the incompatibility tensor. Its four-dimensional analogue or the four-dimensional counterpart of $M_{\mu\lambda}$ is the general-relativistic 'material-energy tensor', as is well known. Tensor $G_{\mu\lambda}$ is obviously the Einstein tensor. It is non-divergent in the sense

$$\overset{*}{\nabla}_{\kappa} G_{\cdot\lambda}^{\kappa} = 0, \tag{17.44}$$

where $\overset{*}{\nabla}_\kappa$ gives the covariant derivative with $\left\{ {\kappa \atop \mu\lambda} \right\}$ as the parameters of connection.

The structure of the virtual teleparallelism permits us to assume that the incompatibility tensor $G_{\mu\lambda}$ can be expressed in terms of the torsion tensor field of true or virtual dislocation distributions as by Eq. (17.42). It is important to recognize this in respect of the analysis of the problem in which all distributed sources of internal stresses, whether dislocational or not, are virtually lumped into an incompatibility tensor.

Sakata (1970) grasped Finslerian imperfections in plastic materials by the metrics $g(x_\kappa, \dot{x}_\kappa)$, where $\dot{x}_\kappa$ is the field of direction variables (the osculation) as a function of x_κ. Then, by averaging microscopic Finslerian imperfections $\overline{g(x_\kappa, \dot{x}_\kappa)}$, he found non-Riemannian manifold (space) in consequence of averaging those features from the macroscopic point of view. From a similar view point, by averaging the micro-gravitational fields, Ikeda (1979) derived macro-gravity fields as a non-Riemannian space. Then under the assumption that this macro-field has a teleparallelism, the Einsteinian formalism can be described by Eq. (17.40).

17.6 High-Order Spaces and Non-Locality of Deformation

In this chapter, we will reconsider the continuum with microstructures taking into account the concepts of non-locality, asymmetry and inner-rotation of deformation from the viewpoint of the differential geometry of high-order spaces.

The high-order space (or the Kawaguchi space: Kawaguchi 1931, 1937, 1962) of order M (= 1, 2, 3, ...) is a metrical space $K_n^{(M)}$ in which the arc lengths along a curve $x^\kappa = x^\kappa(t)$ (t is an arbitrary parameter) are given by the integral

$$S = \int F(\mathbf{x}, \mathbf{x}^{(1)}, \mathbf{x}^{(2)}, \mathbf{x}^{(3)}, \ldots, \mathbf{x}^{(M)})\, dt, \qquad (17.45)$$

where F means the fundamental function satisfying some homogeneity conditions and where $x^{(\alpha)i}$ are the independent internal variables:

$$\mathbf{x}^{(\alpha)} \equiv d^\alpha \mathbf{x} / dt^\alpha; \quad \text{or} \quad x^{(\alpha)i} \equiv d^\alpha x^i / dt^\alpha;$$
$$\alpha = 1, 2, \ldots, M \leq n - 1.$$

Of course, this space is regarded as a generalized Riemannian or Finsler space non-localized by $\mathbf{x}^{(\alpha)}$ (here $\mathbf{x}^{(1)}$ is a vector, but $\mathbf{x}^{(\alpha)}(\alpha \geq 2)$ are the tensors). It turns out that the Riemannian space is a higher-order space of order 0, the Finsler space is a higher-order space of order 1, and the Cartan space is a higher-order space of order $\alpha = n - 1$. Moreover, when we regard the non-Riemannian space and $\mathbf{x}^{(\alpha)}$ as a base space and a fiber, respectively, this high-order space is a fiber bundle space (Kawaguchi 1931, 1937, 1962).

We shall consider a geometrical grasp of the inherent law of the independent internal variables. At first, we shall write the related laws for $x^{(\alpha)\kappa}$ in the form

$$\delta x^{(\alpha)\kappa} = M_{\lambda}^{(\alpha)\kappa} \left(dx^{\lambda} + \sum_{\beta=1}^{\alpha} N_{(\beta)\mu}^{\lambda} \, dx^{(\beta)\mu} \right), \qquad (17.46)$$

which are essentially regarded as the base connections of high order spaces (Kawaguchi 1962); in this equation, $M_{\lambda}^{(\alpha)\kappa}$ and $N_{(\beta)\mu}^{\lambda}$ represent the interactions between each order of the internal variables.

Concerning the concept of “non-locality”, this is carried by the internal variables such as $\mathbf{x}^{(\alpha)}$, so that a “non-local” field (Yukawa 1950) can be obtained by attaching an internal variable to each point of a local (or Riemannian) field. This way of thinking descends from the theory of high order spaces (Kawaguchi 1931, 1962).

Finsler space can be regarded as a generalization of Cosserat continua (Cosserat and Cosserat 1909). When an independent internal variable $\mathbf{x}^{(1)}$ is a vector attached to each point of the continuum, the space can be regarded as an ordinary Cosserat continuum. When an independent internal variable $\mathbf{x}^{(1)}$ is a deformable director attached to each point of the continuum, we can get the theory of continuum mechanics of oriented media (Ericksen and Trusdell 1958). $\mathbf{x}^{(1)}$ is regarded as an osculation in the microscopic Finslerian imperfections (Sakata 1970). Moreover, when an independent internal variable $\mathbf{x}^{(1)}$ is a tensor attached to each point of the continuum, we can derive the multipolar theory (Green and Rivlin 1964). More generally, when an independent internal variable $\mathbf{x}^{(1)}$ is an m-dimensional manifold attached to each point of the continuum, we can derive Capriz’s continuum with microstructures (Capriz 1989) or space-time as a micromorphic continuum (Sławianowski 1990).

17.7 Interaction Between Microscopic and Macroscopic Fields: Comparison Between the Different Approaches

In this section, we shall derive an interaction field (Ikeda 1985) between microscopic and macroscopic fields based on the base connections of high order spaces. This interaction field is the physical interaction field (Ikeda 1972) and the mathematical method is equivalent to the contact tensor calculus proposed by Yano and Davies (1954).

At first, we shall consider an interaction field between microscopic deformation field $\zeta^i (\equiv x^{(1)i})$ and macroscopic deformation field x^λ (in our notation the Latin index letter refers to the microscopic deformation field and the Greek one to the macroscopic deformation field). When microscopic deformation fields ζ^j satisfy the inherent laws, we can put $\delta\zeta^j = 0$. In this case, the base connection becomes

$$dx^\lambda = A_i^\lambda d\zeta^i, \tag{17.47}$$

where A_i^λ is the interaction coefficient between microscopic and macroscopic deformation fields and is equivalent to $N^\lambda_{(1)i}$. The interaction coefficient A_i^λ is non-symmetrical in general.

Next, based on the differential geometry methods (Kondo 1953), we shall consider geometrical backgrounds of the inner-rotation. If small disturbances alone are considered, the deformation state of macroscopic deformation field can be grasped by the metric $g_{\lambda\kappa}$ and the coefficient of connection after deformation in the form:

$$g_{\lambda\kappa} = A_\lambda^j A_\kappa^i \delta_{ji} = \delta_{\lambda\kappa} + 2\varepsilon_{(\lambda\kappa)}, \quad \varepsilon_{\lambda\kappa} = \beta_\lambda^i \delta_{i\kappa}, \tag{17.48}$$

$$\Gamma^\kappa_{\mu\lambda} = A_i^\kappa \partial_\mu A_\lambda^i = \partial_\mu \varepsilon_\lambda^\kappa, \quad \varepsilon_\lambda^\kappa \equiv \beta_\lambda^i \delta_i^\kappa, \tag{17.49}$$

where $\varepsilon_{(\lambda\kappa)}$ is an ordinary strain, β_λ^i represents deformation.

The antisymmetric strain does not represent the metric imperfection, but the rotational characteristic of microelements. Moreover, using the symmetric and antisymmetric strains, the coefficient of connection after deformation can be expressed in the form:

$$\Gamma_{\mu\lambda\kappa} = \partial_\mu \varepsilon_{\lambda\kappa} = \partial_\mu \varepsilon_{(\lambda\kappa)} + \partial_\mu \varepsilon_{[\lambda\kappa]}. \tag{17.50}$$

Thus, antisymmetric strains affect the metric imperfection, but the coefficient of connection after deformation remains unchanged. Let us introduce the inner-rotation $\varphi_{\kappa\lambda}$; the relation between the strain and the inner -rotation is generally given by

$$\varepsilon_{\lambda\kappa} = \partial_\lambda u_\kappa + \varphi_{\kappa\lambda}. \tag{17.51}$$

Therefore, the object of anholonomity in the macroscopic deformation field after deformation can be expressed by

$$\Omega_{\mu\lambda\kappa} = -\Gamma_{[\mu\lambda]}\kappa = -\partial_{[\mu}\varphi_{\lambda]\kappa}. \tag{17.52}$$

This shows the geometrical object for the inner-rotation.

In the so-called Cosserat continuum (Cosserat and Cosserat 1909), the deformation and rotation tensors are given by

$$\begin{aligned} \varepsilon_{(\lambda\kappa)} &= \partial_{(\lambda} u_{\kappa)} + \varphi_{(\kappa\lambda)} = \partial_{(\lambda} u_{\kappa)}, \\ \varepsilon_{[\lambda\kappa]} &= \partial_{[\lambda} u_{\kappa]} + \varphi_{[\lambda\kappa]} = \partial_{[\lambda} u_{\kappa]} + \omega_{\lambda\kappa}, \end{aligned} \tag{17.53}$$

where u_κ is the displacement of a material element and $\omega_{\lambda\kappa}$ is the rotation independent of the rotation $\partial_{[\lambda} u_{\kappa]}$ originating from the displacement. In the micromorphic continuum theory (Suhubi and Eringen 1964), the deformations are represented not only by the displacement vector u_κ, but also by a new tensor that describes deformations and rotations of microelements (e.g., grains, blocks or some internal surface defects). It is a microdisplacement tensor $\phi_{\lambda\kappa}$. The deformation can be now expressed by the following strain measures (e.g. Eringen 1968):

strain tensor

$$e_{\lambda\kappa} = \partial_\lambda u_\kappa, \tag{17.54}$$

microstrain tensor

$$\varepsilon_{\lambda\kappa} = \partial_\lambda u_\kappa + \phi_{\lambda\kappa}, \tag{17.55}$$

microstrain moment tensor

$$\gamma_{\kappa\lambda\mu} = -\partial_\mu \phi_{\lambda\kappa}. \tag{17.56}$$

The microdisplacement tensor $\varphi_{\lambda\kappa}$ represents the relative deformation between microscopic and macroscopic fields. Regarding the microdisplacement tensor $\varphi_{\lambda\kappa}$ as the inner-rotation $\phi_{\lambda\kappa}$, Eq. (17.55) is equivalent to Eq. (17.53). In this case, the order of the microstrain moment is the order of the coefficient of connection, and microdisplacement tensor $\varphi_{\lambda\kappa}$ plays an important role as the inner-rotation or the object of anholonomity in the macroscopic deformation field after deformation.

In Chapter 7 we have presented a new approach to asymmetric continuum with the structural indexes; such an approach is partly equivalent to that used in the Kröner method with the self fields and also that in micropolar theories.

We can write some conditions presenting these equivalences for the stresses, strains and rotations. In the Kröner method we put:

$$S_{ks} = S_{ks}^{0} - S_{ks}^{S}, \quad E_{ks} = E_{ks}^{0} - E_{ks}^{S}, \quad \omega_{ks} = \omega_{ks}^{0} - \omega_{ks}^{S}, \tag{17.57}$$

where we write

- for the elastic (physical) fields: S_{ks}, E_{ks}, ω_{ks};
- for the total fields: $S_{ks}^{T} = S_{ks}^{0}$, $E_{ks}^{T} = E_{ks}^{0}$, $\omega_{ks}^{T} = \omega_{ks}^{0}$;;
- for the self fields: S_{ks}^{S}, E_{ks}^{S}, ω_{ks}^{S}..

In the standard asymmetric theory we have (Chap. 7):

$$S_{ks} = S_{ks}^{0} + S_{[ks]}, \qquad E_{ks} = e^{0} E_{ks}^{0}, \qquad \omega_{ks} = \chi^{0} \omega_{ks}^{0} \tag{17.58}$$

and for the deformation tensor $D_{ks} = E_{ks} + \omega_{ks} = e^{0} E_{ks}^{0} + \chi^{0} \omega_{ks}^{0}$.

In comparison with the self fields we would obtain only the restrained values of these fields:

$$S_{[ks]} \to -S_{ks}^{S}, \qquad (e^{0} - 1) E_{ks}^{0} \to -E_{ks}^{S}, \qquad (\chi^{0} - 1)\omega_{ks}^{0} \to -\omega_{ks}^{S}. \tag{17.59}$$

The comparison with the micromorphic theories brings for the relations (17.54)-(17.56):

$$e_{(ks)} = E_{ks}^{0}, \qquad \varphi_{ks} = \phi_{lk} = \chi^{0} \omega_{ks}^{0}. \tag{17.60}$$

17.8 Asymmetric and Anholonomic Deformation

Based on the base connections of high order spaces and the non-locality of deformation, an interaction field between microscopic and macroscopic deformation fields can be grasped by the $(x^{\lambda}, A_{i}^{\lambda})$-fields as non-local fields. Moreover, based on the differential geometry methods (Kondo

1953), the inner-rotations as asymmetric fields can be derived from the (x^λ, A_i^λ)-fields, and the geometrical backgrounds of the inner-rotation are considered. Therefore, the interaction coefficient A_i^λ can be used to investigate the correlation between the macroscopic deformation field and the microscopic deformation field with microstructures. The metric and the coefficient of connection in strain space are dual to the stress and couple stress in stress space, respectively (Amari and Kagekawa 1964, Yamasakai and Nagahama 1999, 2002). This means that the asymmetrical fields originate from the non-local fields: an interaction field between microscopic and macroscopic deformation fields is a non-local field with internal variables as the inner-rotations $\phi^{\lambda\kappa}$ or $x^{(\alpha)}$. This internal variable $x^{(\alpha)}$-dependence is combined, in general, not only with the concept of "non-locality" but also the concept of "anisotropy" (Takano 1968).

In the micromorphic continuum theory (Suhubi and Eringen 1964), owing to the axiom of affine motion, and by analogy with the deformation gradients, the microdisplacement as the inner-rotation is linked to the relative deformation defined by the position of a material point of the microvolume relative to the center of mass of the macrovolume of the body. In this case, non-linear or irreversible behaviours with the inner-rotation have not been clear and not been linked to anholonomity in an interaction field between microscopic and macroscopic deformation fields. According to the theory of the physical interaction field (Ikeda 1972, 1975, Muto and Nagahama 2004), the inner-rotation plays an important role in the object of anholonomity in the macroscopic deformation field after deformation. The object of anholonomity can describe the non-linear or irreversible behaviour of an interaction field between microscopic and macroscopic fields, because the interaction coefficient A_i^λ is non-symmetric in general.

The Lagrangian of these defects is invariant with respect to three-dimensional rotations $SO(3)$ and spatial translations $T(3)$ (Kadić and Edelen 1983), and the deformation of micromorphic structure (continuum) induces the appearance of dislocations and disclinations, $SO(3) \triangleright T(3)$. These structural defects are related to anholonomity caused by the inner-rotation (microdisplacement or microstrains moment) in the form:

$$\Lambda_{\kappa\lambda} = -\in_{\lambda\mu\nu} \partial_\mu \phi_{\kappa\nu}, \tag{17.61}$$

where $\Lambda_{\kappa\lambda}$ denotes the microdislocation density (Nagahama and Teisseyre 2001b), $\in_{\lambda\mu\nu}$ is Eddington's epsilon (the skew-symmetric tensor; 0, 1, –1).

Therefore, the internal nuclei (dislocations, disclinations, vacancies, thermal nuclei or electric nuclei) are the objects/sources that create internal stresses (self stresses) (Teisseyre 2002).

17.9 Micromorphic Continuum with Defects

In the micromorphic continuum it is assumed that the body possesses a certain microstructure. After Suhubi and Eringen (1964) and Eringen and Claus (1970), we define the following strain measures (see also Teisseyre 1973, 1974, 1995d, Teisseyre and Nagahama 1999, Nagahama and Teisseyre 1998, 2000, 2001a, b): strain tensor e_{nl}, microstrain tensor ε_{nl}, and microstrain moment tensor γ_{klm}. From the compatibility conditions for the strains and microstrains, we obtain

$$e_{nl} = \frac{1}{2}\left(\frac{\partial u_n}{\partial x_l} + \frac{\partial u_l}{\partial x_n}\right), \qquad \varepsilon_{nl} = \frac{\partial u_l}{\partial x_n} + \phi_{nl}, \qquad \gamma_{klm} = -\frac{\partial \phi_{kl}}{\partial x_m}, \tag{17.62}$$

where u_n is a displacement and φ_{kl} is a microdisplacement. Here we will confine ourselves to the linear theory and Cartesian coordinate system. For rotation, we have

$$\omega_{ij} = \frac{1}{2}\left(\frac{\partial u_i}{\partial x_j} - \frac{\partial u_j}{\partial x_i}\right), \tag{17.63}$$

while the bend twist (gradient of rotation) is defined as follows:

$$\kappa_{mi} = \frac{\partial \omega_i}{\partial x_m} = \frac{1}{2} \in_{ikl} \frac{\partial u_l}{\partial x_k \partial x_m}, \qquad \omega_i = \frac{1}{2} \in_{ijk} \omega_{jk}. \tag{17.64}$$

Moreover, from Eq. (17.62), the total strain is obtained by the sum of two strains: macrostrain and microstrain:

$$e^t_{ij} = e_{ij} + \omega_{ij} + \phi_{ij}. \tag{17.65}$$

Now we assume that the microdisplacements φ_{ij} and the microstrain moment tensor γ_{ijk} are independent sources of incompatibilities (Nagahama and Teisseyre 2001a: case II). In such a case, we obtain

$$\alpha_{sn} = -\frac{1}{2} \in_{smk} \left(\frac{\partial \phi_{kn}}{\partial x_m}\right). \tag{17.66}$$

In this case, the disclinations also appear and we shall relate the disclinations to moments of microstrains:

$$\kappa_{kq} = \in_{qst} \gamma_{stk}, \qquad \theta_{pq} = -\in_{pmk} \frac{\partial \kappa_{kq}}{\partial x_m} = -\in_{pmk} \in_{qst} \frac{\partial \gamma_{stk}}{\partial x_m}. \tag{17.67}$$

From Eqs. (17.66) and (17.67), dislocation density α_{sn} and disclination density θ_{pq} are related to the microdisplacement ϕ_{ij}, and the deformation of the micromorphic continuum leads to the appearance of defects (i.e., dislocations and disclinations).

17.10 Taylor-Bishop-Hill Model

The derived formula on the theory of micromorphic continuum with defects is applied to lattice preferred orientation (LPO) of polycrystals described by Taylor-Bishop-Hill (TBH) model. In this chapter, we briefly introduce the TBH model (Taylor 1938, Bishop and Hill 1951, see also Gil Sevillano et al. 1980, Van Houtte and Wagner 1985, Fleck et al. 1994).

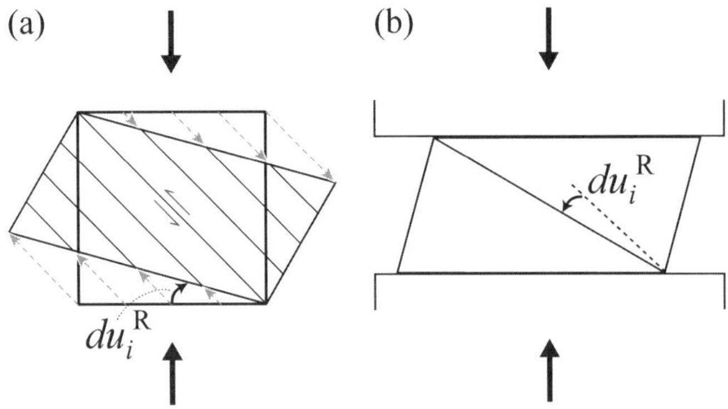

Fig. 17.1 The schematic model for additional crystalline lattice rotation with the intracrystalline slip undergone. (a) The case when a crystal is not constrained by the external displacement. The base of the square rotates clockwise following the torque of the intracrystalline slip. (b) The case when the bases of the crystal are fixed by the piston. In this boundary condition, the crystal rotates counterclockwise against the torque of the intracrystalline slip. In polycrystal, the piston is replaced with other crystals (modified from Wenk et al. 1986)

External rotation must be created in the case of polycrystal deformation, because the deformation of each grain in polycrystals is constrained by the surrounding grains (Fig. 17.1). The simplest model for the constrained deformation in polycrystals is the Taylor homogeneous deformation hypothesis. Thus, the development of LPO is a natural outcome in the TBH model. In the TBH model, we assume that the material deforms through the crystal lattice by intracrystalline slip and the lattice undergoes rotation. The basic relation in TBH model used extensively in plasticity of polycrystals is given by

$$d\tilde{u}_i = d\tilde{u}_i^S + d\tilde{u}_i^R, \tag{17.68}$$

where

$$d\tilde{u}_i^S \equiv \tau_{ij}\, dx_j, \qquad d\tilde{u}_i^R \equiv \phi_{ij}\, d\tilde{x}_j. \tag{17.69}$$

Here, $d\tilde{u}_i^S$ is the relative displacement due to intracrystalline slip and $d\tilde{u}_i^R$ is due to additional lattice rotation for bringing the crystal lattice to rotate. In Eq. (17.69), $d\tilde{u}_i^S$ is linearly related to $d\tilde{x}_j$ via the slip tensor τ_{ij}, and $d\tilde{u}_i^R$ is related to $d\tilde{x}_j$ via the rotation tensor ϕ_{ij}. A particular slip system (α) is specified by the slip vector s_i and the vector of slip plane normal n_j. The slip tensor τ_{ij} is associated with an amount of slip $\tau^{(\alpha)}$ on each of the active slip systems, hence

$$\tau_{ij} = \sum_{\alpha} \tau^{(a)} s_i^{(a)} n_j^{(a)}, \tag{17.70}$$

where the summation is taken over all active slip systems.

The physical meaning of these equations is that the imposed strain in each crystal du_i (i.e. macroscopic strain in the Taylor model) can be accommodated by the strain created by multiple slips in crystals $d\tilde{u}_i^S$ and the additional lattice rotation generating the LPO $d\tilde{u}_i^R$. In other words, strain compatibility leads to the appearance of lattice rotation. From Eqs. (17.65) and (17.68), we have the correspondences of variables, i.e.

$$d\tilde{u}_i \Leftrightarrow e_{ij}^t, \qquad d\tilde{u}_i^S \Leftrightarrow e_{ij} + \omega_{ij}, \qquad d\tilde{u}_i^R \Leftrightarrow \phi_{ij}. \tag{17.71}$$

The one-to-one correspondence (17.67) shows that deformation of micromorphic continuum is related to the TBH model. Moreover, microdisplacements φ_{ij} are equivalent to additional lattice rotation, and the deformation of micromorphic continuum creates anisotropic textures in polycrystals.

The deformation of micromorphic continuum leads to the appearance of defects. In Sects. 17.9 and 17.10, we reconsidered the relation of the

TBH models from the view point of the theory of micromorphic continuum with defects. In the micromorphic theory, the microdisplacement which corresponds to additional lattice rotation is related to dislocation density in Eq. (17.66). Thus, the increase in the deformation, i.e. the increase of the dislocation density, leads to the increases in the additional lattice rotation $d\tilde{u}_i^R$ in Eq. (17.68). As mentioned in Sects. 17.9 and 17.10, microdisplacements φ_{ij} in micromorphic continuum correspond to additional lattice rotation $d\tilde{u}_i^R$ in polycrystals, and deformation of micromorphic continuum in polycrystals.

In the mathematical description of defect fields, gauge theory has played the main role (e.g. Kadić and Edelen 1983, Nagahama 2001, Yamasaki and Nagahama 1999, 2002). In particular, a 45-fold Abelian gauge condition called the Golebiewska gauge (Golebiewska-Lasota and Edelen 1979, Kadić and Edelen 1983, Edelen and Lagoudas 1988) is admitted by the defect field (Yamasaki and Nagahama 1999, 2002). From this point of view, Yamasaki and Nagahama (2002) derived the TBH model from the Golebiewska gauge transformation. In the differential form, the physical quantity ψ^i can be transformed under Golebiewska gauge transformation, as follows;

$$\psi^i = \psi_e^i - \mathrm{H}(\Gamma_j^i \wedge \psi^j), \tag{17.72}$$

where ψ_e^i is called the exact part of ψ^i, H is a linear homotopy operator, Γ_j^i is a connection 1-form and the symbol $\wedge$ denotes the exterior product. From the differential geometrical description of a deformed medium including a defect field (Edelen and Lagoudas 1988), the physical quantities in the strain space-time can be expressed as follows: $\psi^i = B^i$ and $\Gamma_j^i \wedge \psi^j = K^i$ where B^i and K^i represent distortion-velocity 1-form and bend-twist-spin 2-form, respectively. From definitions $\mathrm{H}B^i = u^i$, $\mathrm{H}K^i = r^i$, relation (17.72) can be rewritten:

$$B^i = \mathrm{d}u^i - r^i. \tag{17.73}$$

Relation (17.73) means that total distortions B^i are given by sums of two terms: gradient of displacement $\mathrm{d}u^i$ and internal rotation r^i. From the one-to-one correspondences between Eqs. (17.68) and (17.73), the Golebiewska gauge transformation corresponds to another expression of the TBH model in strain space-time (Yamasaki and Nagahama 1999, 2002).

References

Aifantis EC (1982) Dislocation kinematics and the formation of deformation bands. In: Sih GC, Provan JW (eds) Defect fracture and fatigue. Martiun-Nijhoff, pp 75-84

Aifantis EC (1984) On the microstructural origin of certain inelastic models. Trans ASME **106**: 326-330

Aifantis EC (1985) On dislocation patterns. In: Suzuki H, Ninomiya T, Sumino K, Takeuchi S (eds) Dislocations in solids. Univ Tokyo Press, Yamada Sciences Formation, Tokyo, pp 41-47

Aifantis EC (1986) On the dynamical origin of dislocation patterns. Mater Sci Eng **81**: 563-574

Amari S, Kagekawa K (1964) Dual dislocations and non-Riemannian stress space. RAAG Research Notes Third Ser No.**82**: 1-24

Anthony K, Essmann U, Seeger A, Truble H (1968) Disclinations and Cosserat-continuum with incompatible rotations. In: Kröner E (ed) Mechanics of generalized continua. Springer-Verlag, Berlin, pp 355-358

Bishop JFW, Hill R (1951) A theory of the plastic dislocation of a polycrystalline aggregate under combined stresses. Phil Mag **42**: 414-427

Caglioti G, Botani CE (1988) Nonlinear mechanical properties. In: Lundqvist S, March NH, Tosi MP (eds) Order and chaos in nonlinear physical systems, Plenum, New York, pp 423-446

Capriz G (1989) Continua with microstructure. In: Trusdell C (ed) Springer Tracts in Natural Philosophy Vol 35. Springer-Verlag, Belrin

Cosserat E, Cosserat F (1909) Téorie des corps déformables. Librairie Scientifique A, Hermann, Paris

Czechowski Z, Yamashita T, Teisseyre R (1994) Theory of the earthquake premonitory and fracture processes: evolution of stresses. Acta Geophys Pol **42**: 119-135

Czechowski Z, Yamashita T, Teisseyre R (1995) Stress evolution: creep and dynamic patterns. Acta Geophys Pol **43**: 187-196

DeWitt R (1971) Relation between dislocations and disclination. J Appl Phys **42**: 3304-3308

Edelen DGB, Lagoudas D (1988) Gauge theory and defects in solids. Elsevier, Amsterdam

Ericksen JL, Trusdell C (1958) Exact theory of stress and strain in rods and shells. Arch Rational Mech Anal **1**: 295-323

Eringen AC (1968) Theory of micropolar elasticity. In: Liebowitz H (ed) Fracture, vol 2. Acadmic Press, New York, pp 621-729

Eringen AC, Claus Jr WD (1970) A micromorphic approach to dislocation theory and its relation to several existing theories. In: Simmons JA, de Wit R, Bullough R (eds) Fundamental aspects of dislocation theory, vol 2 (NBS special publication, vol 317). US Nat Bur Stand, Washington DC, pp 1023-1040

Fleck NA, Muller GM, Ashby MF, Hutchinson JW (1994) Strain gradient plasticity: theory and experiment. Acta Metall Mater **42**: 475-487

Gil Sevillano J, Van Houte P, Aernoudt E (1980) Large strain work hardening and textures. Prog Mater Sci **25**: 69-412

Golebiewska-Lasota AA, Edelen DGB (1979) On the gauge transformations admitted by the equations of defect dynamics. Int J Eng Sci **17**: 335-339

Green AE, Rivlin RS (1964) Multipolar continuum mechanics. Arch Rational Mech Anal **17**: 113-147

Harris WF, Scriven LE (1971) Intrinsic disclinations as dislocation sources and sinks in surface crystals. J Appl Phys **42**: 3309-3312

Hehl F and Kröner E (1965) Über den Spin in der allgemeinen Relativitätstheorie: Eine notwendige Erweiterung der Einsteinschen Feldgleichungen. Zeit Phys **187**: 478-489

Ikeda S (1972) A geometrical construction of the physical interaction field and its application to the rheological deformation field. Tensor NS **24**: 60-68

Ikeda S (1975) Prolegomena to applied geometry. Mahā Shobō, Saitama

Ikeda S (1979) Some physical-geometrical remarks on the relationship between the micro- and macro-gravitational fields. Lett Nuovo Cimento **25**: 21-25

Ikada S (1985) The principles of applied geometry. Nippatsu Shuppan, Tokyo

Kadić A, Edelen DG (1983) Gauge theory of dislocations and disclinations. Springer-Verlag, Berlin

Katanaev MO, Volovich IV (1999) Scattering on dislocations and cosmic strings in the geometric theory of defects. Ann Phys **271**: 203-232

Kawaguchi A (1931) Theory of connections in a Kawaguchi space of higher order. Proc Imper Acad Japan **13**: 237-240

Kawaguchi A (1937) Beziehung zwischen einer metrischen linearen Uebertragung und einer nicht-metrischen in einem allgemeinen metrischen Raum. Proc Kon Akad Wet **40**: 596-601

Kawaguchi M (1962) An introduction to the theory of higher order spaces I: The theory of Kawaguchi space. In: Kondo K (ed) RAAG memoirs of the unified study of basic problems in engineering and physical sciences by means of geometry vol III, 3-Div Misc Gakujyutsu-Bunken Fukkyukai, Tokyo, pp 718-734

Kawai T (2000) Energy-momentum and anglar momentaum densities in gauge theories of gravity. Phys Rev D **62**: 104014

Kossecka E, DeWitt R (1977a) Disclination kinetic. Arch Mech **29**: 633-651

Kossecka E, DeWitt R (1977b) Disclination dynamics. Arch Mech **29**: 749-767

Kondo K (1953) On the geometrical and physical foundations of the theory of yielding. Proc 2nd Japan Nat Congr Appl Mech, held 1952, pp 41-47

Kondo K (1955) Geometry of elastic deformation and incompatibility. In: K. Kondo (ed.), RAAG Memoirs of the Unified Study of Basic Problems in Engineering and Physical Sciences by Means of Geometry vol I, C-1, 361-373, Gakujyutsu-Bunken Fukkyukai, Tokyo

Kondo K (1964) On the analytical and physical foundations of the theory of dislocations and yielding by the differential geometry of continua. Int J Engng Sci **2**: 219-251

Kröner E (1958) Kontinuumstheories der Versetzungen und Eingenspannungen. Erg Angew Math **5**: 1-179

Kröner E (1981) Continuum theory of defects. In Balian R (ed) Physics of defects, Proc Les Houches XXXV, North-Holland, Amsterdam

Lardener RW (1969a) Plane strain plasticity of single crystals. Int J Eng Sci **7**: 417-425

Lardener RW (1969b) Dislocation dynamics and the theory of the plasticity of single crystals. ZAMP **20**: 514-529

Mataga PA, Freund LB, Hutchinson JW (1987) Crack tip plasticity in dynamic fracture. J Phys Chem Solid **48**: 985-1005

McConnel JA (1931) Applications of the absolute differential calculus. Balckie and Son, London-Glasgow; reproduced under the title "Applications of tensor analysis" by Dover, New York (1957)

Miguel M-C, Vespignani A. Zapperi S, Weiss J, Grasso J-R (2001) Intermittent dislocation flow in viscoplastic deformation. Nature **410**: 667-671

Minagawa S (1971) Dislocation, disclination, and the source or sink of dislocation-line. RAAG Research Notes, Third Series, No. 177, Res Ass Appl Geomet, Tokyo, 1-10.

Minagawa S (1979) A non-Riemannian geometrical theory of imperfections in a Cosserat continuum. Arch Mech **31**: 783-792

Mura T (1963) On dynamic problems of continuous distribution of dislocations. Int J Eng Sci **1**: 371-381

Muto J, Nagahama H (2004) Dielectric anisotropy and deformation of crustal rocks: physical interaction theory and dielectric mylonites. Phys Earth Planet Inter **141**: 27-35

Nabarro FRN (1967) Theory of crystal dislocations, Claredon Press, Oxford

Nabarro FRN (1951) The synthesis of elastic dislocation field. Philos Mag **42**: 1224-1231

Nagahama H (2001) Gauge theory of dislocational electromagnetic field in earthquake preparation zone. Acta Geophys Pol **49**: 437-448

Nagahama H, Teisseyre R (1998) Micromorphic continuum, rotational wave and fractal properties of earthquakes and faults. Acta Geophys Pol **46**: 277-294

Nagahama H, Teisseyre R (2000) Micromorphic continuum and fractal fracturing in the lithosphere. Pure Appl Geophys **157**: 559-574.

Nagahama H, Teisseyre R (2001a) Seismic rotation waves: dislocations and disclinations in a micromorphic continuum. Acta Geophys Pol **49**: 119-129; Acta Geophys Pol **49**: No. 2 ERRATUM

Nagahama H, Teisseyre R (2001b), Micromorphic continuum and fractal properties of faults and earthquakes. In: Teisseyre R, Majewski E (eds) Earthquake thermodynamics and phase transformations in the Earth's interior. Academic Press, San Diego, pp 425-440

Puntigam RA, Soleng HH (1997) Volterra distortions, spinning strings, and cosmic defects. Class Quantum Grav **14**: 1129-1149

Sabbata V, Gasperini M (1985) Introductin to gravitaion. World Scientific, Singapore

Sabbata V, Sivaram C (1995)Spin and torsion in gravitaion. World Scientific, Singapore

Sakata S (1970) A constructive approach to non-teleparallelism and non-metric representations of plastic materials manifold by generalized diakoptical tearing. RAAG Research Notes, Third Series, No. 149, Res Ass Appl Geomet, Tokyo, 1-27.

Schouten JA (1954) Ricci-Calculus, 2nd ed, Springer-Verlag, Belrin

Sławianowski JJ (1990) Space-time as a micromorphic continuum. Int J Theor Phys **29**: 1177-1184

Suhubi ES, Eringen AC (1964) Nonlinear theory of micro-elastic solids, II. Int J Eng Sci 2: 389-404

Takano Y (1968) Theory of fields in Finsler spaces. I. Prog Theor Phys **40**: 1159-1180

Takeo M, Ito HM (1997) What can be learned from rotational motion excited by earthquakes? Geophys J Int **129**: 319-329

Taylor GI (1938) Plastic strain in metals. J Inst Metal **62**: 307-324

Teisseyre R (1969) Dislocational representation of thermal stresses. Acta Geophys Pol **17**: 3-12

Teisseyre R (1973) Earthquake processes in a micromorphic continuum. Pure Appl Geophys **102**: 15-28

Teisseyre R (1974) Symmetric micromorphic continuum wave propagation, point source solution and some applications to earthquake processes. In: Thoft-Christenen P (ed) Continuum mechanics aspects of geodynamics and rock fracture mechanics, D Riedel Pub, Dordrecht, pp 201-244

Teisseyre R (1990) Earthquake premonitory and rebound theory: synthesis and revision of principles. Acta Geophys Pol **38**: 269-278

Teisseyre R (1995a) Part IV Differential geometry methods in deformation problems, Ch.3 Dislocation density and geometry objects. In: Teisseyre R (ed) Theory of earthquake premonitory and fracture processes. Pol Sci Pub, Warszawa, pp 512-519

Teisseyre R (1995b) Part III Earthquake premonitory and rebound processes, Ch.2 Deformations and defect distribution. In: Teisseyre R (ed) Theory of earthquake premonitory and fracture processes. Pol Sci Pub, Warszawa, pp 324-332

Teisseyre R (1995c) Part II Dislocations and cracks: earthquake and fault models, Ch.2 Dislocations and simple models of earthquake sources. In: Teisseyre R (ed) Theory of earthquake premonitory and fracture processes. Pol Sci Pub, Warszawa, pp 136-168

Teisseyre R (1995d) Micromorphic model of a seismic source zone, 2. Symmetric micromorphic theory; applications to seismology. In: Teisseyre R (ed) Theory of earthquake premonitory and fracture processes. Pol Sci Pub, Warszawa, pp 616-627

Teisseyre R (1996) Motion and flow equation for stresses. Acta Geophys Pol **44**: 19-29

Teisseyre R (1997) Dislocation-stress relations and evolution of dislocation fields. Acta Geophys Pol **45**: 205-214

Teisseyre R (2002) Continuum with defect and self-rotation fields. Acta Geophys Pol **50**: 51-68

Teisseyre R, Czechowski Z (1993) Unifies earthquake premonitory and rebound theory. Acta Geophys Pol **41**: 1-16

Teisseyre R, Nagahama H (1999) Micro-inertia continuum: rotations and semi-waves. Acta Geophys Pol **47**: 259-272

Teodosiu C (1970) A dynamic theory of dislocations and its applications to the theory of the elastic-plastic continuum. In: Simmons JA, DeWitt R, Bullogh R (eds) Fundamental aspect of dislocation theory, Nat Bur Stand (U.S.), Sp Publ 317, II, pp.837-876

Truski Ł (1966) Variational principle for equilibrium and incompatibility equations in dislocation theory. Bull Acad Pol Sci, Sér sci tech **14**: 289294

Utiyama R (1956) Invariant theoretical interpretation of interaction. Phys Rev **101**: 1597-1607

Van Houtte P, Wagner F (1985) Development of textures by slip and twinning. In Wenk HR (ed) Preferred orientation in deformed metals and rocks: an introduction to modern texture analysis. Academic Press, London, pp 233-258

Webster GA (1966a) A widely applicable dislocation model of creep. Phil Mag **14**: 775-783

Webster GA (1966b) In support of a model of creep based on dislocation dynamics. Phil Mag **14**: 1303-1307

Werne RW, Kelly JM (1978) A dislocation theory of isotropic polycrystalline plasticity. Int J Eng Sci **16**: 951-965

Wenk HR, Takeshita T, Van Houtte P, Wagner F (1986) Plastic anisotropy and texture development in calcite polycrystals. J Geophys Res **91** (B3): 3861-3869

Yamasaki K, Nagahama H (1999) Continuum theory of defects and gravity anomaly. Acta Geophys Pol **47**: 239-257

Yamasaki K, Nagahama H (1999) Hodge duality and continuum theory of defects. J Phys A Math Gen **32**: L475-L481

Yamasaki K, Nagahama H (2002) A deformed medium including a defect field and differential forms. J Phys A Math Gen **35**: 3767-3778

Yano K and Davies ET (1954) Contact tensor calculus. Ann Mate Pura Appl **37**: 1-36

Yukawa H (1950) Quantum theory of non-local fields. Part I. Free fields. Phys Rev 77: 219-226; Quantum theory of non-local fields. Part II. Irreducible fields and their interaction. Phys Rev **80**: 1047-1052

18 Spinors and Torsion in a Riemann-Cartan Approach to Elasticity with a Continuous Defect Distribution and Analogies to the Einstein-Cartan Theory of Gravitation

Eugeniusz Majewski

Institute of Geophysics, Polish Academy of Sciences
ul. Księcia Janusza 64, 01-452 Warsaw, Poland
e-mail: emaj@igf.edu.pl

18.1 Introduction

Inspired by the original ideas of Albert Einstein on the relativistic theory of the non-symmetric field (cf., Einstein 1955a), which were at the basis of unification of gravity and electromagnetism theories, we briefly review possible constructive analogies between an elastic solid body with a continuum defect distribution and the Einstein-Cartan Theory (ECT) of gravitation. The aim of this approach is to illuminate the analogies between dynamical localized concentrations of continuous fields of crystalline defects in elastic continua, such as disclinations and dislocations, and the notions of curvature and torsion in the ECT. Beyond reviewing, a new element in this chapter is our attempt to relate the defect densities with spinors defined in spin-spaces. In the result, we identify disclination and dislocation spinors. Moreover, using some analogies between the so-called degenerate continuum and electromagnetism pointed out by Roman Teisseyre, we formulate Maxwell-like equations for spin and twist motions in terms of spinors. Roger Penrose claims that spinors are more fundamental quantities than tensors. The applications of differential geometry in physics have become an extremely enlightening and spectacular intellectual achievement. However, sometimes the geometry remained only a background or visualization. Truly intimate relationships between gravity and geometry have been revealed by Einstein. On April 4, 1955 Albert Einstein wrote: "*the essential achievement of general relativity, namely to overcome 'rigid' space (i.e., the inertial frame), is only indirectly connected with the introduction of a Riemannian metric. The directly relevant conceptual element is the 'displacement field' (Γ^k_{ij}), which expresses the infinitesimal* *displacement of vectors. It is this which replaces the parallelism of spatially arbitrarily separated vectors fixed by the inertial frame (i.e., the equality of corresponding components) by an infinitesimal operation. This makes it possible to construct tensors by differentiations and hence to dispense with the introduction of 'rigid' space (the inertial frame). In the face of this, it seems to be of secondary importance in some sense that some particular* Γ *field can be deduced from a Riemannian metric...*" (cf., Einstein 1955b). His General Relativity Theory (GRT) relates geometry of curvilinear space with gravity field. Any mass generates Riemannian curvature which transforms the Euclidean space into a curvilinear, e.g., the Riemannian space. If the mass is spinning, it causes Cartan's torsion and transforms the Riemannian space into the Riemann-Cartan space. Here we shall take a closer look at this beautiful branch of science, where geometry and physics live in real symbiosis. First of all, we shall consider some aspects of discontinuity fields. An idea of continuous distribution of infinitesimally small discontinuities illuminates the foundations and predictive powers of elasticity theory. This idea becomes more appealing and the obtained results much more attractive and universal when formulating them in the framework of the differential geometry. A mighty machinery of differential geometry allows us to describe elastic continua with continuous fields of disclinations and dislocations permeating the deformed crystal. Disclinations and dislocations are treated as rotational and translational defects, respectively.

Here some basic concepts of Riemann-Cartan (RC) geometry have been briefly recalled. It is worth to mention that this approach is very universal and can be applied in other branches of science. It is only a question of time that new applications will appear. So far, a similar approach was used to deal with fields of errors in measurements of many physical quantities. For example, the discontinuity fields can describe the fields of disclosures of geodetic networks and telecommunication networks.

A Riemann-Cartan approach to elasticity casts new light on some analogies between the theory of elasticity and gravity with torsion. Mathematical formalisms describing the both theories are similar. Simply, the gravity theory describes an elastic crystal under tension. In fact, we can observe here a kind of duality in mathematical theories of elasticity and gravity with torsion. Geometrization of physics initiated by Einstein has profound significance for unification of all fundamental interactions. John Wheeler has coined a new name for this fascinating geometrical approach, i.e., "geometrodynamics."

Sacharov (1967) expressed an interesting idea that "*the geometry does not possess a dynamics of its own, but the stiffness of spacetime could be entirely due to the vacuum fluctuations of the fundamental fields in the universe, e.g., scalar, vector, tensor, and spinor fields.*"

Thus, we can compare the geometry to a flax sack. When it is empty, it is flat and it does not possess any dynamics of its own. However, when it is full of sand grains, it takes a shape determined by rotational and translational degrees of freedom of sand grains in the gravity field. A similar physical interpretation, comparing the geometry to an empty infinitely thin plastic bag and a field to water that can fill it up, was suggested by Kleinert (2008).

Usually, the field equations are derived from appropriate variational principles. A variational approach to an asymmetric elastic continuum was discussed by Majewski (2006b). He considered theoretical aspects of this phenomenon in the framework of the mechanics of a micromorphic continuum. The micromorphic continuum allows for rotational motions of each microvolume of the continuum. Rotational motions are the result of microscopic defects in the form of dislocations and disclinations. Majewski (2006c) applied a Lagrangian formulation to a solid with dislocations in order to describe tectonic solitons propagating along an earthquake fault.

Kondo (1952, 1962) formulated a geometrical theory of defects in the framework of Riemann-Cartan geometry. He revealed the roles played by curvature and torsion in the continuum theory of crystalline defects. It has been known for some time in the physical community that a mathematical structure of linear elasticity in a Riemannian space is identical with Einstein's general relativity theory. It means that the universe can be treated as a gigantic elastic solid crystal under tension. Space curvature in Einstein's theory corresponds to a surface density of disclinations in the solid crystal. A space metric tensor in Einstein's general relativity corresponds to a strain tensor in elasticity. It should be emphasized that both tensors are symmetrical. In order to observe an analogy between the solid crystal and general relativity describing vacuum without matter in a flat spacetime, we need only a pure elastic crystal without defects. Thus, the pure elastic crystal represents the elastic ether in spacetime. In order to observe an analogy in a curved space, it is enough to introduce only one kind of defects in the crystal, namely, disclinations. Thus, the disclinations represent matter. When we add dislocations to the crystal, then we should add torsion to the spacetime and such a gravitational theory with curvature and torsion is called the Einstein-Cartan theory. The second Einstein-Cartan equation relates spin with torsion. Thus, the dislocations represent spinning of matter. Paraphrasing here the famous Wheeler's expression about spacetime and mass, we can say colloquially in our situation that spacetime tells defects how to move and defects tell spacetime how to curve and twist. Deformations of the solid crystal with dislocations correspond to distortions of metric tensor. The connection is no longer symmetrical in the Einstein-Cartan theory.

Rotational motions (spin and twist) are present in many natural phenomena and processes. For instance, they can be observed in earthquakes (cf., Majewski 2006a). It is noteworthy that other tectonic, volcanic, mining, and land sliding events can be sources of rotational motions as well. Rotational motions in solids can excite rotational elastic waves and solitons.

Elasticity theory reproduces only the linear approximation of the Einstein-Cartan theory of gravitation. The equations of asymmetric elasticity theory for the Cosserat media can also be naturally incorporated into the geometric theory as the gauge conditions.

Geometrodynamics relates geometrical properties of space to physical fields, eg., space curvature is related to matter; space torsion is related to spinning of matter. In order to describe all four fundamental physical interactions, we need to enrich the geometry, i.e., to introduce more spatial dimensions. For instance, the recent string theories require 11 dimensions in order to describe all fundamental interactions. The more spatial dimensions have been employed in a geometrical description, the more physical fields or interactions can be included.

18.2 The Riemann-Cartan Geometry

In 1922 Élie Cartan introduced torsion to the Riemannian geometry and formulated foundations of the Riemann-Cartan (RC) geometry. He was inspired by Cosserat brothers (1909). It is noteworthy that electron spin was discovered later, i.e., in 1925.

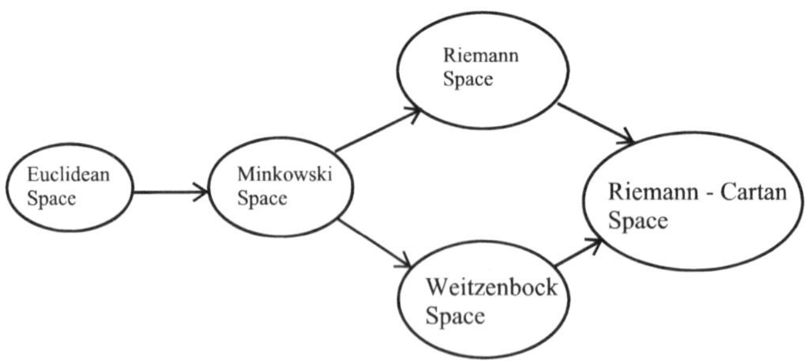

Fig. 18.1 Classification of spaces

The RC spacetime (Fig. 18.1), $\mathbb{R}_4$, is a differentiable 4D material manifold $\mathfrak{M}$ endowed with a metric tensor $g_{\mu\nu}$ and with a metric-compatible connection $\Gamma^{\kappa}_{\mu\nu}$, which is non-symmetrical in its lower indices.

An assumption of parallelism requires that lengths and angles should be preserved by parallel translation. A consequence of this requirement is the following condition for the covariant derivative of the metric tensor $g_{\mu\nu}$ in the form

$$\nabla_{\kappa} g_{\mu\nu} = g_{\mu\nu,\kappa} - \Gamma^{\rho}_{\kappa\mu} g_{\rho\nu} - \Gamma^{\rho}_{\kappa\nu} g_{\rho\mu} = 0, \tag{18.1}$$

where ∇_{κ} and, κ denote a covariant derivative and a partial derivative with respect to x^{κ}, respectively.

The above equation is called a metricity condition for the connection and expresses the fact that the connection is compatible with the metric.

The antisymmetric part of the connection defines a new tensor, the torsion tensor (Fig. 18.2),

$$T^{\kappa}_{\mu\nu} = \Gamma^{\kappa}_{[\mu\nu]} = \frac{1}{2}\left(\Gamma^{\kappa}_{\mu\nu} - \Gamma^{\kappa}_{\nu\mu}\right),, \tag{18.2}$$

where (.) and [.] denote symmetrization and antisymmetrization, respectively. For instance,

$$\Gamma_{(\mu\nu)} = \frac{1}{2}\left(\Gamma_{\mu\nu} + \Gamma_{\nu\mu}\right), \qquad \Gamma_{[\mu\nu]} = \frac{1}{2}\left(\Gamma_{\mu\nu} - \Gamma_{\nu\mu}\right), \tag{18.3}$$

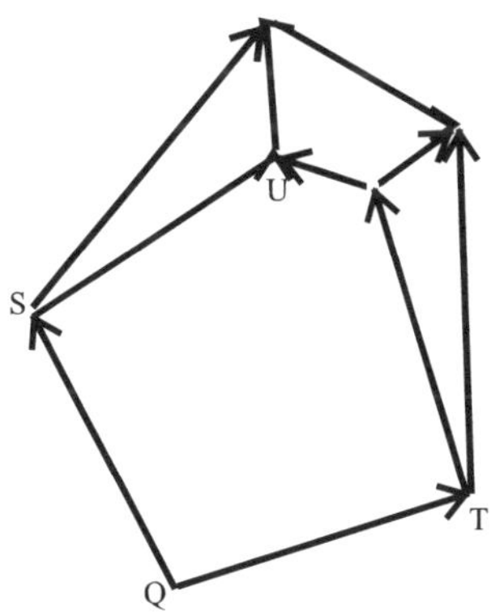

Fig. 18.2 Illustration of torsion. We shift two vectors in parallel directions and find that the obtained parallelogram is unclosed

Exclusion of an index from symmetrization or antisymmetrization is described as follows

$$\Gamma_{(\mu|\nu|\kappa)} = \frac{1}{2}\left(\Gamma_{\mu\nu\kappa} + \Gamma_{\kappa\nu\mu}\right), \qquad \Gamma_{[\mu|\nu|\kappa]} = \frac{1}{2}\left(\Gamma_{\mu\nu\kappa} - \Gamma_{\kappa\nu\mu}\right). \tag{18.3}$$

In order to estimate torsion, Trautman (1973a, b, 2007) introduced the "Cartan radius" $r_{Cart} \approx 10^{-23}$cm that is greater than the Planck length, which is $l \approx 10^{-23}$cm.

The curvature tensor is defined in terms of the connection as follows

$$R^{\kappa}_{\mu\nu\rho} = 2\left(\Gamma^{\kappa}_{[\nu|\rho|,\mu]} + \Gamma^{\kappa}_{[\mu|\lambda|}\Gamma^{\lambda}_{\nu]\rho}\right). \tag{18.5}$$

18.3 Spinors and Spin-Spaces

In addition to vector bases, we shall introduce spinor bases to express geometrical objects in a spin-space that is a 2D complex vector space S with a symplectic form ε or a fundamental spinor, which is an antisymmetric complex bilinear form and is defined as follows (Penrose 1983, Penrose and Rindler 1986a, b)

$$\varepsilon_{AB} = \varepsilon^{AB} = \begin{pmatrix} 0 & 1 \\ -1 & 0 \end{pmatrix}. \tag{18.6}$$

Due to the fact that there are two spin-spaces: (i) unprimed spin-space (S, ε) and (ii) primed spin-space (S', ε'), the primed symplectic form ε or a primed fundamental spinor is defined in the following way

$$\varepsilon_{A'B'} = \varepsilon^{A'B'} = \begin{pmatrix} 0' & 1' \\ -1' & 0' \end{pmatrix}. \tag{18.7}$$

Primed spin-spaces are the complex conjugates of their corresponding unprimed ones. Note that we must pay attention to the ordering of the unprimed indices on a spinor symbol, and also of the primed ones, the relative ordering between primed and unprimed indices is unimportant (Penrose and Rindler 1986a, b).

18.4 Elastic Crystal with a Continuous Defect Distribution

Let us consider an elastic crystal with a continuous distribution of internal defects in the framework of the Riemann-Cartan geometry. For the first time, such a broad and elegant approach was developed by Kondo (1952,

1955, 1962) and later by Bilby et al (1955), and Kröner (1981). Moreover, Amari (1962, 1968, 1981), Holländer (1962), Teisseyre (1964a, b, 1969, 1986, 1995, 2001, 2004), Minagawa (1971, 1979), Kleinert (1988, 1989, 2008), Ruggiero and Tartaglia (2003), Teisseyre and Boratyński (2006), Takeo (2006) and others were extending the Kondo's original idea of the geometrization of the theory of defects. We shall briefly review their main assumptions, concepts and results.

The deformation processes described in the differential manifolds can be introduced by defining the Lagrangian (material) and/or Eulerian (spatial) descriptions. We shall confine our considerations here to an elastic continuum and a spacetime coordinate system. However, a generalization for other spaces is straightforward.

Before deformation and in a defect-free configuration, we shall describe the elastic continuum in the 3D Euclidean space with the coordinates $x^i (i = 1, 2, 3)$. After generating defects and deforming the elastic continuum, we shall describe it using the Riemann-Cartan (RC) space. We assume that $x^\mu (\mu = 0, 1, 2, 3)$ are the coordinates in the 4D Riemann-Cartan space $\mathbb{R}^4$ in which an elastic crystal $\mathbb{C}$ is immersed. Here $x^0 = t$ is the time coordinate. The other coordinates $x^i (i = 1, 2, 3)$ are space coordinates. Greek indices range from 0 to 3 and refer to spacetime, while Roman indices range from 1 to 3 and refer to space. We shall concentrate our considerations on rotational motions (spin and twist).

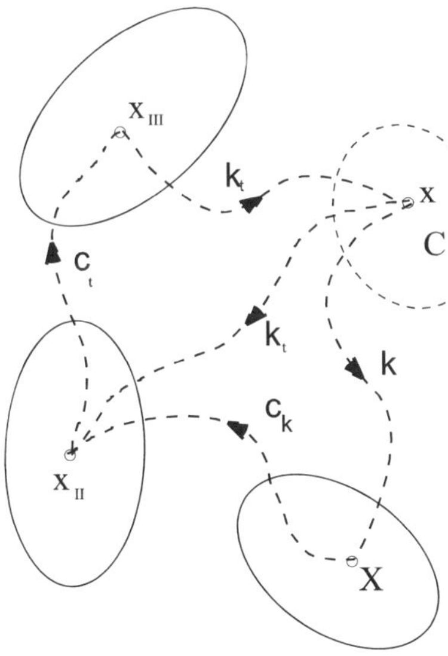

Fig. 18.3 Four material configurations for deformation processes of an elastic crystal with a continuous defect distribution

We assume that $\zeta_i (i = 1, 2, 3)$ are the vector basis at point Q of the material manifold $\mathfrak{M}$. Moreover, we assume the existence of the tangent space $\mathfrak{A}_Q$ as a vector space spanned by the vectors ζ_i. Usually, the elastic crystal $\mathbb{C}$ is deformed and under stress due to an external loading and it has internal stresses due to crystal defects existing inside. If we remove the external loading, then the crystal can be transformed to the strain-free configuration. This process is named unloading. The configurations before and after the unloading are called the deformed and the undeformed configurations, respectively (see Fig. 18.3). We can also consider a defect-free configuration.

We assume that the unloaded or undeformed configuration is realized by an affine transformation of the small material volume. Suppose that in the result of unloading the vectors ζ_μ transform into the vectors $\hat{\zeta}_\mu$. The squared length ds_0^2 of sector ds after unloading can be expressed as follows

$$ds_0^2 = g_{\mu\nu} dx^\mu dx^\nu , \tag{18.8}$$

where $g_{\mu\nu} = (\hat{\zeta}_\mu, \hat{\zeta}_\nu)$ is the translational metric tensor in the tangent space $\mathfrak{A}_Q$ and $(\hat{\zeta}_\mu, \hat{\zeta}_\nu)$ is the inner product. We assume here the Einstein summation convention for repeated indices.

The above metric was expressed in a vector basis. The translational metric in a spinor basis can be expressed as follows

$$g_{\mu\nu} = \varepsilon_{AC} \varepsilon_{B'C'} \sigma_\mu^{AB'} \sigma_\nu^{CD'} , \tag{18.9}$$

where ε_{AB} denotes the symplectic form or the fundamental spinor, and $\sigma_\mu^{AB'}$ denote the Newman-Penrose symbols (Newman and Penrose 1962).

18.5 The Disclination-Curvature Analogy

The total Franck vector Ω^k can be written as follows

$$\Omega^k = \frac{1}{2} \upsilon^l \int_A R^k_{ijl} da^{ij} , \tag{18.10}$$

where the integration is carried out over surface A, and da^{ij} is an infinitesimal surface element of A. The tensor R^k_{ijl} in the Eq. (18.10) is a curvature tensor is defined by Eq. (18.5). We deal with a linear approach to elasticity where second order and higher-order terms of curvature and torsion are discarded. We review the results obtained by Kondo (1952, 1955, 1962), Kossecka and deWit (1977a, b), Teisseyre (1995, 2004), Teisseyre and Boratyński (2006), and Takeo (2006). We express the disclination density tensor θ^{mn} in terms of the curvature tensor R_{pqrs} as follows

$$\theta^{mn} = \frac{1}{4} g^{-1} \varepsilon^{mpq} \varepsilon^{nrs} R_{pqrs}, \tag{18.11}$$

where $g = \det(g_{ij})$ and $\varepsilon_{mnp} = \varepsilon^{mnp} = 0, +1, -1$ is the totally antisymmetric 3D permutation symbol, $\varepsilon_{mnp} = +1, -1$ if mnp is an even and an odd permutation of (1, 2, 3), respectively, otherwise $\varepsilon_{mnp} = 0$.

Eq. (18.11) reveals a kind of analogy between physical and geometrical quantities. The curvature tensor corresponds to defects called the disclinations. Thus, we call it a disclination-curvature duality.

18.6 The Dislocation-Torsion Analogy

The total Burgers vector B^k can be written as follows

$$B^k = \int_A T_{ij}^k da^{ij}, \tag{18.12}$$

Now we are in a position to express the dislocation density tensor α^{mn} in terms of the torsion tensor T_{pq}^n as follows

$$\alpha^{mn} = g^{-1/2} \varepsilon^{mpq} T_{pq}^n. \tag{18.13}$$

Eq. (18.13) illustrates some links between physical and geometrical quantities. The defects that correspond to the torsion tensor are called the dislocations. We observe here a dislocation-torsion duality.

18.7 The Rotational and Translational Strain Tensors

Now, we introduce two strain tensors, a rotational strain tensor ω_{ijk} and a strain tensor E_{ij} defined by

$$\omega_{ijk} = \Xi_{i[jk]} - \Gamma_{i[jk]}, \tag{18.14}$$

$$g_{ij} = \eta_{ij} - 2E_{ij}, \tag{18.15}$$

thus,

$$E_{ij} = \frac{1}{2}(\eta_{ij} - g_{ij}) = \frac{1}{2}h_{ij}, \tag{18.16}$$

where $\eta_{ij} = (\zeta_i, \zeta_j)$ is a metric tensor in the 3D Euclidean space and h_{ij} is proportional to self distortion $\bar{\beta}_{\mu\nu}$.

We adopt a new independent variable, i.e., the vielbein, representing the partial derivatives of displacements $\partial_\mu u^i = u^i_{,\mu}$ in the form

$$\zeta^i_\mu = u^i_{,\mu}. \tag{18.17}$$

The vielbein is a differentiable function. The incompatibility condition resulting from the presence of a dislocation can be expressed in terms of the vielbein as follows

$$\zeta^i_{\nu,\mu} - \zeta^i_{\mu,\nu} \neq 0. \tag{18.18}$$

The vielbein has a physical interpretation as a distortion β^i_μ, i.e.,

$$\zeta^i_\mu = u^i_{,\mu} = \beta^i_\mu. \tag{18.19}$$

Employing the vielbein, we can express the following differentials

$$dx^\mu = \zeta^\nu \zeta^\mu_\nu, \qquad d\zeta^\mu_\kappa = A^\nu_\kappa \zeta^\mu_\nu, \tag{18.20}$$

where

$$\zeta^\nu = \zeta^\nu_\mu dx^\mu, \qquad A^\nu_\kappa = \zeta^\nu_\mu \, d\zeta^\mu_\kappa = A^\nu_{\kappa\rho} \, dx^\rho. \tag{18.21}$$

Now, we are in a position to express the Cartan structural equations as follows

$$\begin{aligned} d\zeta^\mu - \zeta^\nu \wedge A^\mu_\nu &= -T^\mu \\ dA^\kappa_\mu - A^\nu_\mu \wedge A^\kappa_\nu &= -R^\kappa_\mu \end{aligned} \tag{18.22}$$

where T^μ represents the 2-form of the torsion tensor, R^κ_μ represents the 2-form of the curvature tensor, and the sign $\wedge$ denotes an external product that is defined as

$$\zeta^\mu \wedge \zeta^\nu = \zeta^\mu \zeta^\nu - \zeta^\nu \zeta^\mu. \tag{18.23}$$

Using Eq. (18.14), and taking into account that the connection Γ_{ijk} is related to the metric as $\Gamma_{ijk} = g_{pk}\Gamma^{p}_{ij}$, we obtain the relation

$$\Xi_{ijk} = \frac{1}{2}(\eta_{jk,i} + \eta_{ik,j} - \eta_{ij,k}). \tag{18.24}$$

Applying a linear approximation, we can determine the strain tensor E_{ij} in terms of deformed and undeformed metrics in the following way

$$E_{ij}dx^i dx^j = (\eta_{ij}dx^i dx^j)^{1/2} - (g_{ij}dx^i dx^j)^{1/2}. \tag{18.25}$$

The curvature tensor can be expressed in terms of the rotational strain tensors as follows

$$R_{ijlk} = -2(\omega_{[j|lk|,i]} + g^{rs}\omega_{[i|kr|}\omega_{j]ls}). \tag{18.26}$$

This equation can be expressed in a spinor basis as

$$\begin{aligned} X_{IJLK}\varepsilon_{I'J'}\varepsilon_{L'K'} + \Phi_{IJL'K'}\varepsilon_{I'J'}\varepsilon_{LK} + \overline{\Phi}_{I'J'LK}\varepsilon_{IJ}\varepsilon_{L'K'} \\ + \overline{X}_{I'J'L'K'}\varepsilon_{IJ}\varepsilon_{LK} = -2(\omega_{[j|lk|,i]} + g^{rs}\omega_{[i|kr|}\omega_{j]ls}), \end{aligned} \tag{18.27}$$

where X and Φ are curvature spinors; the overbar denotes a complex conjugation, and ε_{AB} denotes the symplectic form or the fundamental spinor. Due to the disclination—curvature analogy, we call the above-mentioned curvature spinors by a new name, i.e., disclination spinors.

The torsion tensor can be expressed in terms of the rotational strain tensor and the strain tensor as follows

$$T^{k}_{ij} = -\omega_{[ij]k} - E_{[j|k|,i]}, \tag{18.28}$$

and in a spinor basis it takes the form

$$\chi^{KK'}_{IJ}\varepsilon_{I'J'} + \tilde{\chi}^{KK'}_{I'J'}\varepsilon_{IJ} = -\omega_{[ij]k} - E_{[j|k|,i]}, \tag{18.29}$$

where χ and $\tilde{\chi}$ are torsion spinors that are symmetric in IJ and $I'J'$, respectively (cf., Penrose and Rindler 1986a).

The torsion spinors, due to the dislocation—torsion analogy, we call the dislocation spinors.

The contortion tensor or the rotational Ricci coefficients can be expressed in the spinorial form as

$$K_{AB'CD'\kappa} = \sigma^{\mu}_{CD'}\nabla_{\kappa}\sigma_{AB'\mu}. \tag{18.30}$$

The compatibility conditions require vanishing of the curvature and torsion tensors, i.e., $R_{ijlk} = T_{ijk} = 0$. After linearization, the compatibility conditions for the displacement vector u_i and the rotational tensor ω_{ij} are as follows

$$E_{ij} = u_{(j,i)} = \frac{1}{2}(u_{i,j} + u_{j,i}), \tag{18.31}$$

$$\omega_{ijk} = \omega_{jk,i} + u_{[k,ji]}. \tag{18.32}$$

18.8 Description of Moving Defects in 4D

So far, we were describing static defects. In order to generalize our approach to describe moving defects, we change the Euclidean space into spacetime, i.e., we supplement the Euclidean space by a time dimension in order to construct a 4D Euclidean spacetime. So, the time factor will be considered from the very beginning of the deformation process. The Euclidean spacetime will be our starting point from the undeformed and defect-free configuration. Next, in the defected and deformed configuration, we shall use the Riemann-Cartan space.

18.9 Rotational Metric

Here, rotational coordinates are denoted by ω_i. A rotational metric can be defined as follows

$$d\psi^2 = d\xi^n_m d\xi^{nm}_n = K^m_{ni} K^n_{mj} dx^i dx^j, \tag{18.33}$$

where $d\xi_{mn} = -d\xi_{nm}$, and K^m_{ni} and K^n_{mj} are contortion tensors.

The contortion tensor K^n_{mi} also known as the rotational Ricci coefficients can be expressed as

$$K^k_{mn} = -T^k_{mn} + g^{kr}\left(g_{ms} T^s_{rn} + g_{ns} K^s_{rm}\right), \tag{18.34}$$

and 1-form of contortion is

$$K^m_n = K^m_{ni} dx^i, \tag{18.35}$$

where $K_{(mn)} = 0$.

The structural equations of a rotational group after discarding the matrix indices can be written as

$$\nabla_{[i}\nabla_{j]}\zeta^k = \frac{1}{2}R_{ij}\zeta^k, \tag{18.36}$$

where

$$R_{ij} = 2\nabla_{[j}T_{i]} + [T_j, T_i]. \tag{18.37}$$

18.10 Complex Vielbein, Rotational Field, and Metric

A complex vielbein ζ^k_μ and its inverse ζ^μ_k are defined by $\delta^i_k = \zeta^i_\mu \zeta^\mu_k$ and $\delta^\nu_\mu = \zeta^\nu_k \zeta^k_\mu$ and their 1-form is defined as $\zeta^\nu = \zeta^\nu_\mu dx^\mu$. We can express the vielbein in the form

$$\zeta^\nu_\mu = \tilde{\zeta}^\nu_\mu + \mathrm{i}\hat{\zeta}^\nu_\mu. \tag{18.38}$$

Teisseyre et al (2006) considered a complex rotational field that can be expressed as

$$\omega_{\mu\nu} = \omega_{[\mu\nu]} + \mathrm{i}\,\omega_{(\mu\nu)}, \tag{18.39}$$

where $\omega_{[\mu\nu]}$ and $\omega_{(\mu\nu)}$ are the spin and twist components, respectively.
In general, we can express a complex translational metric tensor after unloading defined by $g_{\mu\nu} = (\hat{\zeta}_\mu, \hat{\zeta}_\nu)$ in the form

$$g_{\mu\nu} = \tilde{g}_{\mu\nu} + \mathrm{i}\,\hat{g}_{\mu\nu}, \tag{18.40}$$

or as follows

$$g_{\mu\nu} = \begin{bmatrix} 1 & -\tilde{\upsilon}_1 & -\tilde{\upsilon}_2 & -\tilde{\upsilon}_3 \\ -\tilde{\upsilon}_1 & \tilde{g}_{11} & \tilde{g}_{12} & \tilde{g}_{13} \\ -\tilde{\upsilon}_2 & \tilde{g}_{21} & \tilde{g}_{22} & \tilde{g}_{23} \\ -\tilde{\upsilon}_3 & \tilde{g}_{31} & \tilde{g}_{32} & \tilde{g}_{33} \end{bmatrix} + \mathrm{i} \begin{bmatrix} 1 & -\hat{\upsilon}_1 & -\hat{\upsilon}_2 & -\hat{\upsilon}_3 \\ -\hat{\upsilon}_1 & \hat{g}_{11} & \hat{g}_{12} & \hat{g}_{13} \\ -\hat{\upsilon}_2 & \hat{g}_{21} & \hat{g}_{22} & \hat{g}_{23} \\ -\hat{\upsilon}_3 & \hat{g}_{31} & \hat{g}_{32} & \hat{g}_{33} \end{bmatrix}. \tag{18.41}$$

However, we shall use only a real part of the metric in the sequel.

18.11 Disclination Density and Current Tensor

The physical and mathematical interpretations of disclination density tensor θ^{mn} will be elucidated in the sequel. In these calculations here, we use extensively properties of tensor transformations with respect to the expression $\varepsilon^{mnp}g^{-1/2}$ and with respect to the tensor θ^{mn}. From Eq. (18.11) we can obtain the following expression

$$R_{pqrs} = g\varepsilon_{mpq}\varepsilon_{mrs}\theta^{mn}, \tag{18.42}$$

and in a spinor basis (cf., Penrose and Rindler 1986a) it can be expressed as

$$\begin{aligned} X_{PQRS}\varepsilon_{P'Q'}\varepsilon_{R'S'} + \bar{X}_{P'Q'R'S'}\varepsilon_{PQ}\varepsilon_{RS} \\ + \Phi_{PQR'S'}\varepsilon_{P'Q'}\varepsilon_{RS} + \bar{\Phi}_{P'Q'RS}\varepsilon_{PQ}\varepsilon_{R'S'} = g\varepsilon_{mpq}\varepsilon_{mrs}\theta^{mn}. \end{aligned} \tag{18.43}$$

The disclination current tensor Υ_r^s is written as

$$\Upsilon_r^{\ s} = \frac{1}{2}g^{-1/2}\varepsilon^{suv}R_{rtuv}, \tag{18.44}$$

and in a spinor basis it is

$$\begin{aligned} \Upsilon_r^{\ s} = \frac{1}{2}g^{-1/2}\varepsilon^{suv}(X_{RTUV}\varepsilon_{R'T'}\varepsilon_{U'V'} + \bar{X}_{R'T'U'V'}\varepsilon_{RT}\varepsilon_{UV} \\ + \Phi_{RTU'V'}\varepsilon_{R'T'}\varepsilon_{UV} + \bar{\Phi}_{R'T'U'V'}\varepsilon_{RT}\varepsilon_{R'T'}). \end{aligned} \tag{18.45}$$

18.12 Dislocation Density and Current Tensor

From Eq. (18.13) we can obtain the following relation

$$T_{pq}^{n} = \frac{1}{2}g^{1/2}\varepsilon_{mpq}\alpha^{mn}, \tag{18.46}$$

and in terms of spinors it is

$$\chi_{PQ}^{NN'}\varepsilon_{P'Q'} + \tilde{\chi}_{P'Q'}^{NN'}\varepsilon_{PQ} = \frac{1}{2}g^{1/2}\varepsilon_{mpq}\alpha^{mn}. \tag{18.47}$$

The dislocation current tensor J_r^s is defined in the form

$$J_r^{\ s} = T_{rt}^{\ \ s} = -T_{tr}^{\ \ s}, \tag{18.48}$$

and it can be expressed spinorially as

$$J_r^{\ s} = T_{rt}^{\ \ s} = \chi_{RT}^{SS'}\varepsilon_{R'T'} + \tilde{\chi}_{R'T'}^{SS'}\varepsilon_{RT}. \tag{18.49}$$

18.13 Additive Decomposition of the Total Strain Tensors

Let us take now a closer look at the strain tensor E_{mn} and the rotational strain tensor ω_{mnp}. Due to the existence of defects in the elastic crystal, the deformation has elastic and self components (plastic). We may apply here an additive decomposition of the total tensors into elastic and self deformation (plastic) components in the following way

$$E_{mn} = E^{E}_{mn} + \overline{E}_{mn} , \tag{18.50}$$

$$\omega_{mnp} = \omega^{E}_{mnp} + \overline{\omega}_{mnp} , \tag{18.51}$$

$$S_{mn} = S_{(mn)} + S_{[mn]} , \tag{18.52}$$

$$\overline{\omega}_{[mn]} = \frac{1}{2\mu^{*}} \overline{S}_{[mn]} , \tag{18.53}$$

$$\overline{\omega}_{(mn)} = \frac{1}{2\mu^{*}} \overline{S}_{(mn)} , \tag{18.54}$$

where the superscript E and the overbar denote the elastic and the self components (plastic), respectively, and the constant μ^{*} represents the rotation rigidity of bonds and is related to the inner friction (Teisseyre and Boratyński 2006). However, due to the incompatibility in the self deformation field, the displacement components u_j and the rotational strain tensor ω_{jk} become multi-value functions. We should be aware of the existence of multivaluedness here. Due to the fact that the elastic deformations perfectly satisfy the compatibility conditions, then the elastic strain tensor E^{E}_{ij} and the elastic rotational strain tensor ω^{E}_{ijk} have no influence on the curvature and torsion tensors. This fact, after linearization, allows to express the curvature tensor in the simple way

$$R_{ijlk} = -2\overline{\omega}_{[j|lk|,i]} . \tag{18.55}$$

and in a spinorial form (in terms of disclination spinors or curvature spinors) it is as follows

$$\begin{aligned} X_{IJLK}\varepsilon_{I'J'}\varepsilon_{L'K'} + \Phi_{IJL'K'}\varepsilon_{I'J'}\varepsilon_{LK} \\ + \overline{\Phi}_{I'J'LK}\varepsilon_{IJ}\varepsilon_{L'K'} + \overline{X}_{I'J'L'K'}\varepsilon_{IJ}\varepsilon_{LK} = -2\overline{\omega}_{[j|lk|,i]} . \end{aligned} \tag{18.56}$$

Now, the torsion tensor after the linear approximation takes the form

$$T_{ijk} = -\bar{\omega}_{[ij]k} - \bar{E}_{[j|k|,i]}, \tag{18.57}$$

and in a spinor basis it can be expressed as

$$\chi_{IJKK'}\varepsilon_{I'J'} + \tilde{\chi}_{I'J'KK'}\varepsilon_{IJ} = -\bar{\omega}_{[ij]k} - \bar{E}_{[j|k|,i]}. \tag{18.58}$$

Consequently, the disclination and dislocation densities and their current densities Υ^n_m and J_{mn} can be expressed by the following relations

$$\theta^{ij} = -\varepsilon^{ipq}\bar{\wp}^j_{q,p}, \tag{18.59}$$

$$\Upsilon^n_m = \dot{\bar{\wp}}^{Sn}_m - \bar{\wp}^{Sn}_{t,m}. \tag{18.60}$$

$$\alpha^i_j = -\varepsilon^{ipq}(\varepsilon_{qjm}\bar{\wp}^n_p + \bar{E}_{qj,p}), \tag{18.61}$$

$$J_{mn} = \frac{1}{2}(\varepsilon_{mnl}\bar{\wp}^l_t + \dot{E}^t_{mn} - \upsilon_{n,m}), \tag{18.62}$$

where the self part is denoted by the overbar $\bar{\wp}^p_n = \wp^{Sp}_n$ and

$$\bar{\wp}^j_i = \frac{1}{2}\varepsilon^{jpq}\bar{\omega}_{ipq} \tag{18.63}$$

18.14 The Einstein-Cartan Theory

Seven years after creation of the Einstein General Relativity Theory (GRT), Élie Cartan (1922) extended it by including torsion, and created a theory, which now is called the Einstein-Cartan Theory (ECT). He was inspired by Cosserat brothers (1909). Cartan (1922) assumed that the metric connection of spacetime is antisymmetric. According to Cartan, the antisymmetric part of the metric connection should be related to the spin tensor of physical fields which are the source of gravity (Trautman 1973a, b, 2006). The Einstein-Cartan theory is a step closer to unification of general relativity with quantum mechanics. In the Einstein-Cartan theory the spin is the source of torsion of spacetime. Spin is the angular momentum intrinsic to a particle. Spin is a physical quantity but torsion of spacetime is a geometrical quantity. From geometrical point of view, the curvature is connected with Lorentz transformations, but torsion is connected with translations. The Special Relativity Theory admitted the Poincaré group (Lorentz transformations and translations). In Einstein's GRT translations were missing. Thus, by adding torsion and connecting it to spin, Cartan

(1922) reintroduced the Poincaré group to gravitational theories (Trautman 2006). In order to illustrate it, let us consider a position vector field A^i determined on the curve $x^i(t)$, defined by the formula (Trautman 1973a, b)

$$\dot{x}^i \nabla_i A^j = \dot{x}^j . \tag{18.64}$$

If the curve $x^j(t)$ is closed, then after going along it with a full cycle, the position vector gains the increment

$$\Delta A^l \cong (R^l_{jkl} A^i + T^l_{kl}) \Delta\Omega^{kl} , \tag{18.65}$$

where $\Delta\Omega^{kl}$ denotes an oriented surface element spanned on the curve.

In the Einstein-Cartan theory the connection is asymmetric, but it satisfies the metricity condition

$$\nabla_i g_{jk} = 0. \tag{18.66}$$

Equations of gravity field, following from an appropriate variational principle, take the form (Trautman 1973a, b, 2006)

$$R_{ij} - \frac{1}{2} g_{ij} R = \frac{8\pi\kappa}{c^4} P_{ij} , \tag{18.67}$$

$$T^k_{ij} + \delta^k_i T^l_{jk} - \delta^k_j T^j_{il} = -\frac{8\pi\kappa}{c^3} S^k_{ij} , \tag{18.68}$$

where R_{ij} is the curvature tensor, R is the scalar curvature, g_{ij} is the metric tensor, P_{ij} denotes the asymmetric canonical energy-momentum tensor, T^l_{ij} is the torsion tensor, S^k_{ij} denotes the spin tensor, κ is the gravitational constant, and c is the speed of light in vacuum. The second equation in the set of fundamental equations of the Einstein-Cartan theory relates spin with torsion. If there are some regions of spacetime where the spin vanishes, the torsion vanishes there as well and the equations of the Einstein-Cartan theory (ECT) reduce to equations of the Einstein's general relativity theory (GRT). Hehl et al (1995) generalized the Einstein-Cartan theory by assuming non-metric linear connections and dilation and shear currents.

18.15 The Analogy Between the Disclination Density Tensor and the Einstein Tensor

Let us now show a link between the disclination density tensor and the Einstein tensor. Following Schouten (1954), it is necessary to find from

Bianchi's identities the equations describing the torsion and curvature tensors. After some calculations, assuming the value $g = 1$ for the Cartesian coordinates and $R_{\mu\nu(\lambda\kappa)} = 0$, and following Takeo (2006), we arrive at the relation

$$\alpha^{p}_{i,p} = -\varepsilon_{ipq}\theta^{pq}. \tag{18.69}$$

Following Schouten (1954) and Takeo (2006), we can write the following formula for the permutation symbol

$$\varepsilon^{ijk}\varepsilon^{lmn} = gg^{pl}g^{qm}g^{rn}\begin{bmatrix} \zeta^{i}_{h}\zeta^{h}_{p} & \zeta^{i}_{s}\zeta^{s}_{q} & \zeta^{i}_{u}\zeta^{u}_{r} \\ \zeta^{j}_{s}\zeta^{s}_{p} & \zeta^{j}_{h}\zeta^{h}_{q} & \zeta^{j}_{i}\zeta^{i}_{r} \\ \zeta^{k}_{u}\zeta^{u}_{p} & \zeta^{k}_{u}\zeta^{u}_{q} & \zeta^{k}_{h}\zeta^{h}_{r} \end{bmatrix}, \tag{18.70}$$

where the vielbein ζ^{k}_{μ} and its inverse ζ^{μ}_{k} are defined by $\delta^{i}_{k} = \zeta^{i}_{\mu}\zeta^{\mu}_{k}$ and $\delta^{\nu}_{\mu} = \zeta^{\nu}_{k}\zeta^{k}_{\mu}$ and their 1-form is defined as $\zeta^{\nu} = \zeta^{\nu}_{\mu}dx^{\mu}$.

After some calculations (cf., Takeo 2006), one can conclude that the disclination density tensor θ^{ij} defined by Eq (18.11) is identical with the Einstein tensor G^{ji}, thus it satisfies the following equation

$$\theta^{ji} = R^{ji} - \frac{1}{2}g^{ji}R, \tag{18.71}$$

in a 3D space. In a spinor basis it can be expressed as

$$\theta^{ji} = -2\Phi^{JIJ'I'} - 6\Lambda\varepsilon^{JI}\varepsilon^{J'I'}, \tag{18.72}$$

where Φ is the disclination spinor (or curvature spinor), $\Lambda = R/24$ and R is the scalar curvature.

18.16 The Evolution Equation for the Disclination Density

Let us now show a linear version of the evolution equation for disclination density. Following Takeo (2006), after some linearization, we have the relation

$$R_{[\nu\mu|\lambda\kappa|,\omega]} = 0. \tag{18.73}$$

On one hand, taking the time derivative of the curvature, one can obtain

$$\dot{R}_{ijpq} + R_{jtpq,i} + R_{tipq,j} = 0. \tag{18.74}$$

On the other hand, taking the time derivative of the disclination tensor θ^{mn}, we get

$$\dot{\theta}^{mn} = \frac{1}{4}\varepsilon^{mij}\varepsilon^{npq}\dot{R}_{ijpq} = -\varepsilon^{mij}\Upsilon^{n}_{j,i}. \tag{18.75}$$

Comparison of these two derivatives and using the definition of the disclination current tensor, yields the evolution equation for disclination density in terms of the disclination current in the form

$$\dot{\theta}^{mn} = -\varepsilon^{mpq}\Upsilon^{n}_{q,p}. \tag{18.76}$$

18.17 The Evolution Equation for the Dislocation Density

Let us now show a linear version of the evolution equation for dislocation density. Following Takeo (2006), after some calculations, one can obtain

$$\dot{\alpha}_{k}^{\ m} + \varepsilon^{mij}(2J_{jk,i} + \varepsilon_{njk}\Upsilon_{i}^{\ n}) = \frac{1}{2}\varepsilon^{mij}R_{ijtk}. \tag{18.77}$$

After some linearization, we can get the following relation

$$R_{ijtk} = \frac{1}{2}(\upsilon_{k,ij} - \upsilon_{k,ji}) = 0, \tag{18.78}$$

and, after combining these equations, we can arrive at the evolution equation for the dislocation density in the form

$$\dot{\alpha}_{k}^{m} = -\varepsilon^{mij}(J_{jk,i} + \varepsilon_{njk}\Upsilon_{i}^{n}). \tag{18.79}$$

The previous equations allow us to find the following relationship between the dislocation tensor and disclination tensor

$$\alpha^{p}_{i,p} = -\varepsilon_{ipq}\theta^{pq}. \tag{18.80}$$

18.18 Spin Energy Potential

Following Hehl and Obukhov (2007), we can define a spin energy potential $\aleph$, which satisfies the following relation

$$\omega_{[\mu\nu]} = \hat{S}_{\mu\nu} = C\frac{\partial\aleph}{\partial\alpha_{\mu\nu}}, \tag{18.81}$$

where $\omega_{[\mu\nu]} = \hat{S}_{\mu\nu}$ is the spin tensor, C is a constant coefficient, and $\alpha_{\mu\nu}$ is the dislocation density.

18.19 Degenerate Asymmetric Continuum in Terms of Spinors: Analogy to Maxwell's Equations

Using the properties of differential forms on a spacetime, one can express the equation of a superfluid in a form similar to Maxwell's equations. Vorticity current becomes the source term and the Euler equations can be viewed as the extention to the entire spacetime of the known fact that the number of vortex lines crossing any 2D surface spanned on a closed curve can be related with a circulation around this curve. A similar strategy can be applied for the ideal MHD. It appears that by applying this approach some helicity conservation laws may be obtained. Sciama (1962) pointed out the analogy between charge and spin in general relativity. Teisseyre (2005) introduced a concept of a degenerated asymmetric continuum that has only rotational motions but has no translational motions at all. He claimed that for such a continuum (where only spin and twist motions are present) the respective equations appear to have exactly the form of Maxwell-like equations (see also Teisseyre and Białecki 2005, Teisseyre et al 2006). Here, we formulate these equations for spin and twist motions in a spinor basis as follows

$$\omega_{AB} = \nabla_{A'(A}\Theta^{A'}_{B)}, \tag{18.82}$$

$$\nabla^{A}_{B'}\omega_{AB} = 2\pi\ \Upsilon_{BB'}, \tag{18.83}$$

where $\Theta^{A'}_{B}$ is the spin potential and $\Upsilon_{BB'}$ is the disclination current.

18.20 Conclusions

This chapter briefly reviews the main results concerning analogies between the theory of elastic crystal with a continuous defect distribution and the Einstein-Cartan theory of gravitation. The Riemann-Cartan space provides a very general framework for our comparisons because it contains both: curvature and torsion. A defected elastic crystal can be described by the defected metric $g_{\mu\nu} = \eta_{\mu\nu} + h_{\mu\nu}$, where $h_{\mu\nu}$ is proportional to the self distortion $\bar{\beta}_{\mu\nu}$. This defected metric contains all rotational and translational effects which are a consequence of the defects presented in the crystal. We recalled the disclination—curvature and the dislocation—torsion analogies. Moreover, an analogy between the disclination density and the Einstein tensor was recalled. We expressed the curvature and torsion in terms of spinors. Due to the disclination—curvature analogy, we identified the curvature spinors as disclination spinors. Consequently, due to the dislocation—torsion analogy, we identified the torsion spinors as the dislocation spinors. In addition, we defined the spin energy potential as a function of defect density. Finally, we formulated the Maxwell-like equations for spin and twist motions in elasticity in a spinor basis.

The presence of dislocations accompanied by the vanishing of curvature tensor ($R_{ijlk} = 0$) can be very well described in the framework of the Weitzenböck space, which is also the most useful space for the description of the Einstein's teleparallelism theory. The GRT was formulated in the Riemann space, which is a very good framework for describing an elastic crystal with disclinations. The Einstein-Cartan theory is a more general theory of gravitation that is making use of the Riemann-Cartan geometry. A spacetime with torsion and curvature can be generated from a flat spacetime, e.g., the Minkowski spacetime, by employing general coordinate transformations and is similar to an elastic crystal which has undergone irreversible deformation and is filled with disclinations and dislocations. The physical laws in curved spacetime are the direct images of the flat-spacetime laws under nonholonomic (i.e., multivalued) mappings.

To be more specific, we can say that torsion in the Einstein-Cartan theory is proportional to the dislocation density α_{ij}. Disclinations generate spacetime curvature, i.e., an elastic crystal containing disclinations is curved in the differential geometric sense. Curvature is an evidence of existing of disclinations and these are rotational defects. The disclination density θ_{ij} is identical with the Einstein tensor G_{ij}. Using the defect language, it is safe to say that the fundamental Bianchi's identities are a nonlinear generalization of the conservation laws of defect densities. These conservation laws can be formulated as the conservation laws of energy-momentum and angular momentum from an appropriate variational principle. These laws results from the invariance under general coordinate transformations, which may be viewed as the Poincaré group acting on orthonormal frames in the tangent spaces of the material manifold $\mathfrak{M}$ (i.e., local translations, and under local Lorentz transformations, respectively). These transformations correspond to elastic deformations (translational and rotational Lorentz transformations) of the space and the invariance insures that elastic deformations are independent on the defect structure. Thus, curvature and torsion can be viewed as the surface densities of Lorentz transformations and translations, respectively.

References

Amari S (1962) On some primary structures of non-Riemannian plasticity theory, RAAG Mem, **3**, 99-108

Amari S (1968) A geometrical theory of moving dislocations and elasticity, RAAG Mem, **4**, 142-152

Amari S (1981) Dualistic theory of non-Riemannian material manifolds, Int J Engng Sci **19**: 1581-1594

Bilby BA, Bullough R, Smith E (1955) Continuous distributions of dislocations: a new application of the methods of non-Riemannian geometry, Proc Roy Soc London, **A231**: 263-273.

Cartan É (1922) Sur une généralization de la notion de courbure de Riemann et les espaces à torsion, Comptes Rendus de l' Academie des Scieces (Paris), **174**, pp 593

Cosserat E and F (1909) Théorie des corps déformables, Hermann, Paris

Einstein A (1955a) The meaning of relativity, Appendix II of the 5th edition, Princeton University Press, Princeton

Einstein A (1955b) Preface, In: M Pantaleo (ed) Cinquant'anni di relatività 1905-1955. Edizioni Giuntine and Sansoni Editore, Firenze

Hehl FW, McCrea JD, Mielke EW, Ne'eman Y (1995) Metric-affine gauge theory of gravity: field equations, Noether identities, world spinors, and breaking of dilation invariance, Phys Reports **258**: 1-171

Hehl FW, Obukhov YuN (2007) Élie Cartan's torsion in geometry and in field theory, an essay, Annales de la Fondation Louis de Broglie, Manuscript

Holländer EF (1962) The dislocation equation of dislocation dynamics, Czech J Phys **B12**, 35-47

Kleinert H (1988) Lattice defect model with two successive melting transitions. Phys Lett **A130**: 443

Kleinert H (1989) Gauge Fields in Condensed Matter. World Scientific, Singapore

Kleinert H (2008) Gravity with Torsion (in preparation)

Kondo K (1952) The geometry of defects. In: Proc II Japan Nat Congress Appl Mech, Tokyo

Kondo K (1955) Geometry of elastic deformation and incompatibility, RAAG Memoirs 1, Div C, pp 361-373

Kondo K (1962) Continuum theory of defects. In: Kondo K (ed) RAAG Memoirs of the Unified Study of Basic Problems in Engineering and Science by Means of Geometry, vol 3: 148

Kossecka E, deWit R (1977a) Disclination kinematics, Arch Mech **29**: 633-651

Kossecka E, deWit R (1977b) Disclination dynamics, Arch Mech **29**: 749-767

Kröner E (1981) Continuum theory of defects, In: Balian R, Kléman M, Poirer JP (eds) Physics of Defects, Proc Les Houches XXXV, North Holland Publ Comp, Amsterdam

Majewski E (2006a) Rotational energy and angular momentum of earthquakes. In: Teisseyre R, Takeo M, Majewski E (eds) Earthquake source asymmetry, structural media and rotation effects, Springer-Verlag, Berlin Heidelberg, pp 217-225

Majewski E (2006b) Soliton physics. In: Teisseyre R, Takeo M, Majewski E (eds) Earthquake source asymmetry, structural media and rotation effects, Springer-Verlag, Berlin Heidelberg, pp 113-128

Majewski E (2006c) Seismic rotation waves: spin and twist solitons. In: Teisseyre R, Takeo M, Majewski E (eds) Earthquake source asymmetry, structural media and rotation effects, Springer-Verlag, Berlin Heidelberg, pp 255-272

Majewski E (2006d) Tectonic solitons propagating along the fault. In: Teisseyre R, Takeo M, Majewski E (eds) Earthquake source asymmetry, structural media and rotation effects, Springer-Verlag, Berlin Heidelberg, pp 301-309

Minagawa (1971) Dislocations, disclinations, and the source or sink of dislocation-line, RAAG Research Notes, Third Series, No 177, Res Ass Appl Geomet, Tokyo, pp 1-10

Minagawa (1979) A non-Riemannian geometrical theory of imperfections in a Cosserat continuum, Arch Mech **31**: 783-792

Ruggiero ML, Tartaglia A (2003) Einstein-Cartan theory as a theory of defects in space-time, Amer J Phys **71**: 1303-1313

Sacharov AD (1967) DAN SSSR **177**, 70. Reprinted in AD Sacharov (2000) Gen Rel Grav **32**, 365

Schouten JA (1954) Ricci Calculus, Springer, Berlin

Sciama DW (1962) On the analogy between charge and spin in general relativity. In: (Volume dedicated to L Infeld) Recent Developments in General Relativity, Pergamon Press-PWN, Oxford Warszawa, pp 415-439

Takeo M (2006) Rotational motions excited by earthquakes. In: R Teisseyre, M Takeo, E Majewski (eds.), Earthquake source asymmetry, structural media and rotation effects, Springer-Verlag, Berlin Heidelberg, pp 131-156

Teisseyre R (1964a) The method of the continuous dislocation field and its application to the fold theory, Bull Seismol Soc Am **54**: 1059-1072

Teisseyre R (1964b) A method of solving the field equation for local structural anomalies of the medium, Acta Geophys Pol **12**: 13-22

Teisseyre R (1969) Dislocational representation of thermal stresses, Acta Geophys Pol **17**: 3-12

Teisseyre R (1986) Some problems of mechanics of the continuum media and the applications to eartquake studies. In: R Teisseyre (ed) Continuum theories in solid Earth physics, Elsevier-PWN, Amsterdam Warsaw, pp 256-309

Teisseyre R (1995) Differential geometry methods in deformation problems. In: R. Teisseyre (ed.) Theory of earthquake premonitory and fracture processes, PWN – Polish Scientific Publishers, Warszawa, pp 503-544

Teisseyre R (2001) Deformation dynamics: continuum with self-deformation nuclei. In: R Teisseyre and E Majewski (eds) Earthquake thermodynamics and Phase transformations in the Earth's interior, Academic Press, San Diego, pp 143-165

Teisseyre R (2004) Spin and twist motions in a homogeneous elastic continuum and cross-band geometry of fracturing. Acta Geophys Pol **52**: 173-183

Teisseyre R (2005) Asymmetric continuum mechanics: deviations from elasticity and symmetry. Acta Geophys Pol **53**: 115-126

Teisseyre R, Białecki M (2005) Complex relativity: gravity and electromagnetic fields. arXiv:physics/0506033

Teisseyre R, Boratyński W (2006) Deviations from symmetry and elasticity: asymmetric continuum mechanics. In: R Teisseyre, M Takeo, E Majewski (eds.), Earthquake source asymmetry, structural media and rotation effects, Springer-Verlag, Berlin Heidelberg, pp 31-41

Teisseyre R, Białecki M, Górski M (2006) Degenerated asymmetric continuum Theory. In: R Teisseyre, M Takeo, E Majewski (eds.), Earthquake source asymmetry, structural media and rotation effects, Springer-Verlag, Berlin Heidelberg, pp 43-55

Trautman A (1973a) Spin and torsion may avert gravitational singularities, Nature (Phys Sci) **242**: 7

Trautman A (1973b) On the structure of the Einstein-Cartan equations, Symp Math **12**: 139-162

Trautman A (2006) Einstein-Cartan theory, In: Encyclopedia of Math Physics, J-P Francoise et al (eds), Elsevier, Oxford, pp 189-195

19 Twistors as Spin and Twist Solitons

Eugeniusz Majewski

Institute of Geophysics, Polish Academy of Sciences
ul. Księcia Janusza 64, 01-452 Warsaw, Poland
e-mail: emaj@igf.edu.pl

19.1 Introduction

Penrose and Rindler (1986) applied twistors to describe massless spinning particles. They pointed out that twistors can describe twisted photons or charges for massless spin-3/2 fields. Due to the well-known particle-soliton duality, it seems reasonable to relate twistors with spin and twist solitons. We adopt the twistor quantization theory developed by Penrose and Rindler (1986) and employ twistors to describe spin and twist solitons, i.e., quanta of spin and twist energy (cf., Majewski 2006a, b, c, d, e).

19.2 The Twistor Equation

To deal properly with a twistor, one needs a twistor equation. Let us start from the twistor equation introduced by Penrose (1968) in the form

$$\nabla_{A'}^{(A}\omega^{B)} = 0. \tag{19.1}$$

The solutions of the twistor equation establish a vector space over the complex numbers. The solutions ω^A of the twistor equation are determined by the four complex components. These solutions ω^A form a 4D vector space $\mathbb{T}^\alpha$ over the complex numbers called twistor space. The elements of twistor space are called $\begin{bmatrix}1\\0\end{bmatrix}$-twistors. We can denote the solution ω^A as $Z^\alpha = [\omega^A]$. Based on the $\begin{bmatrix}1\\0\end{bmatrix}$-twistors, one can create twistors of arbitrary valence $\begin{bmatrix}p\\q\end{bmatrix}$. Unfortunately, the higher-valence twistors cannot be composed of single fields of spinors. In order to have a more systematic approach to twistors in terms of spinor-field descriptions, it is better to apply the pair of spinor fields ω^A, $\pi_{A'}$ to describe Z^α than to use ω^A alone.

19.3 Twistor Definition

Penrose (1968) introduced a concept of a twistor or an α-plane in flat spaces or α-surface in curved spaces. In fact, he introduced four definitions of twistors in curved spacetime, i.e., local twistors, global null twistors, hypersurface twistors and asymptotic twistors. A local twistor Z^α at $Q \in \Im$ can be described by a pair of spinors ω^A, $\pi_{A'}$ at Q:

$$Z^\alpha \leftrightarrow \left(\omega^A, \pi_{A'}\right), \tag{19.2}$$

with respect to the metric g on $\Im$. After a conformal rescaling of the metric, according to the rule: $\hat{g} = \Lambda^2 g$, the twistor Z^α can be expressed in the form

$$\left(\hat{\omega}^A, \hat{\pi}_{A'}\right) = \left(\omega^A, \pi_{A'} + \mathrm{i}\, W_{AA'}\, \omega^A\right), \tag{19.3}$$

where $W_{AA'} = \nabla_{AA'} \log(\Lambda)$. The comparison of local twistors at different points of $\Im$ makes it necessary to introduce the local twistor transport along a curve ξ in $\Im$ with tangent vector υ. This does not lead to a displacement of the twistor along ξ, but moves the point with respect to which the twistor is defined.

19.4 Twistor Quantization Theory Applied to Spin and Twist Solitons

Following Penrose and Rindler (1986), let us recall some aspects of twistor quantization theory and consider the twistor wave functions f in order to formulate analogical spin and twist soliton wave functions Ψ. It is worthwhile to review here the standard quantization procedure of quantum mechanics. This procedure deals with the position x^α and the linear momentum p_α of a particle. One of these quantities may play a role of an independent variable in the particle complex wave function and at the same time it may play a role of a multiplication operator, while the other quantity may act on the wave function as a differentiation operator. However, these two quantities may exchange their roles and the former quantity may act on the wave function as a differentiation operator, while the latter may become an independent variable of the wave function and at the same time act as a multiplication operator. Thus, in the framework of this procedure, the position x^α and the linear momentum p_α of a particle can be treated as operators in the following commutation law

$$p_\alpha x^\beta - x^\beta p_\alpha = \mathrm{i}\hbar g_\alpha^\beta . \tag{19.4}$$

When we choose the x-space description, then a particle wave function is represented by a complex function $\psi(x^\alpha)$. Note that this function has only one variable—the position x^α. The wave function does not depend on the momentum p_α On one hand, the momentum operator $[p_\alpha]$ takes the form of the differentiation operator,

$$[p_\alpha]\psi \to \mathrm{i}\hbar \frac{\partial \psi}{\partial x^\alpha}. \tag{19.5}$$

On the other hand, the position x^α takes the form of the multiplication operator,

$$[x^\alpha]\psi \to x^\alpha \psi . \tag{19.6}$$

If we choose the p-space description, then the wave function is represented by a complex function $\tilde{\psi}(p_\alpha)$. This function has only one variable—the momentum p_α In this case, the wave function does not depend on the position x^α. Now, the momentum and position operators $[p_\alpha]$ and $[x^\alpha]$, respectively, exchange their roles (cf., Penrose and Rindler 1986)

$$[p_\alpha]\tilde{\psi} \to p_\alpha \tilde{\psi} \tag{19.7}$$

and

$$[x^\alpha]\tilde{\psi} \to -\mathrm{i}\hbar \frac{\partial \tilde{\psi}}{\partial p_\alpha}. \tag{19.8}$$

If we apply the x-space description or the p-space description, then we deal with commuting variables and we have to use only one variable i.e., either the position x^α or the momentum p_α. We cannot use both as our variables at the same time, because there is no such thing as a particle with determined position and determined momentum at the same time. If one is determined, the other is not. Thus, the other should act as a differential operator.

Penrose and Rindler (1986) applied a similar quantization procedure for twistors. They considered the twistors Z^α and $\overline{Z}^\alpha$ and the dual twistors $\overline{W}^\alpha$ and W_α. In the virtue of Eq. (19.4), the twistors Z^α and $\overline{Z}^\alpha$ can be treated as operators in the following commutation law

$$Z^\alpha \overline{Z}_\beta - \overline{Z}_\beta Z^\alpha = \hbar \delta_\beta^\alpha . \tag{19.9}$$

Now, we apply twistors to describe spin and twist solitons. Before, we had the x-space description, now, using twistors, we apply the $\mathbb{T}^{\alpha}$-description. However, in our case, we use the $\mathbb{T}^{\alpha}$-description to describe the massless spin and twist solitons. In this description of spin and twist solitons, we consider a soliton wave function $\hat{\Psi}$ which is a complex function expressed in terms of Z^{α}. Thus, this function has only one variable—the twistor Z^{α}. The soliton wave function does not depend on the complex conjugate $\bar{Z}_{\alpha}$, which can be expressed as the condition

$$\frac{\partial \hat{\Psi}}{\partial \bar{Z}_{\alpha}} = 0. \tag{19.10}$$

So, the complex conjugate—the operator $[\bar{Z}_{\alpha}]$ becomes a differentiation operator acting on the soliton wave function in the form

$$[\bar{Z}_{\alpha}]\,\hat{\Psi} \to -\hbar \frac{\partial \hat{\Psi}}{\partial Z^{\alpha}}. \tag{19.11}$$

At the same time, the twistor Z^{α} acts as a multiplication operator,

$$[Z^{\alpha}]\,\hat{\Psi} \to Z^{\alpha}\,\hat{\Psi}.. \tag{19.12}$$

Now, if we apply the $\mathbb{T}_{\alpha}$-description, then the twistor and its complex conjugate Z^{α} and $\bar{Z}_{\alpha}$, respectively, exchange their roles. The soliton wave function $\hat{\Psi}$ has now only one variable—the complex conjugate $\bar{Z}_{\alpha}$; or if we denote $\bar{Z}_{\alpha}$ as its dual twistor W_{α} and Z^{α} as $\bar{W}^{\alpha}$, then the soliton wave function $\hat{\Psi}$ has only one variable—the dual twistor W_{α}. Correspondingly, the dual multiplication and differentiation operators $[W_{\alpha}]$ and $[\bar{W}^{\alpha}]$, respectively, act on the soliton wave function as follows

$$[W_{\alpha}]\,\hat{\Psi} \to W_{\alpha}\,\hat{\Psi} \tag{19.13}$$

and

$$[\bar{W}^{\alpha}]\,\hat{\Psi} \to \hbar \frac{\partial \hat{\Psi}}{\partial W_{\alpha}}. \tag{19.14}$$

If we apply the $\mathbb{T}^{\alpha}$-description or the $\mathbb{T}_{\alpha}$-description, then we deal with commuting variables and we have to use only one variable i.e., either the twistor Z^{α} or its complex conjugate $\bar{Z}_{\alpha}$ or one of the dual twistors W_{α} or its complex conjugate $\overline{W}^{\alpha}$. We cannot use both as our variables at the same time. Thus, the other should act as a differential operator.

19.5 The Spin Operator

In order to obtain some interesting links between twistors and the space-time variables, one can describe the mechanical system in terms of the linear momentum p_α and angular momentum $M_{\alpha\beta}$. We can employ the transformation properties of linear and angular momenta of a spin soliton under a shift of origin from 0 to a point $Q(x^\alpha)$ (Penrose and Rindler 1986),

$$p_\alpha(Q) = p_\alpha(0), \tag{19.15}$$

$$M_{\alpha\beta}(Q) = M_{\alpha\beta}(0) + p_\alpha x^\beta - p_\beta x^\alpha . \tag{19.16}$$

One can express the Pauli-Lubanski spin vector in the form

$$S_\alpha = \frac{1}{2}\varepsilon_{\alpha\beta\gamma\kappa} p^\beta M^{\gamma\kappa} \tag{19.17}$$

For massless spin solitons, the Pauli-Lubanski vector can be expressed as a multiple of the momentum

$$S_\alpha = s p_\alpha , \tag{19.18}$$

where the real number s is called the helicity of the spin soliton and $|s|$, or $|s|\hbar^{-1}$ is called the spin. For quantum systems s is an integer multiple of ½$\hbar$. In case of the non-commutative variables, the quantum operator expressing the helicity s, can be described as follows

$$[s] = \frac{1}{4}\left(Z^\alpha \overline{Z}_\alpha + \overline{Z}_\alpha Z^\alpha\right) = -\frac{1}{2}\hbar\left(Z^\alpha \frac{\partial}{\partial Z^\alpha} + 2\right), \tag{19.19}$$

where the complex conjugate is determined as $\overline{Z}_\alpha = \left(\overline{\pi}_A, \overline{\omega}^{A'}\right)$.

The spin operator for the commutative case (Penrose and Rindler 1986) can be expressed in terms of twistors in the following form

$$[s] = \frac{1}{2} Z^\alpha \overline{Z}_\alpha . \tag{19.20}$$

19.6 The Twist Operator

A twist operator (Caneschi et al 1969) may take the form

$$\Omega = (-1)^n, \tag{19.21}$$

where the exponent n denotes the n-th oscillation.

Now, the rotational field can be expressed as

$$\omega_{\alpha\beta} = \omega_{[\alpha\beta]} + \mathrm{i}\Omega\ \omega_{(\alpha\beta)}, \tag{19.22}$$

where $\omega_{[\alpha\beta]}$ and $\omega_{(\alpha\beta)}$ are the spin and twist parts, respectively.

19.7 Spin and Twist Solitons Described by the Nonlinear Schrödinger Equation

Spin and twist solitons were described in Majewski (2006a, b, c, d, e) as quanta of spin and twist energy, respectively. He investigated nonlinear waves. The wave nonlinearity introduces a new possibility that the spin soliton speed may depend on the magnitude of a spin vector. In such a case, the soliton width and amplitude are modulated during the process of propagation. Its profile changes as a function of the magnitude of a spin vector. The spin and twist solitons are self-trapped packets of energy that propagate without loss of energy and momentum. The solitons are a result of the balance of nonlinearity (due to finite deformations and elastic properties of the medium) and dispersion (due to the microstructure of the medium).

Nikolaevskiy (1996) started from a set of nonlinear equations for the twist and longitudinal waves in the form

$$V_1^2 \frac{\partial^2 \omega}{\partial x^2} + \frac{\partial u}{\partial x}\left(V_1^2 \frac{\partial^2 \omega}{\partial x^2} - C_1 \omega \right) - C_1 \omega = \frac{\partial^2 \omega}{\partial t^2}, \tag{19.23}$$

$$C_2 \frac{\partial^4 u}{\partial x^4} + C_3 \frac{\partial u}{\partial x}\frac{\partial^2 u}{\partial x^2} + V_2^2 \frac{\partial^2 u}{\partial x^2} - C_4 \frac{\partial \omega^2}{\partial x} = \frac{\partial^2 u}{\partial t^2}, \tag{19.24}$$

where ω is the twist angle, u is the displacement, C_1, C_2, C_3, C_4, are the elastic constants, V_1 and V_1 are the twist and longitudinal waves speeds, respectively, and they are expressed as

$$V_1 = \sqrt{\frac{A}{\rho_0 J}} \quad \text{and} \quad V_2 = \sqrt{\frac{E}{\rho_0}}, \tag{19.25}$$

where A and E are elastic moduli and J is the moment of inertia of a twisting grain, ρ_0 is the reference material density, and as deformation progresses, the material density depends on displacements as

$$\rho = \frac{\rho_0}{1 + \frac{\partial u}{\partial x}}. \tag{19.26}$$

It was assumed that the stress tensor is decomposed into the symmetrical and antisymmetrical parts as follows

$$S_{\alpha\beta} = S_{(\alpha\beta)} + S_{[\alpha\beta]}. \tag{19.27}$$

The antisymmetrical part of the stress tensor was assumed as a function of the twist vector ω_κ in the form

$$S_{[\alpha\beta]} = C_4 \varepsilon_{\alpha\beta\kappa} \omega_\kappa , \tag{19.28}$$

where C_4 is the elastic constant and $\varepsilon_{\alpha\beta\kappa}$ is the permutation symbol.

The total stress is expressed as a function of the square of the twist angle, i.e.,

$$S^T = C_5 \frac{\partial^3 u}{\partial x^3} + C_6 \frac{\partial u}{\partial x}\frac{\partial u}{\partial x} + E\frac{\partial u}{\partial x} - C_7 \omega^2 , \tag{19.29}$$

where C_5, C_6, C_7, are elastic constants.

The couple-stress is assumed as a function of the twist angle, i.e.,

$$\mathbb{S} = A\frac{\partial \omega}{\partial x}. \tag{19.30}$$

Now, we collect the main relations between elastic constants in the form

$$C_1 = \frac{C_4}{\rho J}, \qquad C_2 = \frac{C_8}{\rho_0}, \qquad C_3 = \frac{2C_6}{\rho_0} + V_2^2 , \tag{19.31}$$

where C_8 is the elastic constant.

Following Nikolaevskiy (1996), we use the running coordinate system (Ξ, Θ) for a new length and time scales, as follows

$$\Xi = \Theta(x - V_g t), \qquad \wp = \Theta^2 t , \tag{19.32}$$

where V_g is the group velocity.

Now, we change variables, according to the following substitutions

$$\mathfrak{A} = \frac{\partial u}{\partial \Xi}, \qquad \omega = \Psi e^{i\varphi}. \tag{19.33}$$

where $\mathfrak{A}$ is the amplitude of deformation and Ψ is the amplitude of the twisting angle oscillations.

The above assumptions allow us to transform Eqs. (19.23) and (19.24) into the following set of equations

$$2\,i\,\chi \frac{\partial \Psi}{\partial \Theta} - \frac{\partial^2 \Psi}{\partial \Xi^2}\left(V_2^2 - V_g^2\right) = \left(\chi^2 V_2^2 + 2\omega_0^2\right)\mathfrak{A}\Psi, \tag{19.34}$$

$$\left(V_g^2 - V_2^2\right)\frac{\partial^2 \mathfrak{A}}{\partial \Xi^2} - 2\Theta V_g \frac{\partial^2 \mathfrak{A}}{\partial \wp \partial \Xi} = 2C_7 \Theta^{2(n-1)} \frac{\partial^2 |\Psi|^2}{\partial \Xi^2}, \tag{19.35}$$

where

$$\varphi = \chi \Xi - \phi \wp, \qquad V_g = \frac{d\phi}{d\chi}, \qquad \omega_0 = \frac{\sqrt{C_1}}{2}. \tag{19.36}$$

According to Nikolaevskiy (1969), we can infer that the wavelength of the twisting oscillations is the order of the size of twisting grains in the material. The characteristic internal length $l \approx \sqrt{J}$ corresponding to the grain size is present in the parameters of Eqs. (19.34) and (19.35).

Taking $n = 1$ in the exponent in Eq. (19.35) and assuming that $\Theta \to 0$, Nikolaevskiy (1969) obtained the following nonlinear relation between the amplitudes $\mathfrak{A}$ and Ψ

$$\mathfrak{A} = \frac{2C_7 |\Psi|^2}{V_g^2 - V_2^2}. \tag{19.37}$$

Finally, using Eq. (19.34), Nikolaevskiy (1996) arrived at the nonlinear Schrödinger equation for the amplitude of the twisting angle oscillations Ψ in the following form

$$2\,i\,\chi \frac{\partial \Psi}{\partial \wp} + \left(V_2^2 - V_g^2\right)\frac{\partial^2 \Psi}{\partial \Xi^2} = 2C_4 \frac{\chi^2 V_2^2 + 2\omega_0^2}{V_1^2 - V_g^2} |\Psi|^2 \Psi. \tag{19.38}$$

Nikolaevskiy (1969) found that the envelope of high-frequency twisting oscillations described by the above equation has the soliton form as follows

$$\Psi = \Psi_0 \sqrt{\frac{2\left(V_1^2 - V_g^2\right)}{\chi^2 V_2^2 - 2\omega_0^2}} \exp\left\{ \mathrm{i} \frac{\mathfrak{A}_0 \Xi}{2\sqrt{V_2^2 V_1^2}} - \mathrm{i}\left(\frac{\mathfrak{A}_0^2}{4} - \Psi_0^2 \right) \right\} \operatorname{sech} \Psi_0 \left(\frac{\Xi}{\sqrt{V_2^2 - V_1^2}} - \mathfrak{A}_0 t \right) \tag{19.39}$$

The above equation describes an explicit form of the twist Schrödinger solitons.

19.8 The Fracture Solitons

Teisseyre and Yamashita (1999) split the stress motion equation into seismic wave and fault-related fields (see also Boratyński and Teisseyre 2006, Teisseyre et al 2006). We apply this method to Eq. (19.38) and split the nonlinear Schrödinger equation for the amplitude of the twisting angle oscillations Ψ into the elastic soliton and fracture-zone related soliton equations. Equation (19.38) is nonlinear and superposition methods do not apply here. Nevertheless, the elastic part of the twisting angle is relatively small, because the self-field is dominant in the fracture zone. Thus, in the limit, we may assume that the self-twisting oscillation field along the fracture zone is almost equal to the total twisting field during a fracture process. The equation for self-twisting amplitude obtained as a result of splitting Eq. (19.38) can be viewed as an approximation.

An elastic part of the amplitude of the twisting angle oscillation Ψ can be presented as the difference Ψ^T - Ψ^S between the total Ψ^T field and the self-twist part Ψ^S, which is assumed to rapidly decrease away from the fracture surface. We can identify these parts with an elastic field $\Psi^T = \tilde{\Psi}$ and a fracture-zone related field $\Psi^S = \overline{\Psi}$. After splitting the nonlinear Schrödinger equation (19.38), the first equation for the elastic twist field can be expressed as

$$2\,\mathrm{i}\,\chi \frac{\partial \tilde{\Psi}}{\partial \wp} + \left(V_2^2 - V_g^2\right) \frac{\partial^2 \tilde{\Psi}}{\partial \Xi^2} = 2C_4 \frac{\chi^2 V_2^2 + 2\omega_0^2}{V_1^2 - V_g^2} |\, \tilde{\Psi} \,|^2\, \tilde{\Psi}. \tag{19.40}$$

The above equation describes the elastic twist Schrödinger solitons.

The fracture-zone related nonlinear Schrödinger equation for the self-twist amplitude takes the form

$$2\,\mathrm{i}\,\chi \frac{\partial \overline{\Psi}}{\partial \wp} + \left(V_2^2 - V_g^2\right) \frac{\partial^2 \overline{\Psi}}{\partial \Xi^2} = 2C_4 \frac{\chi^2 V_2^2 + 2\omega_0^2}{V_1^2 - V_g^2} |\, \overline{\Psi} \,|^2\, \overline{\Psi}. \tag{19.41}$$

The above equation describes the fracture twist Schrödinger soliton. These solitons are twist pulses. The latter equation describes the self-twist soliton that can be excited by past fracture processes and may propagate slowly along the fracture zone to trigger new fracture processes.

19.9 The Robinson Congruences

Penrose visualization of Ivor Robinson congruence is based on a geometrical construction that was designed to find solutions of Maxwell's free-space equations. Maxwell's field is geodetic, shear-free, non-singular and twisting. Robinson's idea was to base these solutions on a twisting shear-free congruence of light rays (null lines). Such a family of null straight lines (rays) whose tangent directions constitute this field is called the Robinson congruence (cf., Penrose and Rindler 1984).

Figure 19.1 illustrates a motion of the spin soliton in spacetime. It shows two time-slices of the tangents to a twisting family of circles at time $t = 0$ and at $t = t_1$. The figure is based on Penrose geometrical visualization of a Robinson congruence (Penrose and Rindler 1986).

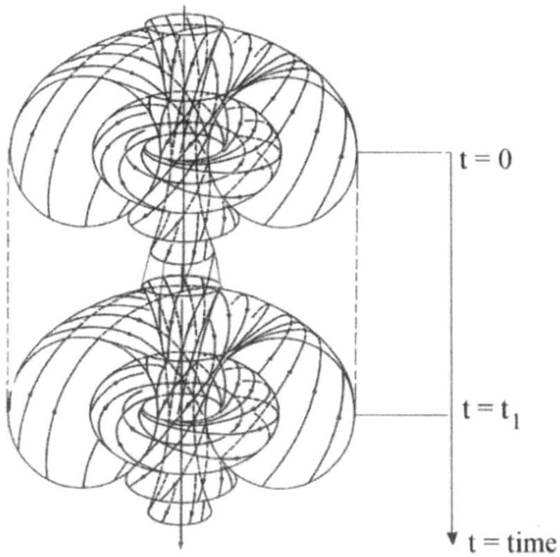

Fig. 19.1 Motion of the spin soliton in spacetime. The first time-slice of the tangents to a twisting family of linked circles is visualized at time $t = 0$; the second time-slice is showed at time $t = t_1$; the entire configuration moves downward along the time axis. The figure is a modification of Penrose geometrical visualization of a Robinson congruence (modified from Penrose and Rindler 1986)

19.10 Conclusions

We summarized briefly a few results of the Penrose twistor theory and attempted to find some links to spin and twist solitons. The twistor wave function was identified as the spin and twist soliton wave function. The Schrödinger soliton equation derived by Nikolaevskiy (1996) for an amplitude of twist waves, was recalled. Here, these results were applied to modeling twist solitons propagating along the fracture surface. The soliton equation was split into the elastic soliton and fracture surface-related soliton equations. The twist soliton can propagate along the fracture surface to trigger new fracture processes. The main conclusion here is that twistors can be used successfully to describe spin and twist solitons. Moreover, the twistor space can be helpful to characterize spin and twist solitons. Using the rich formalism of twistor theory, we can get much deeper insight into the spin and twist solitons. Whether the presented suggestions and links will help us improve the theory of spin and twist soliton propagation in structured media is, for the time being, an unanswered question. However, the presented links certainly reveal the power of twistor theory and the indication that we have essential analogies which hold for massless particles in quantum physics and for twistors in twistor theory but are likely to hold for spin and twist solitons as well. Last but not least, we can view all the above-mentioned links and analogies in a much broader perspective. As time progresses, they may contribute to a remarkable advance toward a possible duality between the physics of continua and twistor theory.

References

Caneschi I, Schwimmer A, Veneziano G (1969) Phys Lett **30B**: 351

Boratyński W, Teisseyre R (2006) Fault dynamics and related radiation. In: Teisseyre R, Takeo M, Majewski E (eds) Earthquake source asymmetry, structural media and rotation effects, Springer-Verlag, Berlin Heidelberg, pp 77-89

Majewski E (2006a) Soliton physics. In: Teisseyre R, Takeo M, Majewski E (eds) Earthquake source asymmetry, structural media and rotation effects, Springer-Verlag, Berlin Heidelberg, pp 113-128

Majewski E (2006b) Seismic rotation waves: spin and twist solitons. In: Teisseyre R, Takeo M, Majewski E (eds) Earthquake source asymmetry, structural media and rotation effects, Springer-Verlag, Berlin Heidelberg, pp 255-272

Majewski E (2006c) Seismic rotation waves in the continuum with nonlinear microstructure. In: Teisseyre R, Takeo M, Majewski E (eds) Earthquake source asymmetry, structural media and rotation effects, Springer-Verlag, Berlin Heidelberg, pp 293-300

Majewski E (2006d) Tectonic solitons propagating along the fault. In: Teisseyre R, Takeo M, Majewski E (eds) Earthquake source asymmetry, structural media and rotation effects, Springer-Verlag, Berlin Heidelberg, pp 301-309

Majewski E (2006e) Complexity of rotation soliton propagation. In: Teisseyre R, Takeo M, Majewski E (eds) Earthquake source asymmetry, structural media and rotation effects, Springer-Verlag, Berlin Heidelberg, pp 255-272

Newman E, Penrose R (1962) J Math Phys **3**: 566-587

Nikolaevskiy VN (1996) Geomechanics and fluidodynamics, Kluwer Academic Publishers, Dordrecht

Penrose R (1968) Twistor quantization and curved space-time, Int J Theor Phys **1**: 61-99

Penrose R (1983) Spinors and torsion in general relativity, Found Phys **13**: 325-339

Penrose R, Rindler W (1984) Spinors and space-time, vol 1: Two-spinor calculus and relativistic fields, Cambridge University Press, Cambridge

Penrose R, Rindler W (1986) Spinors and space-time, vol 2: Spinor and twistor methods in space-time geometry, Cambridge University Press, Cambridge

Teisseyre R, Yamashita T (1999) Splitting stress motion equation into seismic wave and fault-related fields. Acta Geophys Pol **47**: 2, 135-147

Teisseyre R, Takeo M, Majewski E (eds) (2006) Earthquake source asymmetry, structural media and rotation effects, Springer-Verlag, Berlin Heidelberg

20 Potentials in Asymmetric Continuum: Approach to Complex Relativity

Roman Teisseyre

Institute of Geophysics, Polish Academy of Sciences
ul. Księcia Janusza 64, 01-452 Warszawa, Poland
e-mail: rt@igf.edu.pl

20.1 Introduction

In the frame of Standard Asymmetric Continuum Theory (see: Chap. 7) we have introduced the defect density fields; the deformations caused by defects can be adequately presented in the frame of the Riemannian geometry; many papers have been devoted to this kind of presentations; the first ones, that in the Cartan works (Cartan 1923, 1924, 1925), was influenced by work by Cosserat brothers (1909) in which a moment stress tensor is included in a generalized continuum. A gradual development of the Einstein-Cartan Theory (ECT) started by works of Sciama (1962), Kibble (1961) and Trautman (1972a, b, c, 1973); for a review see: Hehl et al. (1995). Kopczyński (1973) has proved that in the ECT the cosmological solutions become free from the singularities, leading to the modified Friedmann equation supplemented with the conservation laws for mass and spin (Trautmann 2006).

The continuum with defect content (dislocation and disclination densities) as described by the Riemannian curvature and torsion was considered by Bilby et al. (1955) and Kondo (1955, 1958) and later by Holländer (1962); Ben-Abraham (1970) and many other authors (for a review see: Teisseyre 1995a,b); the thermal stresses were found to have the same form as that related to dislocation field (Muskhelishvili 1953) and on this basis the thermal effects were included in the continuum with a Riemannian curvature by Kröner (1958), Teisseyre (1963, 1969) and Stojanovic et al. (1964).

In this chapter we approach this problem with the appropriately defined potentials for the spin and twist fields.

We recall the definition of the twist field as the oscillation of the shear off-diagonal axes and the derived equations of spin and twist motions (see:

Chap. 7). The Dirac tensor representation for the symmetric tensor was presented in Chap. 6:

$$\omega_{(\lambda\kappa)} = \omega_{(1)}\varepsilon^1 + \varepsilon_{(2)}\gamma^2 + \varepsilon_{(3)}\gamma^4\gamma^2\gamma^3 = \begin{bmatrix} 0 & -\omega_{(3)} & -\omega_{(2)} & -\omega_{(1)} \\ -\omega_{(3)} & 0 & \omega_{(1)} & -\omega_{(2)} \\ -\omega_{(2)} & \omega_{(1)} & 0 & -\omega_{(3)} \\ -\omega_{(1)} & -\omega_{(2)} & -\omega_{(3)} & 0 \end{bmatrix}$$

where $\omega_{(k)}$ are treated as scalars.

The above-mentioned definition of twist is based on the presented transition from the off-diagonal form of shear tensor to the invariantly defined antisymmetric twist tensor. Therefore, when introducing the potentials for the complex rotation field, spin and twist, we shall try to define such potentials in the adequate tensor forms.

The presented definition can be transferred to 3D as follows:

$$\omega_{(s)} = \frac{1}{2} \in_{skn} \omega_{[(kn)]}, \qquad \omega_{[(kn)]} = \begin{bmatrix} 0 & \omega_{(3)} & -\omega_{(2)} \\ -\omega_{(3)} & 0 & \omega_{(1)} \\ \omega_{(2)} & -\omega_{(1)} & 0 \end{bmatrix},$$

where $\omega_{[(kn)]}$ means the antisymmetric twist tensor (we shall remember that another symbol, $\omega_{[kn]}$ denotes the antisymmetric spin tensor).

We shall also remind that the defined complex rotation field has a close analogy to the electromagnetic field; on this basis we will indicate a new classic approach to the unified Complex Relativity Theory.

20.2 Natural Potentials

For the rotation field, $\omega_{[k]}$, $\omega_{(k)}$, we can introduce the following system of the antisymmetric potentials (cf. Teisseyre and Białecki 2005), in which we introduce the natural vector potentials $\tilde{A}_s$ and $\hat{A}_s$ and the charge and current potentials ψ and ψ_s:

$$\begin{aligned} &\omega_{[k]} = \varepsilon_{kbs}\frac{\partial \tilde{A}_s}{\partial x_b}, \qquad \omega_{(k)} = \varepsilon_{kbs}\frac{\partial \hat{A}_s}{\partial x_b} - \frac{\partial \psi}{\partial x_k}, \\ &\Delta\psi = -4\pi\rho, \qquad 4\pi J_k = \frac{\partial \dot{\psi}}{\partial x_k} + \varepsilon_{kbs}\frac{\partial \psi_s}{\partial x_b}, \end{aligned} \tag{20.1}$$

where we put the conditions

$$\frac{\partial \tilde{A}_n}{\partial x_n} = \frac{1}{c}\psi, \quad \frac{\partial \hat{A}_n}{\partial x_n} = 0, \quad 4\pi\varepsilon_{ndk}\frac{\partial J_k}{\partial x_d} = -\Delta\psi_n, \quad \frac{\partial \psi_n}{\partial x_n} = 0. \tag{20.2}$$

To enable the derivation of equations for $\omega_{[k]}$, and $\omega_{(k)}$, (see: Chap. 7), we demand that

$$\varepsilon_{kms}\frac{\partial \tilde{A}_s}{\partial x_m} + \frac{1}{c}\frac{\partial \hat{A}_k}{\partial t} = \frac{1}{c}\psi_k, \qquad \varepsilon_{kms}\frac{\partial \hat{A}_s}{\partial x_m} + \frac{1}{c}\frac{\partial \tilde{A}_k}{\partial t} = 0. \tag{20.3}$$

For these potentials we come to the wave equations:

$$\square\tilde{A}_n = \tilde{Y}_n, \qquad \square\hat{A}_n = \hat{Y}_n, \qquad \square A_{[\alpha\beta]} = Y_{[\alpha\beta]}, \tag{20.4}$$

where

$$\tilde{Y}_n = -\frac{1}{c}\varepsilon_{ndk}\frac{\partial \psi_k}{\partial x_d}, \qquad \hat{Y}_k = \frac{1}{c^2}\frac{\partial \psi_k}{\partial t},$$

$$A_n = \tilde{A}_n + i\hat{A}_n, \qquad Y_n = \tilde{Y}_n + \mathrm{i}\hat{Y}_n,$$

$$A_{[\alpha\beta]} = \begin{bmatrix} 0 & A_3 & -A_2 & -A_1 \\ -A_3 & 0 & A_1 & -A_2 \\ A_2 & -A_1 & 0 & -A_3 \\ A_1 & A_2 & A_3 & 0 \end{bmatrix}, \qquad Y_{\alpha\beta} = \begin{bmatrix} 0 & Y_3 & -Y_2 & -Y_1 \\ -Y_3 & 0 & \overline{Y}_1 & -Y_2 \\ Y_2 & -Y_1 & 0 & -Y_3 \\ Y_1 & Y_2 & Y_3 & 0 \end{bmatrix}$$

and from (20.2) and the conditions in (20.1) we obtain:

$$A_{[\alpha\beta],\beta} = \frac{1}{c}\psi_\alpha, \tag{20.5}$$

with $\psi_\kappa = \{\psi_k, \psi\}$.

This equation is fulfilled due to the conditions introdued in Eq. (20.1); note that the symmetric tensor of potentials could not satisfy this condition.

The potential tensor, $A_{[\alpha\beta]}$, can be written with the Dirac matrices $\overline{\varepsilon}^s$ (see: Chap. 6):

$$\begin{gathered} A_{[\alpha\beta]} = \mathrm{i}\overline{\varepsilon}^1 A_1 + \mathrm{i}\overline{\varepsilon}^2 A_2 + \overline{\varepsilon}^3 A_3, \\ \overline{\varepsilon}^1 = \varepsilon^1\varepsilon^4, \quad \overline{\varepsilon}^2 = \varepsilon^2\varepsilon^4, \quad \overline{\varepsilon}^3 = \varepsilon^1\varepsilon^2 \end{gathered} \tag{20.6}$$

where

$$\overline{\varepsilon}^1 = \mathrm{i}\begin{bmatrix} 0 & 0 & 0 & 1 \\ 0 & 0 & -1 & 0 \\ 0 & 1 & 0 & 0 \\ -1 & 0 & 0 & 0 \end{bmatrix}, \quad \overline{\varepsilon}^2 = \mathrm{i}\begin{bmatrix} 0 & 0 & 1 & 0 \\ 0 & 0 & 0 & 1 \\ -1 & 0 & 0 & 0 \\ 0 & -1 & 0 & 0 \end{bmatrix},$$

$$\overline{\varepsilon}^3 = \begin{bmatrix} 0 & 1 & 0 & 0 \\ -1 & 0 & 0 & 0 \\ 0 & 0 & 0 & -1 \\ 0 & 0 & 1 & 0 \end{bmatrix}.$$

20.3 Spin and Twist Fields in the Riemannian Space

Any Dirac matrices are defined by the conditions:

$$\overline{\varepsilon}^{\alpha}\overline{\varepsilon}^{\beta} + \overline{\varepsilon}^{\beta}\overline{\varepsilon}^{\alpha} = 2\eta^{\alpha\beta} \tag{20.7}$$

and therefore to generalize the derived relations for the Riemannian space we shall define the following complex perturbations, $h_{[\alpha,\beta]}$, to the metric tensor:

$$g^{\alpha\beta} = \eta^{\alpha\beta} + h^{\alpha\beta}, \quad h_{[\alpha\beta]} = A_{[\alpha\beta]}, \quad \mathrm{d}s^2 = g_{\alpha\beta}\mathrm{d}x^{\alpha}\mathrm{d}x^{\beta} = g_{(\alpha\beta)}\mathrm{d}x^{\alpha}\mathrm{d}x^{\beta} \tag{20.8}$$

Now we may define the complex Einstein-like tensor, $G_{\alpha\beta}$, and the related field relations:

$$\overline{G}_{\alpha\beta} \approx h_{[\mu\beta]}\Big|^{\mu}_{\alpha} + h_{[\mu\alpha]}\Big|^{\mu}_{\beta} - h_{[\alpha\beta]}\Big|^{\mu}_{\mu} = \frac{1}{c}\frac{\partial}{\partial x_{\beta}}\psi_{\alpha} + \frac{1}{c}\frac{\partial}{\partial x_{\alpha}}\psi_{\beta} - Y_{[\alpha\beta]}. \tag{20.9}$$

This tensor is asymmetric due to the introduced perturbations; however, its antisymmetry is not intrinsic as that related to torsion but is similar to that forced by strains in the material continuum: $g_{ks} - \delta_{ks} = 2E_{ks}$, where strains can be asymmetric, as follows from the asymmetric continuum theory (see: Chap. 7).

Splitting expression (20.9) into symmetric and antisymmetric parts we obtain relations as in (20.4) and (20.5):

$$h_{[\alpha\beta]}\Big|^{\mu}_{\mu} = Y_{[\alpha\beta]}, \qquad h_{[\mu\beta]}\Big|^{\mu}_{\alpha} + h_{[\mu\alpha]}\Big|^{\mu}_{\beta} = \frac{1}{c}\frac{\partial}{\partial x_{\beta}}\psi_{\alpha} + \frac{1}{c}\frac{\partial}{\partial x_{\alpha}}\psi_{\beta}. \tag{20.10}$$

20.4 Natural Potentials: Analogy to Electromagnetic Field

The analogy between the electromagnetic and rotation fields permits to introduce the antisymmetric tensor for the electromagnetic potentials; we can follow the sequence of relations (20.1-20.6) in a fully analogical way:

$$B_k = \varepsilon_{kbs}\tilde{A}_{s,b}, \quad E_k = \varepsilon_{kbs}\hat{A}_{s,b} - \psi_{,k}, \quad \tilde{A}_{n,n} = \frac{1}{c}\psi, \quad \hat{A}_{n,n} = 0, \tag{20.11}$$

$$\frac{\partial \tilde{A}_n}{\partial x_n} = \frac{1}{c}\psi, \quad \frac{\partial \hat{A}_n}{\partial x_n} = 0, \quad 4\pi\varepsilon_{ndk}\frac{\partial J_k}{\partial x_d} = -\Delta\psi_n, \quad \frac{\partial \psi_n}{\partial x_n} = 0, \tag{20.12}$$

$$\varepsilon_{kms}\tilde{A}_{s,m} + \frac{1}{c}\frac{\partial \hat{A}_k}{\partial t} = \frac{1}{c}\psi_k, \qquad \varepsilon_{kms}\hat{A}_{s,m} + \frac{1}{c}\frac{\partial \tilde{A}_k}{\partial t} = 0, \tag{20.13}$$

$$\Box\tilde{A}_n = \tilde{Y}_n, \qquad \Box\hat{A}_n = \hat{Y}_n, \qquad \Box A_{[\alpha\beta]} = Y_{[\alpha\beta]}, \tag{20.14}$$

$$\tilde{Y}_n = -\frac{1}{c}\varepsilon_{ndk}\frac{\partial \psi_k}{\partial x_d}, \qquad \hat{Y}_k = \frac{1}{c^2}\frac{\partial \psi_k}{\partial t},$$

$$A_n = \tilde{A}_n + i\hat{A}_n, \qquad Y_n = \tilde{Y}_n + \mathrm{i}\hat{Y}_n,$$

$$A_{[\alpha\beta]} = \begin{bmatrix} 0 & A_3 & -A_2 & -A_1 \\ -A_3 & 0 & A_1 & -A_2 \\ A_2 & -A_1 & 0 & -A_3 \\ A_1 & A_2 & A_3 & 0 \end{bmatrix}, Y_{\alpha\beta} = \begin{bmatrix} 0 & Y_3 & -Y_2 & -Y_1 \\ -Y_3 & 0 & \overline{Y}_1 & -Y_2 \\ Y_2 & -Y_1 & 0 & -Y_3 \\ Y_1 & Y_2 & Y_3 & 0 \end{bmatrix}$$

$$A_{[\alpha\beta],\beta} = \frac{1}{c}\psi_\alpha, \tag{20.15}$$

with $\psi_\kappa = \{\psi_k, \psi\}$

$$A_{[\alpha\beta]} = \mathrm{i}\overline{\varepsilon}^1 A_1 + \mathrm{i}\overline{\varepsilon}^2 A_2 + \overline{\varepsilon}^3 A_3, \quad \overline{\varepsilon}^1 = \varepsilon^1\varepsilon^4, \quad \overline{\varepsilon}^2 = \varepsilon^2\varepsilon^4, \quad \overline{\varepsilon}^3 = \varepsilon^1\varepsilon^2 \tag{20.16}$$

and we repeat that only the antisymmetric tensor of potentials can satisfy relations (20.3a) in the form (20.15).

20.5 Complex Relativity Theory

Again the Riemannian space physics presented for the rotation motions, sequence of relations (20.7-20.10), can be applied to the electromagnetic fields.

From the electromagnetic antisymmetric tensor of potentials, as the metric tensor perturbations, we may arrive at the Complex Relativity:

$$\overline{\varepsilon}^{\alpha}\overline{\varepsilon}^{\beta}+\overline{\varepsilon}^{\beta}\overline{\varepsilon}^{\alpha}=2\eta^{\alpha\beta}, \tag{20.17}$$

$$g^{\alpha\beta}=\eta^{\alpha\beta}+h^{\alpha\beta},\ h_{[\alpha\beta]}=A_{[\alpha\beta]},\ \mathrm{d}s^{2}=g_{\alpha\beta}\mathrm{d}x^{\alpha}\mathrm{d}x^{\beta}=g_{(\alpha\beta)}\mathrm{d}x^{\alpha}\mathrm{d}x^{\beta}, \tag{20.18}$$

$$\overline{G}_{\alpha\beta}\approx h_{[\mu\beta]}\Big|_{\alpha}^{\mu}+h_{[\mu\alpha]}\Big|_{\beta}^{\mu}-h_{[\alpha\beta]}\Big|_{\mu}^{\mu}=\frac{1}{c}\frac{\partial}{\partial x_{\beta}}\psi_{\alpha}+\frac{1}{c}\frac{\partial}{\partial x_{\alpha}}\psi_{\beta}-Y_{[\alpha\beta]}, \tag{20.19}$$

$$h_{[\alpha\beta]}\Big|_{\mu}^{\mu}=Y_{[\alpha\beta]},\quad h_{[\mu\beta]}\Big|_{\alpha}^{\mu}+h_{[\mu\alpha]}\Big|_{\beta}^{\mu}=\frac{1}{c}\frac{\partial}{\partial x_{\beta}}\psi_{\alpha}+\frac{1}{c}\frac{\partial}{\partial x_{\alpha}}\psi_{\beta}, \tag{20.20}$$

This Complex Relativity relations can be directly combined with that for gravity field.

20.6 Concluding Remarks

Our considerations related to physics in a complex Riemannian space include a genaralizaton of spin and twist motions in an asymmetric continuum and unification of fields in the Complex Relativity. We can repeat that an antisymmetry there introduced is not intrinsic as that related to torsion tensor.

- We would like to add some final remarks.

Numerous observations, in the whole spectrum of ranks, kinds and dimensions, show the universal role of rotational motions. The rotational structures and motions appearing in astronomical observations, and observed in geological and tectonic structures and the continent evolution, have become of recent interest in seismology and fracture mechanics and enter into micro-domains and their physics. In the related context, we may mention the two books developing this point of view: "Vortex-Related Events of the Geological Processes" (Vikulin ed 2004) and "Rotational Processes in Geology and Physics" (Milanovsky ed 2007).

Some remarks can touch the Universe evolution:

In the first chapter we have discussed the basic motions and deformations in an asymmetric continuum: the axial motions, displacements, spin fields and twist-oscillations of the off-diagonal shear axes; these fields are a function of time.

Starting with the idea based on the Asymmetric Continuum Theory (see Chap. 7) we present some reflections concerning the Universe origin. Such implications, when do not interfere or deny the essential results of the quantum and relativity theories, may deliver a more simple and intuitional approach to the metric tensor perturbations in relativity and hypotheses related the inflation phase effects in cosmology.

The presented approach to Complex Relativity was inspired by the relations formulated for the rotation motions in the Riemannian space. We believe that the analogies between the physics of the fields defined in the Standard Asymmetric Continuum and the proposed classic approach to relativity has a deep basis; not only the presented analogy between the rotation and electromagnetic field equations but also their extension to Riemannian space relations.

Starting with this point of view, we may present some reflections concerning the hypotheses related the inflation phase effects in cosmology; we may change the usual description to the following one.

The initial explosion (or quantum virtual process) appearing at zero moment (determined with Planck precision) causes the axial spreading in a symmetric system. Such initial expansion runs with the "inflation phase"

Fig. 20.1 Axial phase expansion followed with spiral forming motions

velocity, much greater then the light speed, the initial “condensate” (strings, quarks) could be subjected to motion similar to that for the axial motion in asymmetric continuum. At the end of this phase, rapid expansion slow down; this process is related to the phase transformation from symmetric expansion to a resulting physical state of our Universe with the light velocity limit: the initial “condensate” transforms to our Riemannian 4D space. The related phase transition would mean a transition from symmetric phase to asymmetric one with the possible rotation motions. At this phase change, an enormous expansion energy would be partly transformed to another form of motions with the initial rotations; both the initial rotations and further “slow” expansion process can lead to formation of the spiral structures in the Universe (Fig. 20.1).

Here, we may introduce another hypothesis related to possible appearance of the spin back holes with the energy concentration related to rotations.

References

Ben-Abraham SI (1970) Generalized stress and non-Riemannian geometry, fundamental aspects of dislocation theory. Nat Bur Stand (US) Spec Publ **317, II**: 943-962

Bilby BA, Boullough R, Smith E (1955) Continuous distributions of dislocations: a new application of the methods of non-Riemannian geometry. Proc Roy Soc London **A231**: 263-273

Cartan E (1923) Sur les variétés a connexion affine et la théorie de la relativité généralisée (premiére partie). Ann Sci École Norm Sup **40**: 325-412

Cartan E (1924) Sur les variétés a connexion affine et la théorie de la relativité généralisée (premiére partie suite). Ann Sci École Norm Sup **41**: 1-25

Cartan E (1925) Sur les variétés a connexion affine et la théorie de la relativité généralisée (duexiéme partie). Ann Sci École Norm Sup **42**: 17-88

Cosserat E, Cosserat F (1909) Theorie des corps déformables. A. Hermann et Fils, Paris

Hehl FW, McCrea JD, Mielke EW, Ne’eman Y (1995) Metric-affine gauge theory of gravity: field equations, Noether identities, world spinors, and breaking of dilation invariance. Phys Reports **258**: 1, 1-171

Holländer EF (1962) The geometric equation of dislocation dynamics. Czech J Phys **B12**: 35-47

Kibble TWB (1961) Lorentz invariance and the gravitational field. J Math Phys **2**: 2, 212-221

Kondo K (1955) Geometry of elastic deformation and incompatibility. RAAG Mem **1**:C1, 361-373

Kondo K (1958) Geometry of deformation and stress. RAAG Mem **C2**: 97-164

Kopczyński W (1973) An anisotropic universe with torsion. Phys Lett A **43**: 1, 63-64

Kröner E (1958) Kontinuums Theorie der Versetzungen und Eigenspannungen. Ergeb Angew Math, **5**

Milanovsky EE (ed) (2007) Rotational processes in geology and physics. Izd KomKniga, Moscow, 523 pp (in Russian)

Muskhelishvili NT (1953) Some basic problems of the elasticity. Noordhof, Groningen

Sciama DW (1962) Recent developments in general relativity. Pergamon Press-PWN, Oxford – Warszawa

Stojanovic R, Djuric S, Vujosevic L (1964) On finite thermal deformations. Arch Mech Stos **16**: 103-108

Teisseyre R (1963) Thermo-mechanical model of the earthquake origin and process. Acta Geophys Pol **11**: 4, 229-233

Teisseyre R (1969) Dislocation representation of thermal stresses. Acta Geophys Pol **17**:1, 3-12

Teisseyre R (1995a) Differential geometry methods in deformation problem. In: Teisseyre R (ed) Theory of earthquake premonitory and fracture processes. PWN Warszawa, pp 504-511

Teisseyre R (1995b) Dislocation density and geometrical objects. In Teisseyre R (ed) Theory of earthquake premonitory and fracture processes. PWN Warszawa, pp 512-519

Teisseyre R, Białecki M (2005) Complex relativity: gravity and electromagnetic fields. ArXiv: physics/0506033

Trautman A (1972a) On the Einstein-Cartan equations. I. Bull Acad Pol Sci Ser Sci Math Astronom Phys **20**: 2, 185-190

Trautman A (1972b) On the Einstein-Cartan equations. II. Bull Acad Pol Sci Ser Sci Math Astronom Phys **20**: 6, 503-506

Trautman A (1972c) On the Einstein-Cartan equations. III. Bull Acad Pol Sci Ser Sci Math Astronom Phys **20**: 10, 895-896

Trautman A (1973) Spin and torsion may avert gravitational singularities. Nature **242**: 114, 7-8

Trautman A (2006) Einstein-Cartan theory. In: Françoise J-P, Naber GL, Tsun TS (eds) Encyclopedia of mathematical physics, vol 2, Elsevier, Oxford, pp 189-195

Vikulin AV (ed) (2004) Vortex–related events of the geological processes. Petropavlovsk-Kamchatsky, 297 pp (in Russian)